PHYSICS RESEARCH AND TECHNOLOGY

INVARIANCE PRINCIPLES AND EXTENDED GRAVITY: THEORY AND PROBES

PHYSICS RESEARCH AND TECHNOLOGY

Additional books in this series can be found on Nova's website
under the Series tab.

Additional E-books in this series can be found on Nova's website
under the E-book tab.

PHYSICS RESEARCH AND TECHNOLOGY

INVARIANCE PRINCIPLES AND EXTENDED GRAVITY: THEORY AND PROBES

SALVATORE CAPOZZIELLO
AND
MARIAFELICIA DE LAURENTIS

Nova Science Publishers, Inc.
New York

For permission to use material from this book please contact us:
Telephone 631-231-7269; Fax 631-231-8175
Web Site: http://www.novapublishers.com

LIBRARY OF CONGRESS CATALOGING-IN-PUBLICATION DATA

Capozziello, Salvatore.
Invariance principles and extended gravity : theory and probes / Salvatore Capozziello and Mariafelicia De Laurentis.
p. cm.
Includes bibliographical references and index.
ISBN 978-1-61668-500-3 (hardcover)
1. Gravitation. 2. Gauge invariance. 3. General relativity (Physics) I. Laurentis, Mariafelicia De. II. Title.
QC178.C177 2009
531'.14--dc22

2010016726

Published by Nova Science Publishers, Inc. † New York

Contents

Foreword

This book is devoted to the study of gravitational theories which can be seen as modifications or extensions of General Relativity. The motivation to consider such theories, stemming from Cosmology, High Energy Physics and Astrophysics is thoroughly discussed (cosmological shortcomings, Dark Energy and Dark Matter, the lack in obtaining a successful formulation of Quantum Gravity). The basic principles that a gravitational theory should follow, and their geometrical interpretation, are analysed in a broad perspective which highlights the basic assumptions of General Relativity and suggests possible modifications. In particular, we present the invariance principles that are related to the gravitational field showing how gravity can emerge as a gauge theory. We discuss the bundle approach to gravity and derive geometrical quantities which intervene in the construction of gravitational field. In this context, the role of space-time deformations and conformal transformations is considered. Extensions of General Relativity are presented, focusing on specific classes of theories as scalar-tensor theories and $f(R)$-theories in metric and Palatini approach. The features of these theories are fully explored and attention is payed to issues of dynamical equivalence between them. Also, cosmological phenomenology within the realm of each of these theories is discussed and it is shown as shortcomings can be potentially addressed. A number of viability criteria are presented: observational cosmological tests, stochastic background of gravitational waves, Solar System tests after the discussion of Post-Minkowskian and Post-Newtonian limit. Finally, future perspectives are discussed and the possibility to go beyond a trial and error approach to modified gravity is explored.

Notations

Mathematical Notations

$\partial_\mu = \frac{\partial}{\partial x^\mu}$: Partial derivative with respect to $\{x_\mu\}$

$\{e_\mu\}$: Set with elements e_μ

$\nabla_\mu = \partial_\mu + \Gamma_\mu$ Gauge covariant derivative operator

Γ_μ : Gauge potential 1-form

d : Exterior derivative operator

$\langle V|e\rangle$: Inner multiplication between vector e and 1-form V

$[A, B]$: Commutator of operators A and B

$\{A, B\}$: Anti-commutator of operators A and B

$\wedge$: Exterior multiplication operator

$\rtimes$: Semi-direct product

$\times$: Direct product

$\times_M$: Fibered product over manifold M

$\oplus$: Direct sum

$\otimes$: Tensor product

$A \cup B$: Union of A and B

$A \cap B$: Intersection of A and B

$\mathbb{P}(M, G; \pi)$: Fiber bundle with base space M and G-diffeomorphic fibers

$\pi_{\mathbb{P}M} : \mathbb{P} \to M$: Canonical projection map from $\mathbb{P}$ onto M

$R_h, (L_h)$: Right (left) group action or translation

$\widehat{\mathfrak{R}}\ (\widehat{\mathfrak{L}})$: Right (left) invariant fundamental vector operators

$\Theta\ (\overline{\Theta})$: Right (left) invariant Maurer-Cartan 1-form

$\circ$: Group (element) composition operator

$o_{\alpha\beta} = diag(-1, 1, 1, 1)$ or $\eta_{ij} = diag(-1, 1, 1, 1)$: Lorentz group metric

$A(4, \mathbb{R})$: Group of affine transformations on a real 4-dimensional manifold

$\mathrm{Diff}(4, \mathbb{R})$: Group of diffeomorphisms on a real 4-dimensional manifold

$GL(4, \mathbb{R})$: Group of real 4×4 invertible matrices

$SO(4,2)$: Special conformal group

$SO(3,1)$: Lorentz group

$P(3,1)$: Poincaré group

$\mathfrak{g}$: Lie algebra of group G

$g \in G$: Element g of G

$\{\mathcal{U}\} \subset M$: Set $\mathcal{U}$ is a subset of M

$\boldsymbol{G}$: Algebra generator of group G

$\rho(\boldsymbol{G})$: Representation of G-algebra

C^∞ : Infinitely differentiable (continuous)

${}^*A$: Dual of A with respect to (coordinate) basis indices

${}^\star A$: Dual of A with respect to Lie algebra indices

$\varepsilon_{a_1 \ldots a_n}$ or $\varepsilon_{a_1 \ldots a_n}$: Levi-Civita totally skew tensor density

$\eta_{a_1 \ldots a_n}$: Eta basis volume n-form density

σ^* : Pullback by local section σ

L_{h*} : Differential (pushforward) map induced by L_h

$T_{(a_1 \ldots a_n)}$: Symmetrization of indices

$T_{[a_1 \ldots a_n]}$: Antisymmetrization of indices

$T(M)$: Tangent space to manifold M

$T^*(M)$: Cotangent space to M dual to $T(M)$

${}^\dagger T_{\mu\nu}$: Traceless matrix

$A^\dagger$: Hermitian adjoint of A

$f : A \to B$: Map f taking elements $\{a\} \in A$ to $\{b\} \in B$

$h : C \hookrightarrow D$: Inclusion map, where $C \subset D$

Object	Convention for the gravitational field
Index ranges	$\alpha, \beta = 0,1,2,3;\ i,j = 1,2,3$
Flat metric	$\eta_{\alpha\beta} = \mathrm{diag}(-1,1,1,1)$ or $\eta_{\alpha\beta} = \mathrm{diag}(1,-1,-1,-1)$
Coordinates	$x^\alpha = (x^0, x^1, x^2, x^3) = (ct, x^1, x^2, x^3)$
Vectors	$\underline{v} = (v^1, v^2, v^3);\ \nabla = (\partial/\partial x^1, \partial/\partial x^2, \partial/\partial x^3)$
Symmetrization	$T_{(\alpha\vert\beta\ldots\gamma\vert\delta)} = \frac{1}{2}(T_{\alpha\beta\ldots\gamma\delta} + T_{\delta\beta\ldots\gamma\alpha})$
Kronecker	$\delta^\alpha_\beta = 1$ if $\alpha = \beta$, 0 else
Connection	$\Gamma^\alpha_{\mu\nu} = \frac{1}{2} g^{\alpha\sigma}(g_{\mu\sigma,\nu} + g_{\nu\sigma,\mu} - g_{\mu\nu,\sigma})$
Riemann tensor	$R^\alpha{}_{\beta\mu\nu} = \Gamma^\alpha_{\beta\nu,\mu} - \Gamma^\alpha_{\beta\mu,\nu} + \Gamma^\sigma_{\beta\nu}\Gamma^\alpha_{\sigma\mu} - \Gamma^\sigma_{\beta\mu}\Gamma^\alpha_{\sigma\nu}$
Ricci tensor	$R_{\mu\nu} = R^\sigma{}_{\mu\sigma\nu}$

Dimensions of Physical Quantities	
1	$g_{\alpha\beta}$, $\delta_{\alpha\beta}$, $\mathcal{G}$, a_1, $K_{I,1}$
m	x^{α}, $\Gamma^{\alpha}_{\mu\nu}{}^{-1}$, $R_{\alpha\beta\mu\nu}{}^{-\frac{1}{2}}$, $R_{\alpha\beta}{}^{-\frac{1}{2}}$, $R^{-\frac{1}{2}}$, Y_I, Y_{II}^{-1}, λ_1^{-1}, λ_2^{-1}, $\delta^{-\frac{1}{3}}$, $\tilde{\mathcal{G}}^{\frac{1}{3}}$, ξ, $a_2{}^{\frac{1}{2}}$, $b_1{}^{\frac{1}{2}}$, $K_{I,2}$, $K_{I,3}{}^{-1}$, $K_{I,4}{}^{-1}$, $K_{I,5}{}^{-1}$, $K_{I,6}{}^{-1}$, $K_{I,7}{}^{-2}$, $K_{II,3}$, $K_{II,4}$, $K_{II,5}$, $K_{II,6}$, r
s	$\sigma_I^{-\frac{1}{2}}$, $\sigma_{II}^{-\frac{1}{2}}$, $\tau_{II}^{-\frac{1}{2}}$
Kg	M
$\frac{\mathrm{m}}{\mathrm{s}}$	c, $U^{\frac{1}{2}}$, $V^{\frac{1}{2}}$, $h_{0a}^{\frac{1}{3}}$, v^a, $\Pi^{\frac{1}{2}}$, $A^{\frac{1}{2}}$, $\Phi^{\frac{1}{2}}$, $\tau_I^{\frac{1}{2}}$
$\frac{\mathrm{m}^3}{\mathrm{s}^2}$	U_0, U_1, U_2
$\frac{\mathrm{m}^4}{\mathrm{s}^2}$	B
$\frac{\mathrm{kg}}{\mathrm{m}^3}$	ρ
$\frac{\mathrm{kg}}{\mathrm{s}^2\mathrm{m}}$	$T_{\alpha\beta}$, p
$\frac{\mathrm{m}^3}{\mathrm{s}^2\mathrm{Kg}}$	G_N

Preface

The terms "modified gravity" and "alternative theory of gravity" have become a standard terminology for theories proposed for describing the gravitational interaction which differ from the most conventional one, General Relativity. Modified or alternative theories of gravity have a long history. The first attempts date back to the 1920s, soon after the introduction of Einstein's theory. Interest in this research field, which was initially driven by curiosity or a desire to challenge the newly introduced General Theory of Relativity, has subsequently varied depending on circumstances, responding to the appearance of new motivations. However, there has been more or less continuous activity in this subject over the last 90 years.

When the research presented in this book began, interest in modified gravity was already at a high point and it has continued increasing further until the present day. This recent stimulus has mostly been due to combined motivations coming from the well-known cosmological problems related to the accelerated expansion of the Universe and the feedback from High Energy Physics.

Due to the above motivations, and even the main goal of this book is to present the research conducted by the authors, a significant effort has been made so that this book could be a guide for readers who are interested in this field. To this end, special attention has been paid to give a survey on the foundations of gravitation. Also, an effort has been made to present the theories discussed thoroughly, so that readers less familiar with this subject can be introduced to them before gradually focusing more technical aspects and applications.

The outline of the book is the following: in the Introduction, several open issues related to gravity are discussed, including the cosmological problems related to Dark Matter and Dark Energy, and the search for a theory of Quantum Gravity. Through out the presentation of historical timeline of the passage from Newtonian gravity to General Relativity, and a comparison with the current status of the latter in the light of the problems just mentioned, the motivations for considering alternative theories of gravity are introduced.

In Chapter 2, we discuss how gravity and spin can be obtained as the realization of the local Conformal-Affine group of symmetry transformations. In particular, we show how gravitation is a gauge theory which can be obtained starting from some local invariance as the Poincaré local symmetry. We review previous results where the inhomogeneous connection coefficients, transforming under the Lorentz group, give rise to gravitational gauge potentials which can be used to define covariant derivatives accommodating minimal couplings of matter, gauge fields (and then spin connections). After we show, in a self-contained approach, how the tetrads and the Lorentz group can be used to induce the space-time metric and then the *Invariance Induced Gravity* can be directly obtained both in holonomic and anholonomic pictures. Besides, we show how tensor valued connection forms act as auxiliary dynamical fields associated with the dilation, special conformal and deformation (shear) degrees of freedom, inherent to the bundle manifold. As a result, this allows to determine the bundle curvature of the theory and then to construct boundary topological invariants which give rise to a prototype (source free) gravitational Lagrangian. Finally, the Bianchi identities, the covariant field equations and the gauge currents are obtained determining completely the dynamics. Furthermore starting from the general invariance principle, we discuss the global and the local Poincaré invariance developing the spinor, vector and tetrad formalisms. These tools allow to construct the curvature, torsion and metric tensors by the Fock-Ivanenko covariant derivative. The resulting Einstein-Cartan theory describes a space endowed with non-vanishing curvature and torsion while the gravitational field equations are similar to the Yang-Mills equations of motion with the torsion tensor playing the role of the Yang-Mills field strength.

A definition of space-time metric deformations on an n-dimensional manifolds is given in Chapter 3. We show that such deformations can be regarded as extended conformal transformations. In particular, their features can be related to the perturbation theory giving a natural picture by which gravitational waves are described by small deformations of the metric. As further result, deformations can be related to approximate Killing vectors (approximate symmetries) by which it is possible to parameterize the deformed region of a given manifold.

Chapter 4 is devoted to a survey of what is intended for *Extended Theories of Gravity* in the so called "metric" and "Palatini" approaches. We want to give a survey on the formal and physical aspects of Extended Theories of Gravity, The field equations are derived, specifically, we discuss two interesting cases: $f(R)$ and scalar-tensor gravity considering their relations with General Relativity by conformal transformations. The Palatini approach and its intrinsic conformal structure is discussed giving some peculiar examples. The debate on the physical relevance of conformal transformations can be faced by taking the Palatini approach into account in gravitational theories. We show that conformal transformations are not only a mathematical tool to disentangle gravitational and matter degrees of freedom (passing from the Jordan frame to the Einstein frame) but they acquire a physical meaning considering the bi-metric structure of Palatini formalism which allows to distinguish between space-time structure and geodesic structure. These facts are relevant at least at cosmological scales, while at small scales (i.e. Solar System scales) the conformal factor is slowly varying and its effects are not important. Examples of higher-order and non-minimally coupled theories are worked out and relevant cosmological solutions in Einstein frame and Jordan frames are discussed showing that also the interpretation of cosmological observations can drastically change depending on the adopted frame.

In Chapter 5, we develop the post-Minkowskian limit of Extended Theories Gravity. It is well known that when dealing with General Relativity, such an approach provides massless spin-two waves as propagating degree of freedom of the gravitational field while Extended Theories Gravity imply other additional propagating modes in the gravity spectra. We show that a general analytic $f(R)$-model, together with a standard massless graviton, is characterized by a massive scalar particle with a finite-distance interaction. We briefly discuss how such massive gravitational mode can have relevant consequences on cosmological scales affecting the stochastic background of gravitational waves and representing a valid alternative to Dark Matter of galactic scales. After, we linearize the field equations for higher order theories that contain scalar invariants other than the Ricci scalar. We find that besides a massless spin-2 field (the standard graviton), the theory contains also spin-0 and spin-2 massive modes with the latter which could be ghost modes. Then, we investigate the possible detectability of such additional polarization modes of a stochastic gravitational wave by ground-based and space interferometric detectors. Furtheremore, we extend the formalism of the cross-correlation analysis, including the additional polarization modes, and calculate the detectable energy density of the spectrum for a stochastic background of relic gravitational waves. For the situation considered here, we find that these massive modes are certainly of interest for direct detection by the LISA forthcoming experiment. Finally, we show that the stochastic background of gravitational waves, produced in the early cosmological epochs, strictly depends on the assumed theory of gravity. In particular, the specific form of the function $f(R)$, where R is the Ricci scalar, is related to the evolution and the production mechanism of gravitational waves. On the other hand, detecting the stochastic background by the interferometric experiments (VIRGO, LIGO, LISA) could be a further tool to select the effective theory of gravity.

In Chapter 6, the Newtonian limit of fourth-order gravity is worked out discussing its viability with respect to the standard results of General Relativity. We investigate the limit in the metric approach which, with respect to the Palatini formulation, has been much less studied in the recent literature, due to the higher-order of the field equations. In addition, we refrain from exploiting the formal equivalence of higher-order theories considering the analogy with specific scalar-tensor theories, i.e. we work in the so-called Jordan frame in order to avoid possible misleading interpretations of the results. Explicit solutions are provided for several different types of Lagrangians containing powers of the Ricci scalar as well as combinations of the other curvature invariants. PPN-limit of alternative theories of gravity represents a still controversial matter of debate and no definitive answer has been provided, up to now, about this issue. By using the definition of the PPN-parameters γ and β in term of $f(R)$ of gravity, we show that a family of third-order polynomial theories, in the Ricci scalar R, turns out to be compatible with the PPN-limit and the deviation from General Relativity, theoretically predicted, can agree with experimental data. Viable $f(R)$-gravity models are discussed toward Solar System tests and stochastic background of gravitational waves. The aim is to achieve experimental bounds for the theory at local and cosmological scales in order to select models capable of addressing the accelerating cosmological expansion without cosmological constant but evading the weak field constraints. Beside large scale structure and galactic dynamics, these bounds can be considered complementary in order to select self-consistent theories of gravity working at the infrared limit.

Future perspectives and conclusions are reported in Chapter 7. A wide bibliography on the topics discussed in the book is given at the end.

Finally, we have to acknowledge several collaborators and friends for their invaluable support and fruitful scientific discussions and common work. The list would be too long to be reported here without the risk to forget someone. Last but not least, we thank our families that supported us in the effort for these researches.

Napoli, December 2009.

Salvatore Capozziello & Mariafelicia De Laurentis

Chapter 1

Introduction

1.1. Is General Relativity the Final Theory of Gravity?

It is remarkable that gravity is probably the fundamental interaction which still remains the most enigmatic, even though it is so related with phenomena experienced in everyday life and is the one most easily conceived of without any sophisticated knowledge. As a matter of fact, the gravitational interaction was the first one to have been put under the microscope of experimental investigation, obviously due to the simplicity of constructing a suitable experimental apparatus.

Galileo Galilei was the first to introduce pendula and inclined planes to the study of terrestrial gravity at the end of the 16th century. It seems that gravity played an important role in the development of Galileo's ideas about the necessity of experiment in the study of science, which had a great impact on modern scientific thinking. However, it was not until 1665, when Isaac Newton introduced the now renowned "inverse-square gravitational force law", that terrestrial gravity was actually related to celestial gravity in a single theory. Newton's theory made correct predictions for a variety of phenomena at different scales, including both terrestrial experiments and planetary motion.

Obviously, Newton's contribution to gravity — quite apart from his enormous contribution to physics overall — is not restricted to the expression of the inverse square law. Much attention should be paid to the conceptual basis of his gravitational theory, which incorporates two key ideas:

1. The idea of absolute space, i.e. the view of space as a fixed, unaffected structure; a rigid arena where physical phenomena take place.

2. The idea of what was later called the Weak Equivalence Principle which, expressed in the language of Newtonian theory, states that the inertial and the gravitational mass coincide.

Asking whether Newton's theory, or any other physical theory, is right or wrong, would be ill-posed to begin with, since any consistent theory is apparently "right". A more appropriate way to pose the question would be to ask how suitable is this theory to describe the physical world or, even better, how large a portion of the physical world is sufficiently described by such a theory. Also, one could ask how unique the specific theory is for the description of the relevant phenomena. It was obvious, in the first 20 years after the introduction of Newtonian gravity, that it did manage to explain all of the aspects of gravity known at that time. However, all of the questions above were posed sooner or later.

In 1855, Urbain Le Verrier observed a 35 arc-second excess precession of Mercury's orbit and later on, in 1882, Simon Newcomb measured this precession more accurately to be 43 arc-seconds. This experimental fact was not predicted by Newton's theory. It should be noted that Le Verrier initially tried to explain the precession within the context of Newtonian gravity, attributing it to the existence of another, yet unobserved, planet whose orbit lies within that of Mercury. He was apparently influenced by the fact that examining the distortion of the planetary orbit of Uranus in 1846 had led him, and, independently, John Couch Adams, to the discovery of Neptune and the accurate prediction of its position and momenta. However, this innermost planet was never found.

On the other hand, in 1893, Ernst Mach stated what was later called by Albert Einstein "Mach's principle". This is the first constructive attack on Newton's idea of absolute space after the 18th century debate between Gottfried Wilhelm von Leibniz and Samuel Clarke (Clarke was acting as Newton's spokesman) on the same subject, known as the Leibniz–Clarke Correspondence. Mach's idea can be considered as rather vague in its initial formulation and it was essentially brought into mainstream physics later on by Einstein along the following lines:

"...inertia originates in a kind of interaction between bodies...".

This is obviously in contradiction with Newton's ideas, according to which inertia was always relative to the absolute frame of space. There exists also a later, probably clearer interpretation of Mach's Principle, which, however, also differs in substance. This was given by Dicke [162]:

"The gravitational constant should be a function of the mass distribution in the Universe".

This is different from Newton's idea of the gravitational constant as being universal and unchanging. Now Newton's basic axioms there to be reconsidered.

But it was not until 1905, when Albert Einstein completed Special Relativity, that Newtonian gravity would have to face a serious challenge. Einstein's new theory, which managed to explain a series of phenomena related to non-gravitational physics, appeared to be incompatible with Newtonian gravity. Relative motion and all the linked concepts had gone well beyond Galileo and Newton ideas and it seemed that Special Relativity should somehow be generalised to include non-inertial frames. In 1907, Einstein introduced the equivalence between gravitation and inertia and successfully used it to predict the gravitational redshift. Finally, in 1915, he completed the theory of General Relativity, a generalisation of Special Relativity which included gravity. Remarkably, the theory matched perfectly the experimental result for the precession of Mercury's orbit, as well as other experimental findings like the Lense-Thirring gravitomagnetic precession (1918) and the gravitational deflection of light by the Sun, as measured in 1919 during a Solar eclipse by Arthur Eddington.

General Relativity overthrew Newtonian gravity and continues to be up to now an extremely successful and well-accepted theory for gravitational phenomena. As mentioned before, and as often happens with physical theories, Newtonian gravity did not lose its appeal to scientists. It was realised, of course, that it is of limited validity compared to General Relativity, but it is still sufficient for most applications related to gravity. What is more, at a certain limit of gravitational field strength and velocities, General Relativity inevitably reduces to Newtonian gravity. Newton's equations for gravity might have been generalised and some of the axioms of his theory may have been abandoned, like the notion of an absolute frame, but some of the cornerstones of his theory still exist in the foundations of General Relativity, the most prominent example being the Equivalence Principle, in a more suitable formulation of course.

This brief chronological review, besides its historical interest, is outlined here also for a practical reason. General Relativity is bound to face the same questions as were faced by Newtonian gravity and many would agree that it is actually facing them now. In the forthcoming sections, experimental facts and theoretical problems will be presented which justify that this is indeed the case. Remarkably, there exists a striking similarity to the problems which Newtonian gravity faced, i.e. difficulty in explaining particular observations, incompatibility with other well established theories and lack of uniqueness. This is the reason behind the question mark in the title of this section.

1.2. What a Good Theory of Gravity Has to Do

From a phenomenological point of view, there are some minimal requirements that any relativistic theory of gravity has to match. First of all, it has to explain the astrophysical observations (e.g. the orbits of planets, the potential of self-gravitating structures).

This means that it has to reproduce the Newtonian dynamics in the weak-energy limit. Besides, it has to pass the classical Solar System tests which are all experimentally well founded [436]. As second step, it should reproduce galactic dynamics considering the observed baryonic constituents (e.g. luminous components as stars, sub-luminous components as planets, dust and gas), radiation and Newtonian potential which is, by assumption, extrapolated to galactic scales.

Thirdly, it should address the problem of large scale structure (e.g. clustering of galaxies) and finally cosmological dynamics, which means to reproduce, in a self-consistent way, the cosmological parameters as the expansion rate, the Hubble constant, the density parameter and so on. Observations and experiments, essentially, probe the standard baryonic matter, the radiation and an attractive overall interaction, acting at all scales and depending on distance: the gravity.

The simplest theory which try to satisfies the above requirements is the General Relativity [175]. It is firstly based, as we said above, on the assumption that space and time have to be entangled into a single space-time structure, which, in the limit of no gravitational forces, has to reproduce the Minkowski space-time structure. Einstein profitted also of ideas earlier put forward by Riemann, who stated that the Universe should be a curved manifold and that its curvature should be established on the basis of astronomical observations [178].

In other words, the distribution of matter has to influence point by point the local curvature of the space-time structure. The theory, eventually formulated by Einstein in 1915, was strongly based on three assumptions that the Physics of Gravitation has to satisfy.

The "*Principle of Relativity*", that amounts to require all frames to be good frames for Physics, so that no preferred inertial frame should be chosen a priori (if any exist).

The "*Principle of Equivalence*", that amounts to require inertial effects to be locally indistinguishable from gravitational effects (in a sense, the equivalence between the inertial and the gravitational mass).

The "*Principle of General Covariance*", that requires field equations to be "generally covariant" (today, we would better say to be invariant under the action of the group of all space-time diffeomorphisms) [361].

And - on the top of these three principles - the requirement that causality has to be preserved (the "*Principle of Causality*", i.e. that each point of space-time should admit a universally valid notion of past, present and future).

Let us also recall that the older Newtonian theory of space-time and gravitation - that Einstein wanted to reproduce at least in the limit of weak gravitational forces (what is called today the "post-Newtonian approximation") - required space and time to be absolute entities, particles moving in a preferred inertial frame following curved trajectories, the curvature of which (e.g. the acceleration) had to be determined as a function of the sources (i.e. the "forces").

On these bases, Einstein was led to postulate that the gravitational forces have to be expressed by the curvature of a metric tensor field $ds^2 = g_{\mu\nu}dx^\mu dx^\nu$ on a four-dimensional space-time manifold, having the same signature of Minkowski metric, e.g., the so-called "Lorentzian signature", herewith assumed to be $(+,-,-,-)$. He also postulated that space-time is curved in itself and that its curvature is locally determined by the distribution of the sources, e.g. - being space-time a continuum - by the four-dimensional generalization of what in Continuum Mechanics is called the "matter stress-energy tensor", e.g. a rank-two (symmetric) tensor $T^m_{\mu\nu}$.

Once a metric $g_{\mu\nu}$ is given, its curvature is expressed by the Riemann (curvature) tensor

$$R^\alpha{}_{\beta\mu\nu} = \Gamma^\alpha_{\beta\nu,\mu} - \Gamma^\alpha_{\beta\mu,\nu} + \Gamma^\sigma_{\beta\nu}\Gamma^\alpha_{\sigma\mu} - \Gamma^\sigma_{\beta\mu}\Gamma^\alpha_{\sigma\nu}\,, \tag{1.1}$$

where the comas are partial derivatives. Its contraction

$$R^\alpha{}_{\mu\alpha\nu} = R_{\mu\nu}\,, \tag{1.2}$$

is the "Ricci tensor" and the scalar

$$R = R^\mu{}_\mu = g^{\mu\nu}R_{\mu\nu}\,, \tag{1.3}$$

is called the "scalar curvature" of $g_{\mu\nu}$. Einstein was led to postulate the following equations for the dynamics of gravitational forces

$$R_{\mu\nu} = \frac{\kappa}{2}T^m_{\mu\nu}\,, \tag{1.4}$$

where $\kappa = 8\pi G_N$, with $c = 1$, is a coupling constant. These equations turned out to be physically and mathematically unsatisfactory.

As Hilbert pointed out [361], they were not of a variational origin, i.e. there was no Lagrangian able to reproduce them exactly (this is slightly wrong, but this remark is unessential here). Einstein replied that he knew that the equations were physically unsatisfactory, since they were contrasting with the continuity equation of any reasonable kind of matter. Assuming that matter is given as a perfect fluid, that is

$$T^m_{\mu\nu} = (p+\rho)u_\mu u_\nu - p g_{\mu\nu}\,, \tag{1.5}$$

where $u_\mu u_\nu$ is a comoving observer, p is the pressure and ρ the density of the fluid, then the continuity equation requires $T^m_{\mu\nu}$ to be covariantly constant, i.e. to satisfy the conservation law

$$\nabla^\mu T^m_{\mu\nu} = 0\,, \tag{1.6}$$

where ∇^μ denotes the covariant derivative with respect to the metric.

In fact, it is not true that $\nabla^\mu R_{\mu\nu}$ vanishes (unless $R = 0$). Einstein and Hilbert reached independently the conclusion that the wrong field Eqs. (1.4) had to be replaced by the correct ones

$$G_{\mu\nu} = \kappa T^m_{\mu\nu}\,, \tag{1.7}$$

where

$$G_{\mu\nu} = R_{\mu\nu} - \frac{1}{2} g_{\mu\nu} R\,, \tag{1.8}$$

that is currently called the "Einstein tensor" of $g_{\mu\nu}$. These equations are both variational and satisfy the conservation laws (1.6) since the following relation holds

$$\nabla^\mu G_{\mu\nu} = 0\,, \tag{1.9}$$

as a byproduct of the so-called "Bianchi identities" that the curvature tensor of $g_{\mu\nu}$ has to satisfy [429].

The Lagrangian that allows to obtain the field Eqs. (1.7) is the sum of a "matter Lagrangian" $\mathcal{L}_m$, the variational derivative of which is exactly $T^m_{\mu\nu}$, i.e.

$$T^m_{\mu\nu} = \frac{\delta \mathcal{L}_m}{\delta g^{\mu\nu}}\,, \tag{1.10}$$

and of a "gravitational Lagrangian", currently called the Hilbert-Einstein Lagrangian

$$L_{HE} = g^{\mu\nu} R_{\mu\nu} \sqrt{-g} = R\sqrt{-g}\,, \tag{1.11}$$

where $\sqrt{-g}$ denotes the square root of the value of the determinant of the metric $g_{\mu\nu}$.

The choice of Hilbert and Einstein was completely arbitrary (as it became clear a few years later), but it was certainly the simplest one both from the mathematical and the physical point of view. As it was later clarified by Levi-Civita in 1919, curvature is not a "purely metric notion" but, rather, a notion related to the "linear connection" to which "parallel transport" and "covariant derivation" refer [263].

In a sense, this is the precursor idea of what in the sequel would be called a "gauge theoretical framework" [246], after the pioneering work by Cartan in 1925 [130]. But at the time of Einstein, only metric concepts were at hands and his solution was the only viable.

It was later clarified that the three principles of relativity, equivalence and covariance, together with causality, just require that the space-time structure has to be determined by either one or both of two fields, a Lorentzian metric g and a linear connection Γ, assumed to be torsionless for the sake of simplicity.

The metric g fixes the causal structure of space-time (the light cones) as well as its metric relations (clocks and rods); the connection Γ fixes the free-fall, i.e. the locally inertial observers. They have, of course, to satisfy a number of compatibility relations which amount to require that photons follow null geodesics of Γ, so that Γ and g can be independent, *a priori*, but constrained, *a posteriori*, by some physical restrictions. These, however, do not impose that Γ has necessarily to be the Levi-Civita connection of g [321].

This justifies - at least on a purely theoretical basis - the fact that one can envisage the so-called "alternative theories of gravitation", that we prefer to call "*Extended Theories of Gravitation*" since their starting points are exactly those considered by Einstein and Hilbert: theories in which gravitation is described by either a metric (the so-called "purely metric theories"), or by a linear connection (the so-called "purely affine theories") or by both fields (the so-called "metric-affine theories", also known as "first order formalism theories"). In these theories, the Lagrangian is a scalar density of the curvature invariants constructed out of both g and Γ.

The choice (1.11) is by no means unique and it turns out that the Hilbert-Einstein Lagrangian is in fact the only choice that produces an invariant that is linear in second derivatives of the metric (or first derivatives of the connection). A Lagrangian that, unfortunately, is rather singular from the Hamiltonian point of view, in much than same way as Lagrangians, linear in canonical momenta, are rather singular in Classical Mechanics (see e.g. [27]).

A number of attempts to generalize General Relativity (and unify it to Electromagnetism) along these lines were followed by Einstein himself and many others (Eddington, Weyl, Schrödinger, just to quote the main contributors; see, e.g., [38]) but they were eventually given up in the fifties of XX Century, mainly because of a number of difficulties related to the definitely more complicated structure of a non-linear theory (where by "non-linear" we mean here a theory that is based on non-linear invariants of the curvature tensor), and also because of the new understanding of Physics that is currently based on four fundamental forces and requires the more general "gauge framework" to be adopted (see [242] and the discussion in Chapter 2).

Still a number of sporadic investigations about "alternative theories" continued even after 1960 (see [436] and refs. quoted therein for a short history). The search of a coherent quantum theory of gravitation or the belief that gravity has to be considered as a sort of low-energy limit of string theories [202] - something that we are not willing to enter here in detail - has more or less recently revitalized the idea that there is no reason to follow the simple prescription of Einstein and Hilbert and to assume that gravity should be classically governed by a Lagrangian linear in the curvature.

Further curvature invariants or non-linear functions of them should be also considered, especially in view of the fact that they have to be included in both the semi-classical expansion of a quantum Lagrangian or in the low-energy limit of a string Lagrangian.

Moreover, it is clear from the recent astrophysical observations and from the current cosmological hypotheses that Einstein equations are no longer a good test for gravitation at Solar System, galactic, extra-galactic and cosmic scale, unless one does not admit that the matter side of Eqs. (1.7) contains some kind of exotic matter-energy which is the "dark matter" and "dark energy" side of the Universe.

The idea which we propose here is much simpler. Instead of changing the matter side of Einstein Eqs. (1.7) in order to fit the "missing matter-energy" content of the currently observed Universe (up to the 95% of the total amount!), by adding any sort of inexplicable and strangely behaving matter and energy, we claim that it is simpler and more convenient to change the gravitational side of the equations, admitting corrections coming from non-linearities in the effective Lagrangian. However, this is nothing else but a matter of taste and, since it is possible, such an approach should be explored. Of course, provided that the Lagrangian can be conveniently tuned up (i.e., chosen in a huge family of allowed Lagrangians) on the basis of its best fit with all possible observational tests, at all scales (solar, galactic, extragalactic and cosmic).

Something that - in spite of some commonly accepted but disguised opinion - can and should be done before rejecting a priori a non-linear theory of gravitation (based on a non-singular Lagrangian) and insisting that the Universe has to be necessarily described by a rather singular gravitational Lagrangian (one that does not allow a coherent perturbation theory from a good Hamiltonian point of view) accompanied by matter that does not follow the behavior that standard baryonic matter, probed in our laboratories, usually satisfies.

1.2.1. The Foundation of Metric Theories: The Equivalence Principle

Equivalence principle is the foundation of every metric theory of gravity [80, 436] and the starting point of our considerations.

The first formulation of Equivalence principle comes out from the theory of gravitation formulates by Galileo and Newton; it is called the Weak Equivalence Principle and it states that the "inertial mass" m and the "gravitational mass" M of any object are equivalent. In Newtonian physics, the "inertial mass" m is a coefficient which appears in the second Newton is law: $\vec{F} = m\vec{a}$ where $\vec{F}$ is the force exerted on a mass m with acceleration $\vec{a}$; in Special Relativity (without gravitation) the "inertial mass" of a body appears to be proportional to the rest energy of the body: $E = mc^2$. Considering the Newtonian gravitational attraction, one introduces the "gravitational mass" M: the gravitational attraction force between two bodies of "gravitational mass" M and M' is $F = G_N MM'/r^2$ where G_N is the Newtonian gravitational constant and r the distance between the two bodies. Various experiments [413] demonstrate that $m \equiv M$. The present accuracy of this relation is of the order of 10^{-13}; spatial projects are currently designed to achieve precision of 10^{-15} [220] and 10^{-18} [221].

The Weak Equivalence Principle statement implies that it is impossible to distinguish between the effects of a gravitational field from those experienced in uniformly accelerated frames, using the simple observation of the free-falling particles behavior. The Weak Equivalence Principle can be formulated again in the following statement [436]:

> *If an uncharged test body is placed at an initial event in space-time and given an initial velocity there, then its subsequent trajectory will be independent of*

its internal structure and composition.

A generalization of Weak Equivalence Principle claims that the Special Relativity is only locally valid. It has been achieved by Einstein after the formulation of Special Relativity theory where the concept of mass looses some of its uniqueness: the mass is reduced to a manifestation of energy and momentum. According to Einstein, it is impossible to distinguish between uniform acceleration and an external gravitational field, not only for free-falling particles but whatever is the experiment. This equivalence principle has been called Einstein Equivalence Principle ; its main statements are the following [436]:

- Weak Equivalence Principle is valid;
- the outcome of any local non-gravitational test experiment is independent of velocity of free-falling apparatus;
- the outcome of any local non-gravitational test experiment is independent of where and when in the Universe it is performed.

One defines as "local non-gravitational experiment" an experiment performed in a small-size [1] freely falling laboratory. From the Einstein Equivalence Principle, one gets that the gravitational interaction depends on the curvature of space-time, *i.e.* the postulates of any metric theory of gravity have to be satisfied [436]:

- space-time is endowed with a metric $g_{\mu\nu}$;
- the world lines of test bodies are geodesics of the metric ;
- in local freely falling frames, called local Lorentz frames, the non-gravitational laws of physics are those of Special Relativity.

One of the predictions of this principle is the gravitational red-shift, experimentally verified by Pound and Rebka in 1960 [335].

It is worth stressing that gravitational interactions are specifically excluded from Weak Equivalence Principle and Einstein Equivalence Principle. In order to classify alternative theories of gravity, the Gravitational Weak Equivalence Principle and the Strong Equivalence Principle has to be introduced. The Strong Equivalence Principle extends the Einstein Equivalence Principle by including all the laws of physics in its terms [436]:

- Weak Equivalence Principle is valid for self-gravitating bodies as well as for test bodies (Gravitational Weak Equivalence Principle);
- the outcome of any local test experiment is independent of the velocity of the free-falling apparatus;
- the outcome of any local test experiment is independent of where and when in the Universe it is performed.

Therefore, the Strong Equivalence Principle contains the Einstein Equivalence Principle, when gravitational forces are ignored. Many authors claim that the only theory coherent with the Strong Equivalence Principle is General Relativity.

[1] In order to avoid the inhomogeneities.

1.2.2. The Parametrized Post Newtonian (PPN) Limit

General Relativity is not the only theory of gravitation and, up to now, at least twenty-five alternative theories of gravity have been investigated from the 60's, considering the space-time to be "special relativistic" at a background level and treating gravitation as a Lorentz-invariant field on the background [80].

Two different classes of experiments have been studied: the former ones testing the foundations of gravitation theory – among them the Equivalence Principle – the latter ones testing the metric theories of gravity where space-time is endowed with a metric tensor and where the Einstein Equivalence Principle is valid. However, for several fundamental reasons which we are going to discuss, extra fields might be necessary to describe the gravitation, *e.g.* scalar fields or higher-order corrections in curvature invariants.

Two sets of equations can be distinguished [413]. The first ones couple the gravitational fields to the non–gravitational contents of the Universe, i.e. the matter distribution, the electromagnetic fields, etc.. The second set of equations gives the evolution of non–gravitational fields. Within the framework of metric theories, these laws depend only on the metric: this is a consequence of the Einstein Equivalence Principle and the so-called "minimal coupling". In most theories, including General Relativity, the second set of equations is derived from the first one, which is the only fundamental one; however, in many situations, the two sets are decoupled.

The gravitational field studied in these approaches (without cosmological considerations) is mainly due to the Sun and the Eddington-Robertson expansion gives the corresponding metric. Assuming spherical symmetry and a static gravitational field, one can prove that there exists a coordinate system (called *isotropic*) such as

$$ds^2 = B(r)\,dt^2 - A(r)\,r^2 - (dr^2 \sin^2\theta\, d\phi^2 + d\theta^2)\,. \tag{1.12}$$

dt being the proper time between two neighboring events.

The Newtonian gravitational field does not exceed $G_N M_\odot / R_\odot c^2 \sim 2\times 10^{-6}$, where c is the speed of light, $M_\odot$ is the mass of the Sun and $R_\odot$ its radius. The metric is quasi-Minkowskian, $A(r)$ and $B(r)$ are dimensionless functions which depend only on G_N, M, c and r. Indeed, the only pure number that can be built with these four quantities is $G_N M/rc^2$. The Eddington-Robertson metric is a Taylor expansion of A and B which gives

$$\begin{aligned} ds^2 &= \left(1 - 2\alpha\frac{G_N M}{rc^2} + 2\beta\left(\frac{G_N M}{rc^2}\right)^2 + \cdots\right) dt^2 \\ &- \left(1 - 2\gamma\frac{G_N M}{rc^2} + \cdots\right) r^2\,(dr^2 \sin^2\theta\, d\phi^2 + d\theta^2)\,. \end{aligned} \tag{1.13}$$

The coefficients α, β, γ are called the post Newtonian parameters and their values depend on the considered theory of gravity: in GR, one has $\alpha = \beta = \gamma = 1$. The post Newtonian parameters can be measured not only in the Solar System but also in relativistic binary neutron stars such as *PSR* $1913+16$ [229].

A generalization of the previous formalism is known as *Parametrized Post-Newtonian* (PPN) formalism. Comparing metric theories of gravity among them and with experimental results becomes particularly easy if the PPN approximation is used. The following requirements are needed:

- particles are moving slowly with respect to the speed of light;
- gravitational field is weak and considered as a perturbation of the flat space-time;
- gravitational field is also static, *i.e.* it does not change with time.

The PPN limit of metric theories of gravity is characterized by a set of 10 real-valued parameters; each metric theory of gravity corresponds to particular values of PPN parameters. The PPN framework has been used first for the analysis of Solar System gravitational experiments, then for the definition and the analysis of new tests of gravitational theory and finally for the analysis and the classification of alternative metric theories of gravity.

By the middle 1970's, the Solar System was no more considered as the unique testing ground of gravitation theories. Many alternative theories of gravity agree with GR in the Post-Newtonian limit and thus with Solar System experiments; nevertheless, they do not agree for other predictions (such as cosmology, neutron stars, black holes and gravitational radiation) for which the post-Newtonian limit is not adequate. In addition, the possibility that experimental probes, such as gravitational radiation detectors, would be available in the future to perform extra-Solar System tests led to the abandon of the Solar System as the only arena to test gravitational theories.

The study of the binary pulsar $PSR\ 1913+16$, discovered by R. Hulse and J. Taylor [229], showed that this system combines large post-Newtonian gravitational effects, highly relativistic gravitational fields (associated with the pulsar) with the evidence of an emission of gravitational radiation (by the binary system itself). Relativistic gravitational effects allowed one to do accurate measurements of astrophysical parameters of the system, such as the mass of the two neutron stars. The measurement of the orbital period rate of change agreed with the prediction of the gravitational waves emission in the General Relativity framework, in contradiction with the predictions of most alternative theories, even those with PPN limits identical to General Relativity. However, the evidence was not conclusive to rule out other theories since several shortcomings remain, up to now, unexplained.

1.2.3. Mach's Principle and *G* Variability

Following Bondi [59, 81], there are two, at least in principle, completely different ways for measuring the Earth rotation velocity. The first is a purely terrestrial experiment (e.g. Foucault pendulum), the second is to perform such a measurement astronomically, that is to measure the Earth rotation with respect to the "fixed stars". In the first type experiment, the Earth motion is referred to an idealized inertial frame in which Newton's laws are verified. In the second, the frame of reference is connected to some matter distribution surrounding the Earth, and its motion is considered relative to such a matter distribution. In this way, we are facing the problem concerning Mach principle, which essentially states that the local inertial frame is determined by some average of the motion of the distant astronomical objects. ([59], see also [367,368])[2]. Trying to incorporate the Mach principle into (metric)

[2] A very interesting discussion on this topic, also connected with different theories of space, both in philosophy and in physics, is found in R.H. Dicke "The many faces of Mach", in Gravitation and Relativity" [162,163]. In such a discussion is presented also the problematic position that Einstein had on Mach principle.

gravity theory, Brans and Dicke have constructed a theory of gravity alternative to the standard (Einsteinian) one [61]. Taking into account the influence that the total matter has on each point (then constructing the "inertia"), these two authors introduced a scalar field as a new gravitational variable together with the standard metric tensor (this is the reason why such theory is called scalar–tensor theory; actually, this kind of theories were already proposed years before by Jordan, Fierz and Thiery [47]). An important consequence of such an approach is that the gravitational "constant" is variable, being a function of the total mass distribution, that is of the scalar field. The gravity, in such a picture, is described by the Lagrangian density

$$\mathcal{L} = \sqrt{-g}\left[\varphi R - \omega\left(\frac{\varphi_{\mu}\varphi^{\mu}}{\varphi}\right) + \mathcal{L}_m\right], \tag{1.14}$$

where ω is a dimensionless parameter and $\mathcal{L}_m$ is the Lagrangian of the matter including all the nongravitational fields.

As it has been stressed by Dicke [162, 163], the Lagrangian (1.14) has a property that is very close to higher order theories of gravity as we will discuss in detail below.

Let us start considering the Lagrangian (1.14) and perform the conformal transformation $g_{\mu\nu} \rightarrow \bar{g}_{\mu\nu} = \lambda g_{\mu\nu}$, being λ proportional to the scalar field φ, i.e. $\lambda = G_0\varphi$. Lagrangian (1.14) is transformed into:

$$\bar{\mathcal{L}} = \sqrt{-g}\left(\bar{R} + G_0\bar{\mathcal{L}}_m + G_0\bar{\mathcal{L}}_{\lambda}\right), \tag{1.15}$$

where

$$\bar{\mathcal{L}}_{\lambda} = -\frac{(3+2\omega)}{4\pi G_0\lambda^2}\lambda^{;\mu}\lambda_{,\mu}, \tag{1.16}$$

and $\bar{\mathcal{L}}_m$ is the conformally transformed Lagrangian density of the matter. We see that in this way it has been introduced a sort of total matter density Lagrangian $\bar{\mathcal{L}}_{tot} = \bar{\mathcal{L}}_m + \bar{\mathcal{L}}_{\lambda}$. Einstein equations are written as:

$$\bar{R}_{\mu\nu} - \frac{1}{2}\bar{g}_{\mu\nu}\bar{R} = G_0\bar{\tau}_{\mu\nu}, \tag{1.17}$$

where the stress-energy tensor is the sum of two contributions:

$$\bar{\tau}_{\mu\nu} = T_{\mu\nu}^{(matter)} + \Lambda_{\mu\nu}(\lambda^{(field)}). \tag{1.18}$$

In the same paper where Dicke has presented the results just reported, it is also stressed that this new form of the scalar–tensor gravity theory has some advantages with respect to the theory expressed in the previous untrasformed form; actually the new one is easier to handle, being similar to the Einstein standard description (Einstein's frame); but Dicke has also stressed that, in this new form, his model gives rise to some unpleasant features. If we consider, for example, the motion of a spinless, uncharged massive particle, we get that in the transformed model the trajectory is no more geodesic, only the photons (null) trajectories are unchanged under the above transformation. In fact, as stressed by Dicke, this is connected with the fact that the rest mass is not constant in the (conformal) transformed picture, the photons trajectories are then unaffected, but the massive particle equation of motion are modified.

Such a new approach to gravitation has increased the relevance of theories in which gravitational "constant" is considered variable. They are of particular interest in cosmology since, as we will discuss in much more details in the next sections, they allow to solve many shortcomings of the Standard Cosmological Model. We give a list of Lagrangians describing models of this type which will be relevant for our next analysis.

- String theory (low energy limit [202], [409,410], [359]):

$$\mathcal{L} = \sqrt{-g}e^{-2\phi}\left(R + 4g^{\mu\nu}\phi_\mu\phi_\nu - \Lambda\right) . \tag{1.19}$$

- General scalar-tensor Lagrangian, with completely unspecified gravitational coupling and potential:

$$\mathcal{L} = \sqrt{-g}\left(\phi R + \frac{1}{2}g^{\mu\nu}\phi_\mu\phi_\nu - V(\phi)\right) . \tag{1.20}$$

  An example is the Callan-Coleman-Jackiw Lagrangian [74] in which $\phi = 1/2 - \phi^2/12$.

- Conformal coupled theory [56, 196], [330];

$$\mathcal{L} = \sqrt{-g}\left(-\frac{\phi^2}{6}R + \frac{1}{2}g^{\mu\nu}\phi_\mu\phi_\nu - V(\phi)\right) ; \tag{1.21}$$

  where $V(\phi)$ is generic.

All these theories present a nonconstant gravitational coupling. The standard Newton gravitational constant is replaced by an effective coupling

$$G_{eff} = -\frac{1}{2\phi} , \tag{1.22}$$

which is, in general, different from the constant G_N (we use ϕ as the generic function describing coupling): in string theory or in conformal theory, such functions are specified in (1.19) and (1.21). In particular, in the cosmological scenarios, the coupling is a function of the "epoch" that is of the cosmological time.

We stress that all these scalar–tensor theories of gravity do not satisfy Strong Equivalence Principle. The reason is the above mentioned feature: the variation of G_{eff} implies that local gravitational physics depends (via ϕ) on the scalar field. We have answered, then, to the question of why introducing a stronger form of equivalence principle like Strong Equivalence Principle. Anyway, theories which have such a peculiar aspect are called *non-minimally coupled theories*.

Let us consider, as in (1.20), a general scalar-tensor theory in presence of "standard" matter (see, for example, [147, 149]). That is we consider the total Lagrangian (density) $\phi R + \mathcal{L}_\phi + \mathcal{L}_m$, where $\mathcal{L}_m$ describes "standard" matter. The dynamical equations of such matter are given by $T^{\nu}_{\mu;\nu}(m) = 0$, where $T_{\mu\nu}(m)$ is the matter stress-energy tensor, which is derived from the total Lagrangian, varying it with respect to the metric. In other words: concerning the standard matter, everything goes as in General Relativity (i.e. $\eta_{\mu\nu} \to g_{\mu\nu}$, $\partial_\mu \to \nabla_\mu$), that is everything follows the minimal coupling prescriptions. What is new in these theories is the way in which the scalar part and the metric part (that is the two ingredients describing gravity) are present: now there is a coupling between the scalar part with a function of the tensor part (the metric) and its derivative (actually with the curvature scalar connected with the metric, that is with $R = R^{\alpha}_{\beta\alpha\gamma} g^{\beta\gamma} = R(g, \partial g, \partial^2 g)$). As we already said, this so called nonminimal coupling describes a variable gravitational "constant". This is what we mean with the expression "nonminimally coupled theories" of gravity. Then we have to face two different physical possibilities (still continuing to confine our analysis to the cosmological arena):

- the first is

$$G_{eff}(\phi(t))_{t\to\infty} \longrightarrow G_N\,; \tag{1.23}$$

  this is the case in which the standard relativity (actually, cosmology) is recovered nowadays.

- The second case concerns the possibility that the gravitational coupling is not constant at the present days, that is G_{eff} is still varying with the "epoch", that is $\dot{G}_{eff}/G_{eff}|$now (in brief $\dot{G}/G$) is not zero.

In many theories of gravity, then, it is perfectly conceivable the G-variability with time: in most of them, it is not even obtained the today observed constant value. What we know about this variability from an observational point of view? Essentially there are three main sources for the G-variability analysis: the first are the observations called *Lunar-Laser-Ranging*, the second is connected with the information we can get studying the sun, the third source are the data coming from binary pulsars. The Lunar-Laser-Ranging consists in measuring the round trip travel time and thus the distances between trasmitter and reflector. The change of round trip time contains many informations on the Earth-Moon system. This round trip travel time has been challenged for many years (more than 25, in connection with Apollo 11, 14, 15 and Lunakhod 2 lunar explorations). Combining these data with those coming from the sun evolution and the Earth-Mars radar ranging, the more update estimation of $\dot{G}/G$ ranges from 0.4×10^{-11} to 1.0×10^{-11} per year (see for example [165]). The third source of information on G–variability is given by the binary pulsars systems. In order to extract data from this type of system (the prototype is the binary pulsar PSR 1913+16, see [402]), it has been necessary to extend the post–Newtonian approximation, which can be applied only in presence of a weakly gravitationally interacting n-body system, to strong gravitationally interacting systems. From these strongly interacting systems, the estimation of $\dot{G}/G$ is 2×10^{-11} per year (see [148], [165]).

We have not presented the specific contents of the models we are going to study: we will do this in the forthcoming sections. We want here just to give an idea of some general features that generalized theories of gravity have in connection with GR. In conclusion, it seems reasonable to enlarge GR to more general schemes, since in this way it is possible to explain several theoretical and observational facts, which, otherways, are misleading. As we shall see below, cosmology is the field of many fruitful applications of these generalizations of Einstein's gravity.

1.2.4. Gravity at High and Low Energies

Many people will agree that modern physics is based on two main pillars: General Relativity and Quantum Field Theory. Each of these two theories has been very successful in its own arena of physical phenomena: General Relativity in describing gravitating systems and non-inertial frames from a classical point of view or on large enough scales, and Quantum Field Theory in revealing the mysteries of high energy or small scale regimes where a classical description breaks down. However, Quantum Field Theory assumes that space-time is flat and even its extensions, such as Quantum Field Theory in curved space time, consider space-time as a rigid arena inhabited by quantum fields. General Relativity, on the other hand, does not take into account the quantum nature of matter. Therefore, it comes naturally to ask what happens if a strong gravitational field is present at small, essentially quantum, scales? How do quantum fields behave in the presence of gravity? To what extent are these amazing theories compatible?

Let us try to pose the problem more rigorously. Firstly, what needs to be clarified is that there is no precise proof that gravity should have some quantum representation at high energies or small scales, or even that it will retain its nature as an interaction. The gravitational interaction is so weak compared with other interactions that the characteristic scale under which one would expect to experience non-classical effects relevant to gravity, the Planck scale, is 10^{-33} cm. Such a scale is not of course accessible by any current experiment and it is doubtful whether it will ever be accessible to future experiments either[3]. However, there are a number of reasons for which one would prefer to fit together General Relativity and Quantum Field Theory [62, 231]. Let us list some of the most prominent ones here and leave the discussion about how to address them for the next section.

1.3. The Problem of Quantum Gravity

1.3.1. Intrinsic Limits in General Relativity and Quantum Field Theory

The predictions of a theory can place limits on the extent of its ability to describe the physical world. General Relativity is believed by some one to be no exception to this rule. Surprisingly, this problem is related to one of the most standard processes in a gravitational theory: gravitational collapse. Studying gravitational collapse is not easy since generating solutions to Einstein's field equations can be a tedious procedure. We only have a few exact solutions to hand and numerical or approximate solutions are often the only resort. However, fortunately, this does not prevent one from making general arguments about the ultimate fate of a collapsing object.

This was made possible after the proof of the Penrose–Hawking singularity theorems [208, 209]. These theorems state that a generic space-time cannot remain regular beyond a finite proper time, since gravitational collapse (or time reversal of cosmological expansion)

[3] This fact does not imply, of course, that imprints of Quantum Gravity phenomenology cannot be found in lower energy experiments.

will inevitably lead to space-time singularities. In a strict interpretation, the presence of a singularity is inferred by geodesic incompleteness, i.e. the inability of an observer travelling along a geodesic to extend this geodesic for an infinite time as measured by his clock. In practical terms this can be loosely interpreted to mean that an observer free-falling in a gravitational field will "hit" a singularity in a finite time and Einstein's equations cannot then predict what happens next. Such singularities seem to be present in the centre of black holes. In the Big Bang scenario, the Universe itself emerges out of such a singularity.

Wheeler has compared the problem of gravitational collapse in General Relativity with the collapse of the classical Rutherford atom due to radiation [433]. This analogy raises hopes that principles of quantum mechanics may resolve the problem of singularities in General Relativity, as happened for the Rutherford model. In a more general perspective, it is reasonable to hope that quantization can help to overcome these intrinsic limits of General Relativity.

On the other hand, it is not only General Relativity that has an intrinsic limit. Quantum Field Theory presents disturbing ultraviolet divergences. Such divergences, caused by the fact that integrals corresponding to the Feynman diagrams diverge due to very high energy contributions — hence the name "ultraviolet" — are discretely removed by a process called renormalization. These divergences are attributed to the perturbative nature of the quantization process and the renormalization procedure is somehow unappealing and probably not so fundamental, since it appears to cure them in a way that can easily be considered as non-rigorous from a mathematical point of view. A non-perturbative approach is believed to be free of such divergences and there is hope that Quantum Gravity may allow that (for early results see [157, 159, 232, 245, 324]).

1.3.2. The Perturbative Covariant Approach and the Canonical Approach

One of the main issues of modern physics is to construct a theory able to describe the fundamental interactions of Nature as different aspects of that theory. This goal has led, in the past decades, to the formulation of several unification schemes which, *inter alia*, try to describe gravity on the same ground of the other interactions.

All these schemes describe the fundamental fields in terms of the conceptual apparatus of Quantum Mechanics. It is based on the fact that the states of systems are described by vectors in Hilbert space H and the physical fields are represented by linear operator defined in H. Till now, any attempt to incorporate gravity in this scheme is failed and has not led to satisfying results. The main conceptual problem arises because the gravitational field describes, at the same time, the dynamical degrees of freedom of gravity and the background space-time.

Owing to the difficulties to build up a self-consistent theory of unification of interactions and particles, during last fifthy years, the two fundamental theories of modern physics, General Relativity and Quantum Mechanics, have been critically re-analyzed. From one side, one supposes that the material fields (bosons and fermions) come out from superstructures (i.e., Higgs bosons or superstrings) that passing through phase transitions have generated the known particles. On the other hand, one supposes that the geometry (e.g., the Ricci scalar) interacts with quantum material fields and generates back-reactions. This implies to modify the standard gravitational theory, i.e. Lagrangian of effective fields is connected with respect to the Hilbert-Einstein one, and lead to *Generalized Theories of Gravitation*. Such theories take into account non–trivial terms, as, for example, higher order terms in the invariant of curvature and non–minimal coupling between matter and gravity.

The modifications of the standard gravity imply inflationary scenarios that are of remarkable interest from a cosmological point of view. In any case, the condition in order that such theories are physically acceptable is that one has to recover General Relativity in the weak energy limit.

Despite the remarkable conceptual progresses due to the introduction of Generalized Theories of Gravitation, the mathematical difficulties are highly increased. The correcting terms into Lagrangian increase the (intrinsic) non–linearity of the Einstein equations, making them more difficult to deal with because one gets differential equations of higher order than the second and the non-separability of the geometric and material degrees of freedom. To overcome these difficulties, one tries to look for symmetries of Lagrangian and selects the conserved quantities in order to simplify the equations of motion. In such way, the exact solutions of the dynamics can be derived. The main point to realize this program is to pass from Lagrangian of fields to point-like Lagrangian, that is to pass from infinite degrees of freedom to finite degrees of freedom as cosmology.

The necessity to build up a quantum theory of gravity goes back to the end of the 1950's, when physicists tried, to deal with all interactions at a fundamental level and describe them in terms of Quantum Field Theory.

The original idea to quantize gravity was to use the *canonical approach* and the *covariant approach*, that had been applied with remarkable success to the Electromagnetism.

In the first approach, one takes into account the electric and magnetic fields that satisfies the Heisenberg uncertainty principle and the quantum states are invariant gauge functional generated by vector potential defined on 3–surfaces.

In the second approach, one quantizes two degrees of freedom of Maxwell field, without any decomposition (3+1) of metric, and the quantum states are elements of Fock space [238].

These procedure are in principle equivalent. The former allows to well define the measure, whereas the latter is more convenient for perturbative calculations (as the S matrix in Quantum Electrodynamics).

Such methods have been applied to General Relativity, but many difficulties arose due to the fact that it cannot be formulate as a quantum filed theory on a Minkowski space-time. To be more specific, *in General Relativity, we do not have a geometry, given a priori, for the background space-time*. Space-time is the dynamical variable itself. So that, in order to introduce the notion of causality, time and evolution, one has to solve at first the equations of motion and then to obtain the space-time as a result. For example, in order to know whether a particular initial condition gives rise to a black hole, it is necessary to derive its evolution by solving the Einstein equations. Then, taking into account the casual structure induced by such a solution, one has to study the asymptotic metric in the future in order to see now it is related, in the past, to the initial conditions. This problem become not handling in the Quantum Theory. Due to the uncertainty principle, in non–relativistic quantum mechanics, particles do not move along well defined trajectories. We can only calculate the probability amplitude $\psi(x,t)$ to realize a measurement at time t, in the point x of space-time. Similarly, in Quantum Gravity, the evolution of a initial state do not provide a specific space-time. In absence of a space-time, how is it possible to introduce the basic concepts as causality, time, matrix elements of scattering or black holes?

The canonical and covariant approaches give different answers at these problems. In the canonical approach it is used the Hamiltonian formulation of General Relativity to achieve the canonical quantization. The canonical commutation relations are the same that lead to the Uncertainty Principle. In fact, it is required the commutation of some operators on a spatial three–manifold, fixed for satisfy the notion of causality. In the limit of asymptotically flat space–time, the motion generated by the Hamiltonian has to be interpreted as temporal evolution (in other words, when the background tends to Minkowski space-time, the Hamiltonian operator takes again its role of generator of the translations). The canonical approach preserves the geometric feature of General Relativity without introducing perturbative methods [433].

The covariant approach uses the technique of Quantum Field Theory. The basic idea is that the above mentioned problems can be easily solved by splitting the metric $g_{\mu\nu}$ in a kinematics part $\eta_{\mu\nu}$, usually flat, and a dynamics part $h_{\mu\nu}$

$$g_{\mu\nu} = \eta_{\mu\nu} + h_{\mu\nu}\,.$$

The geometry of the background is given by the flat metric tensor, which allows to define the concepts of causality, time and *scattering*. The quantization procedure is applied to the dynamical field considered as a deviation from the background. Quanta are particles with spin 2, called *gravitons*, which propagates in space–time defined by $h_{\mu\nu}$. Substituting $g_{\mu\nu}$ into the Hilbert-Einstein action, it follows that the Lagrangian of gravity contains a sum whose terms represent, at different orders of approximation, the interaction of gravitons (and eventually terms of interaction matter–graviton, if matter is present). Such terms are analyzed by using the technique of the perturbative theory.

During the sixty and seventy years, both these program of quantization have been followed. Arnowitt, Deser and Misner, using the methods proposed by Dirac and Bergamann, derived the Hamiltonian formulation of the General Relativity in the canonical approaches. The canonical variables are the three-metric on the spatial sub-manifolds, obtained by the foliation of the four-dimensional manifolds. Einstein equation give the constrains between the three–metric and their conjugate momenta, and the equation of evolution of these fields (the Wheeler-DeWitt equation). In this way, General Relativity can be interpreted as the dynamical theory of the three–geometries (*geometrodynamics*). The difficulties arising from this approach are that the quantum equations imply products of operators defined in the same point of the space–time and, in addition, they imply the construction of distributions whose physical meaning is not clear. However, the main problem is the absence of a Hilbert space and, as consequence, the absence of a probabilistic interpretation of the calculated quantities.

The covariant quantization of gravity is more close to the physics of particles and fields, in the sense that it has been possible to extend the perturbative methods of Quantum Electrodynamics to the gravitation. This has allowed to analyze the reciprocal interactions between gravitons and between graviton and matter. The formulation of Feynman rules for gravitons and the demonstration that the theory might be unitary at every order in the expansion, was achieved by DeWitt.

A further progress has been pursued by the Yang-Mills theories, which describe the strong, weak and electromagnetic interactions of quarks and leptons by means of the symmetries. Such theories are renormalizable because it is possible to gives masses to the fermions owing to the spontaneous symmetry breaking. The natural attempt is then to consider gravitation as a Yang–Mills Theory (in the covariant perturbative approach) and check if it is renormalizable. But gravity does not satisfy this scheme. It turns out to be non–renormalizable when one considers the interaction graviton–graviton (diagrams at two–loops), and interaction graviton–matter (diagrams at one–loop)[4]. The covariant method allows to construct a theory of gravitation renormalizable at one–loop in the perturbative series [56, 196].

Due to the non-renormalizability of gravity at different orders, its validity is restricted only at low energy, i.e. at large scales, while it fails at high energy, i.e. at small scales. This implies that the *correct* theory of gravity has to be invoked at the Planck era and far from this epoch General Relativity (and one-loop corrections) are enough to describe the gravitational interactions. In this contest it make sense to add higher order terms to the Hilbert-Einstein action. Besides, if the constants are opportunely chosen, the theory has a better ultraviolet behaviour and it is asymptotically free. Nevertheless, the Hamiltonian of these theories is not limited from below and they result unstable. In particular, the unitarity is violated and the probability is not preserved.

An alternative approach to construct a self-consistent quantum gravity comes from the Electro-Weak interaction. In all these approaches, gravity is treated without taking into account the other fundamental interactions. The unification of the Electromagnetism with the weak interaction suggests that a consistent theory can be obtained when gravity is coupled to some kind of matter. This is the basic idea of *Supergravity* theory. In this theory, the divergences due to the bosons (gravitons in this case) are cancelled out by these ones due to the fermions, leading to a renormalized theory of gravity. Unfortunately, this scheme works only up to two-loop order and also for coupling between matter and gravity. Besides, Hamiltonian is defined positive and the theory results unitary. For higher order loops, the renormalizabilty of the theory cannot be achieved.

The perturbative methods are also used in string theories. In this case, the approach is completely different with respect to the previous ones because the concept of particle is replaced by that one of extended object. The standard usual physical particles, including the graviton with spin 2, are due to the excitations of string. The theory has only a free parameter and the interaction couplings are uniquely determined. It follows that string theory includes all fundamental physics and it is considered the *Theory of Everything*. String theory seems to be unitary and the perturbative series converges implying finite terms. This property follows from the fact that strings are objects intrinsically extended, so that ultraviolet divergencies, coming from small distances, or from large transferred momenta, are naturally cured. In other words, the natural cut–off is given by their length, that is of the Planck length l_P. At scale greater than l_P, the effective string action can be rewritten in terms of non–massive vibrational modes, i.e. in terms of scalar and tensorial fields (*tree–level ef-*

[4]Higher order terms in the perturbative series imply an infinite number of undetermined parameters. At one-loop level, it is enough to renormalize only the effective constants G_{eff} and Λ_{eff}, which, at low energy, reduces to Newton constant G_N and the cosmological constant Λ.

fective action). This eventuality leads to an effective theory of gravitation non–minimally coupled with a scalar fields, the so–called *dilaton*).

To conclude, let us resume the previous considerations:

- a consistent theory of gravity, i.e. unitary and renormalizable, does not yet exists;
- in the program to quantize gravity two approaches can be used: the covariant approach and the perturbative approach. Moreover, these approaches does not lead to self-consistent theory of Quantum Gravity,
- in the regime of low energy with respect to the Planck energy (at large scales) General Relativity can be generalized by introducing into the Hilbert-Einstein action terms of higher orders in the curvature invariants or non-minimal couplings between matter fields and gravity. They lead, at least at one-loop, to self-consistent and renormalizable theory.

1.3.3. A Conceptual Clash: Disagreement between the Approaches

Every theory of physics is based, as we know, on a series of conceptual assumption and General Relativity and Quantum Field Theory are no exceptions. On the other hand, the two theories to work in a complementary way, between their conceptual bases. This is not necessarily the case here.

There are two main points of tension between General Relativity and Quantum Field Theory. The first has to do with the concept of time: Time is given and is not a dynamical quantity in Quantum Field Theory; this assumption is closely related to the fact that space-time is considered a fixed arena where phenomena take place, much like the Newtonian mechanics. On the other hand, General Relativity considers space-time as being dynamical, with time alone not being such a relevant concept. It is of a theory describing relations between different events in space-time than a theory that describes evolution over some running parameter. One could go further and seek for the connection between what is mentioned here and the differences between gauge invariance as a symmetry of Quantum Field Theory and diffeomorphism invariance as a symmetry of General Relativity.

The second conceptual issue has to do with Heisenberg's uncertainty principle in Quantum Theory which is absent in General Relativity, being a classical theory. It is interesting to note that General Relativity, a theory in which background independence is a key concept, actually introduces space-time as an exact and fully detailed record of the past, the present and the future. Everything would be fixed for a super-observer that could look at this 4-dimensional space from a fifth dimension. On the other hand, Quantum Field Theory, a background dependent theory, manages to include a degree of uncertainty for the position of any event in space-time.

Having a precise mathematical structure for a physical theory is always important, but getting answers to conceptual issues is always the main motivation for studying physics in the first place. Trying to attain a quantum theory of gravity could lead to such answers.

1.3.4. Matching General Relativity with Quantum Fields: The Semiclassical Limit Approach

The above considerations can be partially realized by taking into account the Quantum Field Theory in curved space-time. Specifically at high temperature, a description of the matter as a perfect fluid of particles is not adequate: a more accurate description requires Quantum Field Theory formulated in the framework of General Relativity. Since at scales greater than Planck scale, matter has to be treated with quantum theory, one can assume a semiclassical description of gravitation where Einstein equations are of the form

$$G_{\mu\nu} = R_{\mu\nu} - \frac{1}{2}g_{\mu\nu}R = < T_{\mu\nu} > , \tag{1.24}$$

On the left hand side (l.h.s.) there is the standard Einstein tensor and on right hand side (r.h.s.) there is the expectation value of the quantum stress-energy tensor, source of gravitational field. More precisely, if $|\psi>$ is the quantum state describing the early universe, then

$$< T_{\mu\nu} > = < \psi | T^{op}_{\mu\nu} | \psi > , \tag{1.25}$$

where $T^{op}_{\mu\nu}$ is the quantum representation of the energy-momentum tensor. In general, the matter field $\hat{\phi}$ is subject to self-interactions, it interacts with other fields and with the gravitational background. Such terms are included in the definition of an effective potential[5] given by

$$V_{eff}(\phi) = < a | \hat{\mathcal{H}} | a > , \tag{1.26}$$

with

$$\phi = < a | \hat{\phi} | a > , \tag{1.27}$$

where $|a>$ represents normalized states of the theory under consideration, i.e. $< a|a > = 1$. $\hat{\mathcal{H}}$ is the Hamiltonian operator. It satisfies the condition $\delta < a|\hat{\mathcal{H}}|a > = 0$, where δ is the variation on the average of $\mathcal{H}$ states. This condition corresponds to the energy conservation.

In a curved space-time, also in absence of classical matter and radiation, quantum fluctuations of matter fields give to non vanishing contributions $< T_{\mu\nu} >$. The effect is similar to the vacuum polarization of Quantum Electrodynamics [56, 196, 260]. These corrections assume the form

$$< T_{\mu\nu} > = k_1\,{}^{(1)}H_{\mu\nu} + k_3\,{}^{(3)}H_{\mu\nu} , \tag{1.28}$$

when matter fields are free, massless and conformally invariants. k_1 and k_3 are numerical coefficients, while

$${}^{(1)}H_{\mu\nu} = 2R_{;\mu\nu} - 2g_{\mu\nu}\Box R + 2RR_{\mu\nu} - \frac{1}{2}g_{\mu\nu}R^2 , \tag{1.29}$$

$${}^{(3)}H_{\mu\nu} = R^{\sigma}_{\mu}R_{\nu\sigma} - \frac{2}{3}RR_{\mu\nu} - \frac{1}{2}g_{\mu\nu}R^{\sigma\tau}R_{\sigma\tau} + \frac{1}{4}g_{\mu\nu}R^2 , \tag{1.30}$$

[5] Hereafter scalar fields and potential are understood as their effective values, obtained averaging on quantum states. In this sense, fields and classical potentials are nothing else but the expectation values of quantum fields and potentials.

are constructed by curvature invariants The divergence of the tensor ${}^{(1)}H_{\mu\nu}$ is zero

$$ {}^{(1)}H^{\nu}_{\mu;\nu} = 0 \,. \tag{1.31} $$

It can be obtained by varying the local action with the Lagrangian quadratic in R

$$ {}^{(1)}H_{\mu\nu} = \frac{2}{\sqrt{-g}} \frac{\delta}{\delta g_{\mu\nu}} d^4x \sqrt{-g} R^2 \,. \tag{1.32} $$

In order to remove the infinities coming from $< T_{\mu\nu} >$ and obtain a renormalizable theory, one has to introduce infinite counterterms in the Lagrangian of gravitation. One of these terms is $CR^2\sqrt{-g}$ where C is a parameter that diverges logarithmically[6]. Because one can add to C an arbitrary finite constant, the coefficient k_1 can assume any value. Its choice should be determined experimentally [56, 196].

The tensor ${}^{(3)}H_{\mu\nu}$ is conserved only in space-time conformally flat (for example, Friedmann-Robertson-Walker). It cannot be obtained by varying a local action. Finally, one has

$$ k_3 = \frac{1}{1440\pi^2} \left(N_0 + \frac{11}{2} N_{1/2} + 31 N_1 \right) , \tag{1.33} $$

where the coefficients N are due to the number of quantum fields with spin $0, 1/2, 1$. Vector fields contributes more to k_3 due to the multiplier numerical factor of N_1. These massless fields. as well as the spinorial ones, are described by equations conformally invariants and appear in $< T_{\mu\nu} >$ in the form (1.28).

The trace of the stress-energy tensor vanishes for conformally invariants classical fields, while, owing to the term in k_3, one finds that the expectation value of the tensor has non-vanishing trace. This fact generates the so-called *trace anomaly*.

Let us shortly discuss how the conformal anomalies are generated when the origin of the tensor $T_{\mu\nu}$ is not classical, i.e. when quantum field theories are formulated in curved space-time.

As we will see in more details later, if a theory is conformally invariant one can writes

[6]Eqs. (1.24) cannot be generated by a finite action, because of gravitational field should be completely renormalizable, i.e. it should be enough to eliminate a finite number of divergences for making gravitation similar to Quantum Electrodynamics. One can only construct a truncated Quantum Theory of Gravity. The expansion in *loops* is done in terms of $\hbar$, so the truncated theory at the $one-loop$ level contains all terms of the order $\hbar$. In this sense, it is the first quantum correction to General Relativity. It assumes that matter fields are *free* and, due to the equivalence principle, all forms of matter couple in the same way to gravity. It implies a *intrinsic* non-linearity of gravitation, so that the number of loops arise in order to take into account self-interactions or mutual-interactions among matter fields and gravitational filed. At one–loop level, divergences can be removed by renormalizing the cosmological constant Λ_{eff} and the gravitational constant G_{eff}. The contributions to $< T_{\mu\nu} >$ are the above quantities, ${}^{(1)}H_{\mu\nu}$ and ${}^{(3)}H_{\mu\nu}$. Besides, one has to consider

$$ {}^{(2)}H_{\mu\nu} = 2R^{\sigma}_{\alpha;\beta\sigma} - \Box R_{\mu\nu} - \frac{1}{2} g_{\mu\nu} \Box R + R^{\sigma}_{\mu} R_{\sigma\nu} - \frac{1}{2} R^{\sigma\tau} R_{\sigma\tau} g_{\mu\nu} \,. $$

It is shown [56, 196] that, in space-time conformally flat, the relation

$$ {}^{(2)}H_{\mu\nu} = \frac{1}{3} {}^{(1)}H_{\mu\nu} $$

holds. Hence, in Eq. (1.28), only the first and third terms of $H_{\mu\nu}$ are presents.

the relation

$$g_{\mu\nu}(x) \to \Omega^2(x)g_{\mu\nu}(x) = \tilde{g}_{\mu\nu}(x)\,. \tag{1.34}$$

It implies, for an action in $(n+1)$-dimensions, the following functional equation

$$\mathcal{A}\left[\tilde{g}_{\mu\nu}\right] = \mathcal{A}\left[g_{\mu\nu}\right] + \int \frac{\delta \mathcal{A}\left[\tilde{g}_{\mu\nu}\right]}{\delta \tilde{g}^{\rho\sigma}}\delta \tilde{g}^{\rho\sigma} d^{n+1}x\,, \tag{1.35}$$

where, putting

$$\delta \tilde{g}^{\mu\nu}(x) = -2\tilde{g}^{\mu\nu}(x)\Omega^{-1}(x)\delta\Omega(x)\,, \tag{1.36}$$

and using the classical variational principle

$$\frac{2}{\sqrt{-g}}\frac{\delta \mathcal{A}}{\delta g^{\mu\nu}} = T_{\mu\nu}\,, \tag{1.37}$$

one gets

$$\mathcal{A}\left[\tilde{g}_{\mu\nu}\right] = \mathcal{A}\left[g_{\mu\nu}\right] - \int \sqrt{-\tilde{g}}T^{\rho}_{\rho}(\tilde{g}_{\mu\nu})\Omega^{-1}\delta\Omega d^{n+1}x\,. \tag{1.38}$$

From it follows

$$T^{\rho}_{\rho}\left[g_{\mu\nu}(x)\right] = -\left.\frac{\Omega(x)}{\sqrt{-g}}\frac{\delta \mathcal{A}\left[\tilde{g}_{\mu\nu}\right]}{\delta\Omega(x)}\right|_{\Omega=1} \tag{1.39}$$

so that if classical action is invariant under conformal transformations, the trace of the stress-energy tensor is zero. At quantum level, this situation could not manifest for the following reasons. The conformal transformations are essentially, a re-scaling of the lengths in each point x of the space-time. The presence of masses in the theory, and hence of fixed length scales, breaks the conformal invariance, generating the trace anomaly. The requirement that the invariance be preserved forces to consider massless fields, as done in (1.28). In this case we get the condition

$$< T^{\rho}_{\rho} > \equiv 0\,, \tag{1.40}$$

that allows to recover a conformally invariant theory. It should be noted that gravity is not renormalizable in the usual way. Owing to this fact, divergences appear as soon as one considers quantum effects. Performing a loop expansion, one derives

$$< T^{\rho}_{\rho} > = < T^{\rho}_{\rho} >_{div} + < T^{\rho}_{\rho} >_{rin} = 0\,, \tag{1.41}$$

that confirms the validity of Eq. (1.40). In this case, the conformal invariance is preserved if the divergent tensor is equal (up to the minus) to the renormalized tensor. An anomalous trace term appears in Eqs. (1.24). In the *one-loop* case and in the field zero mass limit, it is given by

$$< T^{\rho}_{\rho} >_{div} = \left[\tilde{k}_1\left(M - \frac{2}{3}\Box R\right) + \tilde{k}_3 E\right] = - < T^{\rho}_{\rho} >_{rin}\,, \tag{1.42}$$

where $\tilde{k}_1$ and $\tilde{k}_3$ are proportional to k_1 and k_3, while M and E are obtained from ${}^{(1)}H_{\mu\nu}$ and ${}^{(3)}H_{\mu\nu}$, that is

$$M = R^{\alpha\beta\gamma\delta}R_{\alpha\beta\gamma\delta} - 2R^{\alpha\beta}R_{\alpha\beta} + \frac{1}{3}R^2\,, \tag{1.43}$$

$$E = R^{\alpha\beta\gamma\delta}R_{\alpha\beta\gamma\delta} - 4R^{\alpha\beta}R_{\alpha\beta} + R^2 . \tag{1.44}$$

E is the *Gauss – Bonnet* term. In four dimensions, the integral

$$d^4x\sqrt{-g}E , \tag{1.45}$$

is an invariant which provides information on the topology of the space-time manifold on which the theory is formulated. In Friedmann-Robertson-Walker metric, M vanishes identically, but E gives non-vanishing contributions, even if the variation of (1.45) is zero (in four dimensions).

In general, by summing up all geometrical terms deduced from the Riemann tensor and of the same order as $< T^{\rho}_{\rho} >_{rin}$, one derives the r.h.s. of (1.28). It can be expressed by means of (1.29) and (1.30), if metrics are conformally flat. Then one can conclude that the trace anomaly (due to the geometric terms) arises because the *one-loop* approach is an attempt to formulate quantum field theories on curved space-time[7].

The matter field masses and the interaction among fields can be neglected in the limit of high curvature because $R \gg m^2$. The matter-graviton interactions generate, in the effective Lagrangian, terms of non-minimal coupling. The *one-loop* contributions of such terms are comparable to the ones due to the trace anomaly and generate, from a conformal point of view, the same effects on gravitation.

The simplest effective Lagrangian that takes into account these corrections is

$$\mathcal{L} = \frac{1}{2}g^{\mu\nu}\phi_{;\mu}\phi_{;\nu} + \frac{1}{2}\xi R\phi^2 - V(\phi) , \tag{1.46}$$

where ξ is the coupling constant between scalar and gravitational fields. Also the stress-energy tensor will be modified, but one can find a conformal transformation such that the modifications due to curvature terms can be cast in the form of matter-curvature interaction. The same argument holds for the trace anomaly.

Some Grand Unification Theories lead to a polynomial coupling of the form $\phi = 1 + \xi\phi^2 + \zeta\phi^4$, generalizing that in (1.46), while an exponential-like coupling comes from the effective string-dilaton Lagrangian.

To conclude, any attempt to formulate Quantum Field Theory on curved space-time needs a modification of the Hilbert-Einstein action. It means to add terms in the curvature invariants or non-minimal couplings between matter and curvature. Their origin is due to the perturbative expansion. All these modifications, from a cosmological point of view imply inflationary scenarios, that is the possibilities to eliminate some of the shortcomings of the standard cosmological model.

1.3.5. Induced Gravity and Emergent Gravity

Induced Gravity (or Emergent Gravity) is an idea in quantum gravity that space-time background emerges as a mean field approximation of underlying microscopic degrees of free-

[7]It should be noted that (1.29) and (1.30) can include higher order terms than fourth order (i.e. R^2) in the derivative of metric if all possible Feynman diagrams are considered. For example, corrections like $R\Box R$ or $R^2\Box R$ can be present in ${}^{(3)}H_{\mu\nu}$ implying equations of motion of six order in the derivative. Also these terms can be treated by making use of conformal transformations [18, 48, 68]

dom, similar to the fluid mechanics approximation of BoseEinstein condensates. The concept was originally proposed by Andrei Sakharov in 1967. Sakharov observed that many condensed matter systems give rise to emergent phenomena which are identical to general relativity quantitatively. Crystal defects can look like torsion, for example. His idea was to start with an arbitrary background pseudo-Riemannian manifold (in modern treatments, possibly with torsion) and introduce quantum fields (matter) on it but not introduce any gravitational dynamics explicitly. This gives rise to an effective action which to one-loop order contains the Einstein-Hilbert action with a cosmological constant. In other words, General Relativity arises as an emergent property of matter fields and is not put in by hand. On the other hand, such models typically predict huge cosmological constants. The particular models proposed by Sakharov and others have been proven impossible by the Weinberg-Witten theorem. However, models with emergent gravity are possible as long as other things, such as space-time dimensions, emerge together with gravity. Developments in AdS/CFT correspondence after 1997 suggest that the microphysical degrees of freedom in induced gravity might be radically different. The bulk space-time arises as an emergent phenomenon of the quantum degrees of freedom that live in the boundary of the space-time.

1.3.6. Towards Quantum Gravity

We discussed some of prominent motivations for seeking high energy theory of gravity which would allow a matching between General Relativity and Quantum Field Theory. These triggered research in this direction at a very early stage and already in the 1950s serious efforts were made towards what is referred to as Quantum Gravity. Early attempts followed the conventional approach of trying to quantize the gravitational field in ways similar to the quantization of Electromagnetism, which resulted in Quantum Electrodynamics (QED). This led to influential papers about the canonical formulation of General Relativity [28, 29]. However, it was soon realized that the obvious quantization techniques could not work, since General Relativity is not renormalizable as is the case with Quantum Electrodynamics [444]. In simple terms, this means that if one attempts to treat gravity as another particle field and to assign a gravity particle to it (graviton) then the sum of the interactions of the graviton diverges. This would not be a problem if these divergences were few enough to be removable via the technique called renormalization and this is indeed what happens in Quantum Electrodynamics. Unfortunately, this is not the case for General Relativity and renormalization cannot lead to sensible and finite results.

It was later shown that a renormalizable gravitation theory — although not a unitary one — could be constructed, but only at the price of admitting corrections to General Relativity [394, 444]. Views on renormalization have changed since then and more modern ideas have been introduced such as the concept of effective field theories. These are approximate theories with the following characteristic: according to the length-scale, they take into account only the relevant degrees of freedom. Degrees of freedom which are only relevant to shorter length-scales and higher energies and are, therefore, responsible for divergences, are ignored. A systematic way to integrate out short-distance degrees of freedom is given by the renormalization group (see [156] for an introduction to these concepts).

In any case, quantizing gravity has proved to be a more difficult task than initially expected and quantum corrections appear, introducing deviations away from General Relativity [56,67,196,440]. Contemporary research is mainly focused on two directions: String Theory and Loop Quantum Gravity. Analysing the basis of either of these two approaches would go beyond the scope of this introduction and so we will only make a short mention of them. We refer the reader to [167,202,333] and [30,31,345,346,405] for text books and topical reviews in String Theory and Loop Quantum Gravity respectively.

String Theory attempts to explain fundamental physics and unify all interactions under the key assumption that the building blocks are not point particles but one dimensional objects called strings. There are five different versions of String Theory, namely Type I, Type IIA, Type IIB and two types of Heterotic String Theory. M-Theory is a proposed theory under development that attempts to unify all of the above types. A simplified version of the idea behind String Theory would be that its fundamental constituents, strings, vibrate at resonant frequencies. Different strings have different resonances and this is what determines their nature and results in the discrimination between different forces.

Loop Quantum Gravity follows a more direct approach to the quantization of gravity. It is close to the picture of canonical quantization and relies on a non-perturbative method called loop quantization. One of its main disadvantages is that it is not yet clear whether it can become a theory that can include the description of matter as well or whether it is just a quantum theory of gravitation.

It is worth mentioning that a common problem with these two approaches is that, at the moment, they do not make any experimentally testable predictions which are different from those already know from the standard model of particle physics. As far as gravity is concerned, String Theory appears to introduce deviations from General Relativity (see for example [197,309,439]), whereas, the classical limit of Loop Quantum Gravity is still under investigation.

1.4. Cosmological and Astrophysical Riddles

We started by discussing the possible shortcomings of General Relativity on very small scales, as those which appeared first in literature. However, if there is a scale for which gravity is of utmost importance, this is surely the cosmic scale. Given the fact that other interactions are short-range and that, at cosmological scales, we expect matter's characteristics, related to them, to be "averaged out" — for example we do not expect that the Universe has an overall charge — gravity should be the force which rules cosmic evolution. Let us see briefly how this comes about by considering Einstein's equations combined with our more obvious assumptions about the main characteristics of the observable Universe.

Even though matter is not equally distributed through space and by simple browsing through the sky, one can observe distinct structures such as stars and galaxies. If attention is focused on larger scales the Universe appears as if it was made by patching together multiple copies of the same pattern, i.e. a suitably large elementary volume around the Earth and another elementary volume of the same size elsewhere will have little difference. This suitable scale is actually $\approx 10^8$ light years, larger than the typical size of a cluster of galaxies. In Cosmology, one wants to deal with scales larger than that and to describe the Universe as a whole. Therefore, as far as Cosmology is concerned, the Universe can be very well described as homogeneous and isotropic.

To make the above statement useful from a quantitative point of view, we have to turn it into an idealized assumption about the matter and geometry of the Universe. Note that the Universe is assumed to be spatially homogeneous and isotropic at each instant of cosmic time. In more rigorous terms, we are talking about homogeneity on each one of a set of 3-dimensional space-like hypersurfaces. Concerning, we assume a perfect fluid description; these spacelike hypersurfaces are defined in terms of a family of fundamental observers who are comoving with this perfect fluid and who can synchronise their comoving clocks so as to measure the universal cosmic time. The matter content of the Universe is then just described by two parameters, a uniform density ρ and a uniform pressure p, as if the matter in stars and atoms is scattered through space. Concerning geometry we idealize the curvature of space to be everywhere the same.

Let us proceed by imposing these assumption on the equations describing gravity and very briefly review the dynamics of the Universe, namely the Friedmann equations. We refer the reader to standard textbooks for a more detailed discussion on the precise geometric definitions of homogeneity and isotropy and their implications the form of the metric (e.g. [429]). Einstein's equations have the form (1.7) and (1.8). Under the assumptions of homogeneity and isotropy, the metric is

$$ds^2 = dt^2 - a^2(t)\left[\frac{dr^2}{1-kr^2} + r^2 d\theta^2 + r^2\sin^2(\theta)d\phi^2\right], \tag{1.47}$$

known as the Friedmann-Lemaître-Robertson-Walker metric; $k = -1, 0, 1$ according to whether the Universe is hyperspherical ("closed"), spatially flat, or hyperbolic ("open") and $a(t)$ is called the scale factor[8]. Inserting this metric into Eq. (1.7) and taking into account a perfect fluid like (1.5) one gets the following equations

$$\left(\frac{\dot{a}}{a}\right)^2 = \frac{8\pi G_N \rho}{3} - \frac{k}{a^2}, \tag{1.48}$$

$$\frac{\ddot{a}}{a} = -\frac{4\pi G_N}{3}(\rho + 3p), \tag{1.49}$$

where an overdot denotes differentiation with respect to coordinate time t.

Eqs. (1.48) and (1.49) are called the Friedmann equations. By imposing homogeneity and isotropy as characteristics of the Universe that remain unchanged with time on suitably large scales, we have implicitly restricted any evolution to affect only one remaining characteristic: its size. This is the reason why the Friedmann equations are equations for the scale factor, $a(t)$, which is a measure of the evolution of the size of any length scale in the Universe. Eq. (1.48), being an equation in $\dot{a}$, tells us about the velocity of the expansion or contraction, whereas Eq. (1.49), which involves $\ddot{a}$, tells us about the acceleration of the expansion or the contraction. According to the Big Bang scenario, the Universe starts

[8] The traditional cosmological language of "closed, flat, open" is inaccurate and quite misleading and, therefore, should be avoided. Even if one ignores the possibility of nonstandard topologies, the $k = 0$ spatially flat 3-manifold is, in any sensible use of the word, "open". If one allows non-standard topologies, then there are, in any sensible use of the word, "closed" $k = 0$ spatially flat 3-manifolds (tori), and also "closed" $k = -1$ hyperbolic 3-manifolds. Finally the distinction between flat and spatially flat is important, and obscuring this distinction is dangerous.

expanding with some initial velocity. Setting aside the contribution of the k-term for the moment, Eq. (1.48) implies that the Universe will continue to expand as long as there is matter in it. Let us also take into consideration the contribution of the k-term, which measures the spatial curvature and in which k takes the values $-1, 0, 1$. If $k = 0$ the spatial part of the metric (1.47) reduces to a flat metric expressed in spherical coordinates. Therefore, the Universe is spatially flat and Eq. (1.48) implies that it has to become infinite, with ρ approaching zero, in order for the expansion to halt. On the other hand, if $k = 1$ the expansion can halt at a finite density at which the matter contribution is balanced by the k-term. Therefore, at a finite time the Universe will stop expanding and will re-collapse. Finally, for $k = -1$ one can see that even if matter is completely dissolved, the k-term will continue to "pump" the expansion which means that the latter can never halt and the Universe will expand forever.

Let us now focus on Eq. (1.49) which, as already mentioned, governs the acceleration of the expansion. Notice that k does not appear in this equation, i.e. the acceleration does not depend on the spatial curvature. Eq. (1.49) reveals what would be expected by simple intuition: that gravity is always an attractive force. Let us see this in detail. The Newtonian analogue of Eq. (1.49) would be

$$\frac{\ddot{a}}{a} = -\frac{4\pi G}{3}\rho, \tag{1.50}$$

where ρ denotes the matter density. Due to the minus sign on the right hand side and the positivity of the density, this equation implies that the expansion will always be slowed by gravity.

The presence of the pressure term in Eq. (1.49) is simply due to the fact that in General Relativity, it is not simply matter that gravitates but actually energy and therefore the pressure should be included. For what could be called ordinary matter (e.g. radiation, dust, perfect fluids, etc.) the pressure can be expected to be positive,as the density. More precisely, one could ask that the matter satisfies the four energy conditions [422]:

1. Null Energy Condition: $\rho + p \geq 0$,
2. Weak Energy Condition: $\rho \geq 0, \quad \rho + p \geq 0$,
3. Strong Energy Condition: $\rho + p \geq 0, \quad \rho + 3p \geq 0$,
4. Dominant Energy Condition: $\rho \geq |p|$.

We give these conditions here in terms of the components of the stress-energy tensor of a perfect fluid but they can be found in a more general form in [422]. Therefore, once positivity of the pressure or the validity of the Strong Energy Condition is assumed, gravity remains always an attractive force also in General Relativity [9].

To sum up, even without attempting to solve the Friedmann equations, we have already arrived to a well-established conclusion: Once we assume, according to the Big Bang scenario, that the Universe is expanding, then, according to General Relativity and with ordinary matter acting as source, this expansion should always be decelerated. Is this what actually happens though?

[9] When quantum effects are taken into account, one or more of the energy conditions can be violated, even though a suitably averaged version may still be satisfied. However, there are even classical fields that can violate the energy conditions, as we will see later on.

1.4.1. The First Need for Acceleration

We derived the Friedmann equations using two assumptions: homogeneity and isotropy of the Universe. Both assumptions seem very reasonable considering how the Universe appears to be today. However, there are always the questions of why does the Universe appear to be in this way and how did it arrive at its present state through its evolution. More importantly though, one has to consider whether the description of the Universe by the Big Bang model and the Friedmann equations is self-consistent and agrees not only with a rough picture of the Universe but also with the more precise current picture of it.

Let us put the problem in more rigorous terms. First of all one needs to clarify what is meant by "Universe". Given that the speed of light (and consequently of any signal carrying information) is finite and adopting the Big Bang scenario, not every region of space-time is accessible to us. The age of the Universe sets an upper limit for the largest distance from which a point in space may have received information. This is what is called a "particle horizon" and its size changes with time. What we refer to as the Universe is the part of the Universe causally connected to us – the part inside our particle horizon. What happens outside this region is inaccessible to us but more importantly, it does not affect us, at least not directly. However, it is possible to have two regions that are both accessible and causally connected to us, or to some other observer, but are not causally connected with each other. They just have to be "inside" our particle horizon without being inside other particle horizons. It is intuitive that regions that are causally connected can be homogeneous — they have had the time to interact. However, homogeneity of regions which are not causally connected would have to be attributed to some initial homogeneity of the Universe since local interactions cannot be effective for producing this.

The picture of the Universe that we observe is indeed homogeneous and isotropic on scales larger than we would expect based on our calculations regarding its age and causality. This problem was first posed in the late 1960s and has been known as the "horizon problem" [298, 429]. One could look to solve it by assuming that the Universe is perhaps much older and this is why, in the past, the horizon problem has also been reformulated in the form of a question: how did the Universe grow to be so old? However, this would require the age of the Universe to differ by orders of magnitude from the value estimated by observations. So the homogeneity of the Universe, at least at first sight and as long as we believe in the cosmological model at hand, appears to be built into the initial conditions.

Another problem, which is similar and appeared at the same time, is the "flatness" problem. To pose it rigorously let us return to the Friedmann equations and more specifically to Eq. (1.48). The Hubble parameter H is defined as $H = \dot{a}/a$. We can use it to define what is called the critical density

$$\rho_c = \frac{3H^2}{8\pi G}, \tag{1.51}$$

that is the density which would make the 3-geometry flat. Finally, we can use the critical density in order to create the dimensionless quantities

$$\Omega = \frac{\rho}{\rho_c}, \tag{1.52}$$

$$\Omega_k = -\frac{k}{a^2H^2}. \tag{1.53}$$

It is easy to verify from Eq. (1.48) that

$$\Omega + \Omega_k = 1. \tag{1.54}$$

As dimensionless quantities, Ω and Ω_k are measurable, and by the 1970s it was already known that the current value of Ω appears to be very close to 1 (see for example [164]). Extrapolating into the past reveals that Ω would have had to be even closer to 1, making the contribution of Ω_k, and consequently of the k-term in Eq. (1.48), exponentially small.

The name "flatness problem" can be slightly misleading and therefore it needs to be clarified that the value of k obviously remains unaffected by the evolution. To avoid misconceptions, it is therefore better to formulate the flatness problems in terms of Ω itself. The fact that Ω seems to be taking a value so close to the critical one, at early times, is not a consequence of the evolution and once more, as happened with the horizon problem, it appears as a strange coincidence which can only be attributed to some fine tuning of the initial conditions.

But is it reasonable to assume that the Universe started in such a homogeneous state, even at scales that where not causally connected, or that its density was dramatically close to its critical value without any apparent reason? Even if the Universe started with extremely small inhomogeneities, it would still not present such a homogeneous picture currently. Even if shortcomings like the horizon and flatness problems do not constitute logical inconsistencies of the standard cosmological Model but rather indicate that the present state of the Universe depends critically on some initial state, this is definitely a feature that many people consider undesirable.

So, by the 1970s, Cosmology was facing new challenges. Early attempts to address these problems involved implementing a recurring or oscillatory behaviour for the Universe and therefore were departing from the standard ideas of cosmological evolution [261, 262, 412]. This problem also triggered Charles W. Misner to propose the "Mixmaster Universe" (Bianchi type IX metric), in which a chaotic behaviour was supposed to ultimately lead to statistical homogeneity and isotropy [299]. However, all of these ideas have proved to be non-viable descriptions of the observed Universe.

A possible solution came in the early 1980s when Alan Guth proposed that a period of exponential expansion could be the answer [206]. The main idea is quite simple: an exponential increase of the scale factor $a(t)$ implies that the Hubble parameter H remains constant. On the other hand, one can define the Hubble radius $c/H(t)$ which, roughly speaking, is a measure of the radius of the observable Universe at a certain time t. Then, when $a(t)$ increases exponentially, the Hubble radius remains constant, whereas any physical length scale increases exponentially in size. This implies that in a short period of time, any length-scale which could, for example, be the distance between two initially causally connected observers, can become larger than the Hubble radius. So, if the Universe passed through a phase of very rapid expansion, then the part of it that we can observe today may have been significantly smaller at early times than what one would naively calculate using the Friedmann equations. If this period lasted long enough, then the observed Universe could have been small enough to be causally connected at the very early stage of its evolution. This rapid expansion would also drive Ω_k to zero and consequently Ω to 1 today, due to the very large value that the scale factor $a(t)$ would currently have, compared to its initial value.

Additionally, such a procedure is very efficient in smoothing out inhomogeneities, since the physical wavelength of a perturbation can rapidly grow to be larger than the Hubble radius. Thus, both of the problems mentioned above seem to be effectively addressed.

Guth was not the only person who proposed the idea of an accelerated phase and some will argue he was not even the first. Contemporaneously with him, Alexei Starobinsky had proposed that an exponential expansion could be triggered by quantum corrections to gravity and provide a mechanism to replace the initial singularity [84, 389]. There are also earlier proposals whose spirit is very similar to that of Guth, such as those by Demosthenes Kazanas [244], Katsuhiko Sato [358] and Robert Brout *et al.* [66]. However, Guth's name is the one most related with these idea since he was the first to provide a coherent and complete picture on how an exponential expansion could address the cosmological problems mentioned above. This period of accelerated expansion is known as "inflation", a terminology borrowed from economics due to the apparent similarity between the growth of the scale factor in Cosmology and the growth of prices during an inflationary period. To be more precise, one defines as inflation any period in the cosmic evolution for which

$$\ddot{a} > 0. \tag{1.55}$$

However, a more detailed discussion reveals that an exponential expansion, or at least quasi-exponential (since what is really needed is that the physical scales increase much more rapidly than the Hubble radius increases), is not something trivial to achieve. As discussed in the previous section, it does not appear to be easy to trigger such an era in the evolution of the Universe, since accelerated expansion seems impossible according to Eq. (1.49), as long as both the density and the pressure remain positive. In other words, satisfying Eq. (1.55) requires

$$(\rho + 3p) < 0 \Rightarrow \rho < -3p, \tag{1.56}$$

and assuming that the energy density cannot be negative, inflation can only be achieved if the overall pressure of the ideal fluid which we are using to describe the Universe, becomes negative. In more technical terms, Eq. (1.56) implies the violation of the Strong Energy Condition [422].

It does not seem possible for any kind of baryonic matter to satisfy Eq. (1.56), which directly implies that a period of accelerated expansion in the Universe evolution can only be achieved within the framework of General Relativity if some new form of matter field with special characteristics is introduced. Before presenting any scenario of this sort though, let us resort to observations to convince ourselves about whether such a cosmological era is indeed necessary.

1.4.2. Observations, Precision Cosmology and Cosmological Constant

In reviewing the early theoretical shortcomings of the Big Bang evolutionary model of the Universe we have seen indications for an inflationary era. The best way to confirm those indications is probably to resort to the observational data at hand for having a verification. Fortunately, there are currently very powerful and precise observations that allow us to look back to very early times.

A typical example of such observations is the Cosmic Microwave Background Radiation. In the early Universe, baryons, photons and electrons formed a hot plasma, in which the mean free path of a photon was very short due to constant interactions of the photons with the plasma through Thomson scattering. However, due to the expansion of the Universe and the subsequent decrease of temperature, it subsequently became energetically favourable for electrons to combine with protons to form hydrogen atoms (recombination). This allowed photons to travel freely through space. This decoupling of photons from matter is believed to have taken place at a redshift of $z \sim 1088$, when the age of the Universe was about 380,000 years old or approximately 13.7 billion years ago. The photons which left the last scattering surface at that time, then travelled freely through space and have continued cooling since then. In 1965 Penzias and Wilson noticed that a Dicke radiometer which they were intending to use for radio astronomy observations and satellite communication experiments had an excess 3.5K antenna temperature which they could not account for. They had, in fact, detected the Cosmic Microwave Background Radiation, which actually had already been theoretically predicted in 1948 by George Gamow. The measurement of the Cosmic Microwave Background Radiation, apart from giving Penzias and Wilson a Nobel prize publication [331], was also to become the number one verification of the Big Bang model.

Later measurements showed that the Cosmic Microwave Background Radiation has a black body spectrum corresponding to approximately 2.7 K and verifies the high degree of isotropy of the Universe. However, it was soon realized that attention should be focused not on the overall isotropy, but on the small anisotropies present in the Cosmic Microwave Background Radiation, which reveal density fluctuations [327, 399]. This triggered a numbered of experiments, such as COBE, BOOMERanG and MAXIMA [46, 154, 207, 227, 296, 376]. The most recent one is the Wilkinson Microwave Anisotropy Probe (WMAP) [216] and there are also new experiments planned for the near future, such as the PLANCK which now is flying [217].

The density fluctuations indicated by the small anisotropies in the temperature of Cosmic Microwave Background Radiation are believed to act as seeds for gravitational collapse, leading to gravitationally bound objects which constitute the large scale structures currently present in the Universe [272]. This allows us to build up a coherent scenario about how these structures were formed and to explain the current small scale inhomogeneities and anisotropies. Besides the Cosmic Microwave Background Radiation, which gives information about the initial anisotropies, one can resort to galaxy surveys for complementary information. Current surveys determining the distribution of galaxies include the 2 degree Field Galaxy Redshift Survey (2dF GRS) [218] and the Sloan Digital Sky Survey (SDSS) [219]. There are also other methods used to measure the density variations such as gravitational lensing [340] and X-ray measurements [12].

Besides the Cosmic Microwave Background Radiation and large scale structure surveys, another class of observations that appears to be of special interest in Cosmology are those of type Ia supernovae (SNeIa). These exploding stellar objects are believed to be approximately standard candles, i.e. astronomical objects with known luminosity and absolute magnitude. Therefore, they can be used to reveal distances, leading to the possibility of construction a reliable redshift-distance relation and thereby measuring the expansion of the Universe at different redshifts. For this purpose, there are a number of supernova surveys [32, 179, 343].

But let us return to how we can use the outcome of the experimental measurements mentioned above in order to infer whether a period of accelerated expansion has occurred. The most recent Cosmic Microwave Background Radiation dataset is that of the Three-Year WMAP Observations [377] and results are derived using combined WMAP data and data from supernova and galaxy surveys in many cases. To begin with, let us focus on the value of Ω_k. The WMAP data (combined with Supernova Legacy Survey data [32]) indicates that

$$\Omega_k = -0.015^{+0.020}_{-0.016}, \tag{1.57}$$

i.e. that Ω is very close to unity and the Universe appears to be spatially flat, while the power spectrum of the Cosmic Microwave Background Radiation appears to be consistent with gaussianity and adiabaticity [254, 329]. Both of these facts are in perfect agreement with the predictions of the inflationary paradigm.

In fact, even though the theoretical issues mentioned in the previous paragraph (i.e. the horizon and the flatness problem) were the motivations for introducing the inflationary paradigm, it is the possibility of relating large scale structure formation with initial quantum fluctuations that appears today as the major advantage of inflation [302]. Even if one would choose to dismiss, or find another way to address, problems related to the initial conditions, it is very difficult to construct any other theory which could successfully explain the presence of over-densities with values suitable for leading to the present picture of our Universe at smaller scales [272]. Therefore, even though it might be premature to claim that the inflationary paradigm has been experimentally verified, it seems that the evidence for there having been a period of accelerated expansion of the Universe in the past is very compelling.

However, observational data hold more surprises. Even though Ω is measured to be very close to unity, the contribution of matter to it, Ω_m, is only of the order of 24% A very little fraction of it is baryonic matter. Therefore, there seems to be some unknown form of energy in the Universe, often called *dark energy*. What is more, observations indicate that, if one tries to model dark energy as a perfect fluid with equation of state $p = w\rho$ then

$$w_{de} = -1.06^{+0.13}_{-0.08}, \tag{1.58}$$

so that dark energy appears to satisfy Eq. (1.56). Since it is the dominant energy component today, this implies that the Universe should be undergoing an accelerated expansion currently as well. This is also what was found earlier using supernova surveys [343].

As is well known, between the two periods of acceleration (inflation and the current era) the other conventional eras of evolutionary Cosmology should take place. This means that inflation should be followed by Big Bang Nucleosynthesis, referring to the production of nuclei other than hydrogen. There are very strict bounds on the abundances of primordial light elements, such as deuterium, helium and lithium, coming from observations [72] which do not seem to allow significant deviations from the Standard Cosmological Model [121]. This implies that Big Bang Nucleosynthesis almost certainly took place during an era of radiation domination, i.e. a period in which radiation was the most important contribution to the energy density. On the other hand, the formation of matter structures requires that the radiation dominated era is followed by a matter dominated era.

The transition, from radiation domination to matter domination, comes naturally since the matter energy density is inversely proportional to the volume and, therefore, proportional to a^{-3}, whereas the radiation energy density is proportional to a^{-4} and so it decreases faster than the matter energy density as the Universe expands.

To sum up, our current picture of the evolution of the Universe as inferred from observations comprises a pre-inflationary (probably quantum gravitational) era followed by an inflationary era, a radiation dominated era, a matter dominated era and then a second era of accelerated expansion which is currently taking place. Such an evolution departs seriously from the one expected if one just takes into account General Relativity and conventional matter and therefore appears to be quite unorthodox.

But puzzling observations do not seem to stop here. As mentioned before, Ω_m accounts for approximately 24% of the energy density of the Universe. However, one also has to ask how much of this 24% is actually ordinary baryonic matter. Observations indicate that the contribution of baryons to that, Ω_b, is of the order of $\Omega_b \sim 0.04$ leaving some 20% of the total matter content of the Universe to be accounted for by some unknown unobserved form of matter, called *dark matter*. Differently from dark energy, dark matter has the gravitational characteristics of ordinary matter (hence the name) and does not violate the Strong Energy Condition. However, it is not directly observed since it appears to interact very weakly if at all.

The first indications for the existence of dark matter did not come from Cosmology. Historically, it was Fritz Zwicky who first posed the "missing mass" question for the Coma cluster of galaxies in 1933 [447,448]. After applying the virial theorem in order to compute the mass of the cluster needed to account for the motion of the galaxies near to its edges, he compared this with the mass obtained from galaxy counts and the total brightness of the cluster. The virial mass turned out to be larger by a factor of almost 400.

Later, in 1959, Kahn and Waltjer were the first to propose the presence of dark matter in individual galaxies [241]. However, it was in the 1970s that the first compelling evidence for the existence of dark matter came about: the rotation curves of galaxies, i.e. the velocity curves of stars as functions of the radius, did not appear to have the expected shapes. The velocities, instead of decreasing at large distances as expected from Keplerian dynamics and the fact that most of the visible mass of a galaxy is located near to its centre, appeared to be flat [64, 348, 349]. As long as Keplerian dynamics is considered correct, this implies that there should be more matter than just the luminous matter, and this additional matter should have a different distribution within the galaxy (dark matter halo).

Much work has been done in the last 40 years to analyse the problem of dark matter in astrophysical environments (for reviews see [65, 174, 301]) and there are also recent findings, such as the observations related to the Bullet Cluster, that deserve a special mention[10]. The main conclusion that can be drawn is that some form of dark matter is present in galaxies and clusters of galaxies. What is more, taking also into account the fact that dark matter appears to greatly dominate over ordinary baryonic matter at cosmological scales, it is not surprising that current models of structure formation consider it as a main ingredient (e.g. [381]).

We have just seen some of the main characteristics of the Universe as inferred from

[10] Weak lensing observations of the Bullet cluster (1E0657-558), which is actually a unique cluster merger, appear to provide direct evidence for the existence of dark matter [138].

observations. Let us now set aside for the moment the discussion of the earlier epochs of the Universe and inflation and concentrate on the characteristic of the Universe as it appears today: it is probably spatially flat ($\Omega_k \sim 0$), expanding in an accelerated behaviour as confirmed both from supernova surveys [341] and WMAP, and its matter energy composition consists of approximately 76% dark energy, 20% dark matter and only 4% ordinary baryonic matter. One has to admit that this picture is not only surprising but maybe even embarrassing, since it is not at all close to what one would have expected based on the Standard Cosmological Model and what is more it reveals that almost 96% of the energy content of the Universe has a composition which is unknown to us.

In any case, let us see which is the simplest model that agrees with the observational data. To begin with, we need to find a simple explanation for the accelerated expansion. The first physicist to consider a Universe which exhibits an accelerated expansion was probably Willem de Sitter [158]. A de Sitter space is the maximally symmetric, simply-connected, Lorentzian manifold with constant positive curvature. It may be regarded as the Lorentzian analogue of an n-sphere in n dimensions. However, the de Sitter space-time is not a solution of the Einstein equations, unless one adds a cosmological constant Λ to them, i.e. adds on the left hand side of Eq. (1.7) the term $\Lambda g_{\mu\nu}$.

Such a term was not included initially by Einstein, even though this is technically possible since, according to the reasoning which he gave for arriving at the gravitational field equations, the left hand side has to be a second rank tensor constructed from the Ricci tensor and the metric, which is divergence free. Clearly, the presence of a cosmological constant does not contradict these requirements. In fact, Einstein was the first to introduce the cosmological constant, thinking that it would allow him to derive a solution of the field equations describing a static Universe [176]. The idea of a static Universe was then rapidly abandoned however when Hubble discovered that the Universe is expanding and Einstein appears to have changed his mind about the cosmological constant: Gamow quotes in his autobiography, My World Line (1970): "Much later, when I was discussing cosmological problems with Einstein, he remarked that the introduction of the cosmological term was the biggest blunder of his life" and Pais quotes a 1923 letter of Einstein to Weyl with his reaction to the discovery of the expansion of the Universe: "If there is no quasi-static world, then away with the cosmological term!" [320].

In any case, once the cosmological term is included in the Einstein equations, de Sitter space becomes a solution. Actually, the de Sitter metric can be brought into the form of the Friedmann-Robertson-Walker metric in Eq. (1.47) with the scale factor and the Hubble parameter given by[11].

$$a(t) = e^{Ht}, \tag{1.59}$$

$$H^2 = \frac{8\pi G}{3}\Lambda. \tag{1.60}$$

This is sometimes referred to as the de Sitter Universe and it can be seen that it is expanding exponentially.

[11]Note that de Sitter space is an example of a manifold that can be sliced in 3 ways — $k=+1, k=0, k=-1$ — with each coordinate patch covering a different portions of space-time. We are referring here just to the $k=0$ slicing for simplicity.

The de Sitter solution is a vacuum solution. However, if we allow the cosmological term to be present in the field equations, the Friedmann Eqs (1.48) and (1.49) will be modified so as to include the de Sitter space-time as a solution:

$$\left(\frac{\dot{a}}{a}\right)^2 = \frac{8\pi G\rho + \Lambda}{3} - \frac{k}{a^2}, \tag{1.61}$$

$$\frac{\ddot{a}}{a} = \frac{\Lambda}{3} - \frac{4\pi G}{3}(\rho + 3p). \tag{1.62}$$

From Eq. (1.62) one infers that the Universe can now enter a phase of accelerated expansion once the cosmological constant term dominates over the matter term on the right hand side. This is bound to happen since the value of the cosmological constant stays unchanged during the evolution, whereas the matter density decreases like a^3. In other words, the Universe is bound to approach a de Sitter space asymptotically in time.

On the other hand Ω in Eq. (1.54) can now be split in two different contributions, $\Omega_\Lambda = \Lambda/(3H^2)$ and Ω_m, so that Eq. (1.54) takes the form

$$\Omega_m + \Omega_\Lambda + \Omega_k = 1. \tag{1.63}$$

In this sense, the observations presented previously can be interpreted to mean that $\Omega_\Lambda \sim 0.72$ and the cosmological constant can account for the mysterious dark energy responsible for the current accelerated expansion. One should not fail to notice that Ω_m does not only refer to baryons. As mentioned before, it also includes dark matter, which is actually the dominant contribution. Currently, dark matter is mostly treated as being cold and not baryonic, since these characteristics appear to be in good accordance with the data. This implies that, apart from the gravitational interaction, it does not have other interactions — or at least that it interacts extremely weakly — and can be regarded as collisionless dust, with an effective equation of state $p = 0$ (we will return to the distinction between cold and hot dark matter shortly).

We have sketched our way to what is referred to as the Λ Cold Dark Matter or ΛCDM model. This is a phenomenological model which is sometimes also called the concordance model of Big Bang Cosmology, since it is more of an empirical fit to the data. It is the simplest model that can fit the cosmic microwave background observations as well as large scale structure observations and supernova observations of the accelerating expansion of the Universe with a remarkable agreement (see for instance [377]). As a phenomenological model, however, it gives no insight about the nature of dark matter, or the reason for the presence of the cosmological constant, neither does it justify the value of the latter.

While it seems easy to convince someone that an answer is indeed required to the question "what exactly is dark matter and why is it almost 9 times more abundant than ordinary matter", the presence of the cosmological constant in the field equations might not be so disturbing for some. Therefore, let us for the moment put aside the dark matter problem — we will return to it shortly — and consider how natural it is to try to explain the dark energy problem by a cosmological constant (see [122, 123, 325, 430] for reviews).

It has already been mentioned that there is absolutely no reason to discard the presence of a cosmological constant in the field equations from a gravitational and a mathematical perspectives. Nonetheless, it is also reasonable to assume that there should be a theoretical motivation for including it — after all there are numerous modifications that could be made to the left hand side of the gravitational field equations and still lead to a consistent theory from a mathematical perspective and we are not aware of any other theory that includes more than one fundamental constant. On the other hand, it is easy to see that the cosmological term can be moved to the right hand side of the field equations with the opposite sign and can be regarded as some sort of matter term. It can then be put into the form of a stress-energy tensor $T^{\mu}_{\nu} = \mathrm{diag}(\Lambda, -\Lambda, -\Lambda, -\Lambda)$, i.e. resembling a perfect fluid with equation of state $p = -\rho$ or $w = -1$. Notice the very good agreement with the value of w_{de} inferred from observations (Eq. (1.58)), which explains the success of the ΛCDM model.

Once the cosmological constant term is considered to be a matter term, a natural explanation for it seems to arise: the cosmological constant can represent the vacuum energy associated with the matter fields. One should not be surprised that empty space has a non-zero energy density once, apart from General Relativity, field theory is also taken into consideration. Actually, Local Lorentz Invariance implies that the expectation value of the stress energy tensor in vacuum is

$$\langle T_{\mu\nu}\rangle = -\langle\rho\rangle g_{\mu\nu}, \tag{1.64}$$

and $\langle\rho\rangle$ is generically non-zero. To demonstrate this, we can take the simple example of a scalar field [124]. Its energy density will be

$$\rho_\phi = \frac{1}{2}\dot{\phi}^2 + \frac{1}{2}(\nabla_{\rm sp}\phi)^2 + V(\phi), \tag{1.65}$$

where $\nabla_{\rm sp}$ denotes the spatial gradient and $V(\phi)$ the potential. The energy density becomes constant for any constant value $\phi = \phi_0$ and there is no reason to believe that for $\phi = \phi_0$, $V(\phi_0)$ should be zero. One could in general assume that there is some principle or symmetry that dictates it. So in general one should expect that matter fields have a non-vanishing vacuum energy, i.e. that $\langle\rho\rangle$ is non-zero.

Within this perspective, effectively there should be a cosmological constant in the field equations, given by

$$\Lambda = 8\pi G\langle\rho\rangle. \tag{1.66}$$

One could, therefore, think to use the Standard Model of particle physics in order to estimate its value. Unfortunately, however, $\langle\rho\rangle$ actually diverges due to the contribution of very high-frequency modes. No reliable exact calculation can be made but it is easy to make a rough estimate once a cutoff is considered (see for instance [124, 429]). Taking the cutoff to be the Planck scale ($M_{\rm Plank} = 10^{19}$ GeV), which is the scale where the classical gravity is completely questionable, the outcome is

$$\rho_\Lambda \sim (10^{27}\,{\rm eV})^4. \tag{1.67}$$

On the other hand, observations indicate that

$$\rho_\Lambda \sim (10^{-3}\,{\rm eV})^4. \tag{1.68}$$

Obviously the discrepancy between these two estimates is very large for being attributed to any rough approximation. There is a difference of 120-orders-of-magnitude, which is large enough to be considered embarrassing. One could validly claim that we should not be comparing energy densities but mass scales by considering a mass scale for the vacuum implicitly defined through $\rho_\Lambda = M_\Lambda^4$. However, this will not really make a difference, since a 30-orders-of-magnitude discrepancy in mass scale hardly makes a good estimate. This constitutes the so-called *cosmological constant problem*.

Unfortunately, this is not the only problem related to the cosmological constant. The other known problem goes under the name of *the coincidence problem*. It is apparent from the data that $\Omega_\Lambda \sim 0.72$ and $\Omega_m \sim 0.28$ have comparable values today in order of magnitude. However, as the Universe expands their fractional contributions change rapidly since

$$\frac{\Omega_\Lambda}{\Omega_m} = \frac{\rho_\Lambda}{\rho_m} \propto a^3. \tag{1.69}$$

Since Λ is a constant, ρ_Λ should once have been negligible compared to the energy densities of both matter and radiation and, as dictated by Eq. (1.69), it will come to dominate completely at some point in the late time Universe. However, the striking fact is that the period of transition between matter domination and cosmological constant domination is very short compared to cosmological time scales[12]. The puzzle is, therefore, why we live precisely in this very special era [124]. Obviously, the transition from matter domination to cosmological constant domination, or, alternatively stated, from deceleration to acceleration, would happen eventually. The question is, why now?

To sum up, including a cosmological constant in the field equations appears as an easy way to address issues like the late time accelerated expansion but unfortunately it comes with a price: the cosmological constant and coincidence problems. We will return to this discussion from this point later on but for the moment let us close the present section with an overall comment about the ΛCDM model. Its validity should definitely not be underestimated. In spite of any potential problems that it may have, it is still a remarkable fit to observational data while, at the same time being elegantly simple. One should always bear in mind how useful a simple empirical fit to the data may be. On the other hand, the ΛCDM model should also not be over-estimated. Being a phenomenological model, with poor theoretical motivation, one should not necessarily expect to discover in it some fundamental secrets of nature.

1.4.3. Scalar Fields in Early Universe

We have already discussed the need for an inflationary period in the early Universe. However, we have not yet attempted to trace the cause of such an accelerated expansion. Since the presence of a cosmological constant could in principle account for that, one is tempted to explore this possibility, as in the case of late time acceleration. Unfortunately, this simple solution is bound not to work for a very simple reason: once the cosmological constant dominates over matter there is no way for matter to dominate again. Inflation has to end at some point, as already mentioned, so that Big Bang Nucleosynthesis and structure formation can take place. Our presence in the Universe is all the evidence one needs for that.

[12]Note that in the presence of a positive cosmological constant there is an infinite future in which Λ is dominating.

Therefore, one is forced to seek other, dynamical solutions to this problem.

As long as one is convinced that gravity is well described by General Relativity, the only option left is to assume that it is a matter field that is responsible for inflation. However, this matter field should have a rather unusual property: its effective equation of state should satisfy Eq. (1.56), i.e. it should have a negative pressure and actually violate the Strong Energy Condition. Fortunately, matter fields with this property do exist if we give up the idea that cosmological matter can be only a perfect fluid. A typical example is a scalar field ϕ.

A scalar field minimally coupled to gravity, satisfies the Klein-Gordon equation

$$\Box\phi + V'(\phi) = 0, \tag{1.70}$$

where $\Box$ denotes the d'Alembert operator, $\nabla^{\mu}\nabla_{\mu}$, $V(\phi)$ is the potential and the prime denotes differentiation with respect to the argument. Assuming that the scalar field is homogeneous, and therefore $\phi \equiv \phi(t)$, we can write its energy density and pressure as

$$\rho_\phi = \frac{1}{2}\dot{\phi}^2 + V(\phi), \tag{1.71}$$

$$p_\phi = \frac{1}{2}\dot{\phi}^2 - V(\phi), \tag{1.72}$$

while, in a Friedmann-Robertson-Walker space-time, Eq. (1.70) takes the following form:

$$\ddot{\phi} + 3H\dot{\phi} + V'(\phi) = 0. \tag{1.73}$$

It is now apparent that if $\dot{\phi}^2 < V(\phi)$ then the pressure is indeed negative. In fact $w_\phi = p_\phi/\rho_\phi$ approaches -1 when $\dot{\phi}^2 \ll V(\phi)$.

In general a scalar field that leads to inflation is referred to as the *inflaton*. Since we invoked such a field instead of a cosmological constant, claiming that in this way we can successfully end inflation, let us see how this is achieved. Assuming that the scalar dominates over both matter and radiation and neglecting for the moment the spatial curvature term for simplicity, Eq. (1.48) takes the form

$$H^2 \approx \frac{8\pi G_N}{3}\left(\frac{1}{2}\dot{\phi}^2 + V(\phi)\right). \tag{1.74}$$

If, together with the condition $\dot{\phi}^2 < V(\phi)$, we require that $\ddot{\phi}$ is negligible in Eq. (1.73) then Eqs. (1.74) and (1.73) reduce to

$$H^2 \approx \frac{8\pi G_N}{3}V(\phi), \tag{1.75}$$

$$3H\dot{\phi} \approx -V'(\phi). \tag{1.76}$$

This constitutes the *slow-roll approximation* since the potential terms are dominant with respect to the kinetic terms, causing the scalar to roll slowly from one value to another. To be more rigorous, one can define two slow-roll parameters

$$\varepsilon(\phi) = 4\pi G_N \left(\frac{V'}{V}\right)^2, \tag{1.77}$$

$$\eta(\phi) = 8\pi G_N \frac{V''}{V}, \tag{1.78}$$

for which the conditions $\varepsilon(\phi) \ll 1$ and $\eta(\phi) \ll 1$ are necessary in order for the slow-roll approximation to hold [270, 271]. Note that these are not sufficient conditions since they only restrict the form of the potential. One also has to make sure that Eq. (1.76) is satisfied. In any case, what we want to focus on at this point is that one can start with a scalar that initially satisfies the slow-roll conditions but, after some period, ϕ can be driven to such a value so as to violate them. A typical example is that of $V(\phi) = m^2\phi^2/2$, where these conditions are satisfied as long as $\phi^2 > 16\pi G_N$ but, as ϕ approaches the minimum of the potential, a point will be reached where $\phi^2 > 16\pi G_N$ will cease to hold. Once the slow-roll conditions are violated, inflation can be naturally driven to an end since $\dot{\phi}^2$ can begin to dominate again in Eq. (1.72).

However, just ending inflation is not enough. After such an era, the Universe would be a cold and empty place unable to evolve dynamically to anything close to the picture which we observe today. A viable model for inflation should include a mechanism that will allow the Universe to return to the standard Big Bang scenario. This mechanism is called "reheating" and consists mainly of three processes: a period of non-inflationary scalar field dynamics, after the slow-roll approximation has ceased to be valid, the creation and decay of inflaton particles and the thermalization of the products of this decay [272]. Reheating is an extensive and intricate subject and analyzing it goes beyond the scope of this introduction. We refer the reader to [1, 5, 250, 251, 253, 274–278, 371] for more information.

1.4.4. Inflationary Models

In order to see how scalar field specifically works, let us now outline the principal inflationary models reviewing both the major advantages and problems. It is important to make two remarks before starting.

We can get inflation in several ways but we have to distinguish two main classes of models: firstly there are models where it is changed only matter sector, that is the gravity is described essentially by Einstein equations and the scalar field(s) acts as a source yielding inflation and primordial perturbations; secondly there are models where both the Einstein theory of gravity and matter are altered. The generalized theories of gravity (nonminimally coupled and higher order theories) are included in this second family. As it has been proved, it seems that this last approach is better since it allows to solve a larger number of shortcomings specifically connected to the choice of gravitational sector.

The second remark concerns the nature of scalar field. In some models, like the "old inflation", it has to be a quantum field coming from some fundamental unification theory; in other models, like "chaotic inflation", such a hypothesis is not essential and scalar field can be also a classical field.[13].

[13]In fact, from a cosmological point of view, we are working with the average values $\phi = <\chi|\hat{\phi}|\chi>$ and

Having clarified these two questions, we briefly go to outline the main scenarios currently used to describe inflationary phase.

The so called *old inflation* [206] takes into account a Higgs field minimally coupled to gravitation and an effective potential of a Georgi-Glashow $SU(5)$ theory. When the system reaches a critical temperature T_c for the spontaneous symmetry breaking, we have the bubble nucleation (first order phase transition). In this situation, the system undergoes a quantum tunnelling enucleating bubbles from the regions of space-time with $\phi = 0$ (false vacuum) to the regions under the potential barrier, towards the most favourable energetic state with $\phi = \phi_0 (\neq 0)$ (true vacuum). Such a feature solves the horizon problem but introduces an inhomogeneity problem, since bubbles, carrying a large fraction of initial vacuum energy density and remaining trapped inside the observed horizon, create too large perturbations which should generate a lot of topological defects (domain walls, cosmic strings and magnetic monopoles) not observed at all[14].

A further problem is due to the fact that the phase transition never ends, that is when time passes, a larger and larger part of space reaches the new phase and the regions which remain in false vacuum phase continue to expand exponentially without percolating into each others, but going away more and more; in other words, there is no mechanism able to tie bubbles in the new phase and able to contain the observed universe (this is the "graceful exit problem" [206]).

The scenario called *new inflation* [6] avoids such problems by using a Coleman-Weinberg effective potential [140,428]. In other words, the starting point is a scalar field theory with a double well potential which undergoes a second order phase transition. Another important feature of the model is that the potential has to have a substantially flat slope for small values of ϕ. At the beginning of the evolution, such a field rolls down slowly towards the true vacuum phase (*slow rolling*). In such a situation, the system exponentially expands. Inflation ends when slow rolling stops and scalar field rapidly evolves towards the absolute minimum. The oscillations around such a minimum yield the *reheating* of the universe and the crossing to the radiation-dominated Friedmann-Robertson-Walker phase with the consequent raising of temperature. This model predicts a density perturbation spectrum useful for the formation of galaxies and clusters; but such perturbations, from the fundamental physics side, do not match with parameters of GUT theories (they are too large) and, from the cosmological side, disagree with micro-wave background isotropy [35]. In order to satisfy these constraints, the self-interactions and the couplings with other fields (e.g. fermionic fields of ordinary matter) should be too small: in such a case, thermal equilibrium could not be reached to average at zero the initial value of scalar field and to begin inflation[15]. Hence, new inflation has essentially a *fine tuning* problem for initial data.

$V_{eff}(\phi) = <\chi|\hat{\mathcal{H}}|\chi>$ so that we do not care what is the primordial origin of scalar field. Here $|\chi>$ represents the normalized state of any fundamental quantum theory, where the condition $<\chi|\chi> = 1$ has to hold; $\hat{\mathcal{H}}$ is the Hamilton operator for which $\delta <\chi|\mathcal{H}|\chi> = 0$, that means the variation on the states has to imply the conservation of energy.

[14]Such defects are the 2-dimensional (domain walls), 1-dimensional (cosmic strings), and zero-dimensional (magnetic monopoles) remnants of spontaneous symmetry breaking. They represent the evidence that the system passes from a symmetric state (no defects) to an asymmetric state. It is worthwhile to note that cosmic strings are objects which form at GUT scales while fundamental (bosonic or fermionic) strings are objects which live at Planck's epoch.

[15]A finite temperature effective potential is used to localize the scalar field at small expectation values [338],

Chaotic inflation solves the question by showing that inflationary expansion does not depend on particular field theories. In fact, for large classes of potentials, we can obtain a slow evolution of scalar field: "slow evolution" is intended here in comparison to the variation of Hubble parameter. If such an approach allows to build models with satisfactory parameters, in this case also the density fluctuations force the coupling to be excessively small and also here a fine tuning problem appears.

We have to note that a subclass of the mainly used potentials, in particular the exponential potentials of the form $V(\phi) = V_0 \exp \lambda\phi$, is particularly relevant since it allows the so called *power-law* inflation (mentioned above) where the scale factor of the universe evolves as $a(t) \sim t^q$, with $q > 1$ [283,285,286]. In this case, the initial singularity problem is again present (even if it is obviously avoided in a de Sitter expansion), but one is able to control the background so that density fluctuations can be useful to form large scale structures.

Old, new and chaotic inflations are all based on the use of some fundamental scalar fields which can or cannot be the Higgs fields of some grand unified theory. Anyway, instead of introducing new physics to give inflation via scalar fields, as we said above it is also possible to modify Einstein gravity. The original attempt to obtain inflation from an appproach not using scalar fields is the Starobinsky model [389] in which the Lagrangian for gravity is modified by including a term proportional to R^2. In the initial model, this term was introduced as a contribution to the effective action from quantum corrections [389]. Later, models with bare Lagrangians including αR^2 and, in general βR^N, with N positive integer, were studied [85,297]. Consider the theory with Lagrangian for gravity

$$\mathcal{L} = R + \alpha R^2 , \tag{1.79}$$

where R, as usual, is the Ricci scalar of a metric $g_{\mu\nu}$. Referring to the discussion we made in Sect. 2, a conformal transformation is able to give rise again to the situation of a scalar field minimally coupled to the Einstein gravity (and then the "usual" chaotic inflation).

These results can also be derived by directly solving the equations of motions which follow from (1.79) for a homogeneous universe [36,297].

Adding more than one higher order term (e.g. $R\Box R$ where as above $\Box$ is the d'Alembert operator) can be read as considering more than one (chaotic) scalar fields. Such a feature leads to consider multiple inflationary models. As an example, take the so called *double inflation* [11,17,201] which, yielding two successive inflationary epochs, gives two perturbation spectra useful to form first *very* large scale structure (first inflationary perturbation spectrum for clusters of galaxies) and then *small* large scale structures (second inflationary perturbation spectrum for galaxies), in a *top–down* structure formation scenario.

Some models have been formulated where scalar fields are nonminimally coupled with geometry in a Brans–Dicke theory of gravity like the above (1.14) or, in general, in a nonminimally coupled theory. However, by the simple transformations $\phi \to \varphi$ and $\phi/2F'(\phi)^2 \to \omega(\varphi)$, Brans–Dicke theory and general nonminimally coupled gravity (induced gravity) are equivalent except for the presence of scalar field potential $V(\phi)$. Furthermore, in the original Brans-Dicke approach, ω is a constant while here it can be a function of φ. This fact allows to avoid original shortcomings essentially due to the big ω values requested for the agreement with solar tests. Such models are referred to as *extended* and

such a feature makes inflation start.

hyperextended inflation. Here we outline the main feature of original extended inflationary model which is, essentially, a revival of old inflation. Actually, in the nonminimal coupling context, it is also possible to reformulate both new and chaotic inflations.

Let us take into account the action (1.14). There, φ is the Brans-Dicke scalar field (which induces a time and space dependent gravitational coupling constant), ω is the Brans-Dicke parameter and $\mathcal{L}_m$ is the Lagrangian for matter fields. $\mathcal{L}_m$ is chosen to give rise to a first order phase transition which proceeds by nucleation of true vacuum bubbles in a surrounding sea of false vacuum.

The crucial point is that in Brans-Dicke theory, an equation of state like $p_m = -\rho_m$ leads to power law rather than exponential expansion of the Universe. In Friedmann-Robertson-Walker–case, the evolution equations are (for $k = 0$)

$$H^2 = \frac{1\rho_m}{3\varphi} + \frac{\omega}{6}\left(\frac{\dot{\phi}}{\phi}\right)^2 - H\frac{\dot{\phi}}{\phi}, \tag{1.80}$$

and

$$\ddot{\phi} + 3H\dot{\phi} = \frac{\rho_m - 3p_m}{3+2\omega}, \tag{1.81}$$

for $p_m = -\rho_m$, particular solutions are

$$\phi(t) = (1 + Ht/\alpha)^2, \tag{1.82}$$

and

$$a(t) = (1 + Ht/\alpha)^{\omega+1/2}, \tag{1.83}$$

with $\alpha \sim \omega^2$ and

$$H^2 = \frac{\tilde{\rho}}{3}, \tag{1.84}$$

being $\tilde{\rho}$ the r.h.s. of Eq. (1.80). Since the false vacuum only expands as a power of t in Brans-Dicke theory, true vacuum bubbles are able to percolate. For a point, the probability of remaining in the false vacuum is

$$\mathcal{P}(t) = \exp\left[-\int_{t_c}^{t'} dt'\lambda(t')a(t')^3\frac{4\pi}{3}\left(\int_{t'}^{t}\frac{dt''}{a(t'')}\right)^3\right], \tag{1.85}$$

where $\lambda(t)$ is the nucleation rate per unit time and t_c is the critical time of the phase transition (when nucleation begins). With $\varepsilon = \lambda/H^4$ taken to be time independent [200, 252], we obtain

$$\mathcal{P}(t) \sim \exp\left[-\frac{4\pi}{3}\varepsilon H\right]. \tag{1.86}$$

In the old inflationary Universe, $\mathcal{P}(t)a^3(t)$ diverges since $\varepsilon \ll 1$ and $a(t)$ increases exponentially. This implies that true vacuum bubbles do not percolate and we can never produce a region of the size of our Universe with true vacuum. Instead of this unsatisfactory situation, in extended inflation, $a(t)$ grows as a power of t, then

$$\lim_{t\to\infty} a^3(t)\mathcal{P}(t) = 0\,, \tag{1.87}$$

and true vacuum bubbles percolate.

However, percolation of true vacuum bubbles is not sufficient to give a viable cosmological model. The bubbles must also thermalize in order to produce not too large density perturbations. According to the analysis in [431], thermalization occurs only if $\omega < 10$, a bound which conflicts with the lower bound $\omega > 500$ stemming from time delay measurements. Hence, in order to save extended inflation, new variants of the models are required: for example a model with a non trivial potential $V(\phi)$ for the Brans-Dicke scalar.

As it appears from the above remarks, inflation is an important idea which solves many problems of standard Big Bang cosmology. However, no completely convincing realization of inflation, which does not involve unexplained shortcomings, has emerged til now (for a general discussion of this point see [11]).

Furthermore, it is important to distinguish between models of inflation which are "self consistent" and those which are not. For example, new inflation is not self consistent, whereas chaotic and improved versions of extended inflation are. One of the key issues involves initial conditions. In new inflation, the initial conditions required can only be obtained if the inflaton field is in thermal equilibrium above the critical temperature, which however is not possible because of the density fluctuation constraints on coupling constants. In chaotic inflation, it can be shown [186, 257] that a large portion of phase space of initial conditions gives that kind of inflation, whereas the probability to obtain field configurations which give new inflation (this possibility is only available in double well potentials) is negligibly small [200]. The situation is similar in extended inflation.

There are other points which have to be stressed. In alternative models (e.g. fourth order or nonminimal coupling models), there is no general criterion to choose higher order terms in Einstein-Hilbert action or to choose the couplings in scalar-tensor theories and, in many cases, due to the difficulties connected with dynamics, it is not possible to have a reasonable control of the cosmological behaviour. *Nöther Symmetry Approach*, addresses both issues and provides several physically interesting models (which, moreover, can be exactly solved) where the higher order terms, the coupling ϕ and the potential $V(\phi)$ are selected by the existence of that symmetry. The models derived from Noether symmetry yield realistic cosmological backgrounds, and the conserved quantities (due to the symmetries) have, in many cases, a physical interpretation [81].

As concluding remarks, we have to say that in inflationary cosmology, it is of fundamental importance to control the models and the connected evolutionary behaviours, since only in this way one can check whether the shortcomings of Standard Model are solved or not and whether it is possible to produce perturbation spectra coherent with observations of large scale structure.

1.4.5. The Dark Energy Problem

We have already outlined that there are compelling observational evidence that the Universe is currently undergoing an accelerated expansion and we have also discussed the problems that arise if a cosmological constant is considered to be responsible for this acceleration within the framework of the ΛCDM model. Based on that, one can classify the attempts to address the problem of finding a mechanism that will account for the late-time accelerated expansion in two categories: those that try to find direct solutions to the cosmological constant and the coincidence problems and consequently attempt to provide an appealing theoretical explanation for the presence and the value of the cosmological constant, and those that abandon the idea of the cosmological constant altogether and attempt to find alternative ways to explain the acceleration.

Let us state two of the main approaches followed to solve the cosmological constant problem directly:

The first approach resorts to High Energy Physics. The general idea is simple and can be summed up in the question: Are we counting properly? This refers to the quite naive calculation mentioned previously, according to which the energy density of the cosmological constant as calculated theoretically should be 10^{120} times larger than its observed value. Even though the question is simple and reasonable, giving a precise answer to it is actually very complicated since, as mentioned already, little is known about how to make an exact calculation of the vacuum energy of quantum fields. There are indications coming from contemporary particle physics theories, such as supersymmetry (SUSY), which imply that one can be led to different values for the energy density of vacuum from the one mentioned before (Eq. (1.67)). For instance, since no superpartners of known particles have been discovered up to now in accelerators [16], one can assume that supersymmetry was broken at some scale of the order of 10^3GeV or higher. If this is the case, one would expect that

$$\rho_\Lambda \sim M_{\rm SUSY}^4 \geq (10^{12}{\rm eV})^4. \tag{1.88}$$

This calculation gives an estimate for the energy density of the vacuum which is 60 orders of magnitude smaller than the one presented previously in Eq. (1.67). However, the value estimated here is still 60 orders of magnitude larger than the one inferred from observations (Eq. (1.68)). Other estimates with or without a reference to supersymmetry or based on string theory or loop quantum gravity exist. An example is the approach of Ref. [53, 54, 76, 441] where an attempt is made to use our knowledge from condensed matter systems in order to explain the value of the cosmological constant. We will not, however, list further examples here but refer the reader to [123, 430] and references therein for more details. In any case, the general flavour is that it is very difficult to avoid the cosmological constant problem by following such approaches without making some fine tuning within the fundamental theory used to perform the calculation for the energy density of vacuum. Also, such approaches mostly fail to address the second problem related to the cosmological constant: the coincidence problem.

The second direct approach for solving problems related to the cosmological constant has a long history and Brandon Carter gave it the name of "anthropic principle" by [37, 131,

[16] We are waiting for the results at large hadron Collider (LHC) at CERN [223].

132]. Unfortunately, the anthropic principle leaves a lot of room for different formulations or even misinterpretations. Following [430] we can identify at least three versions, starting from a very mild one, that probably no one really disagrees with but is not very useful for answering questions, stating essentially that our mere existence can potentially serve as an experimental tool. The second version on the other hand is a rather strong one, stating that the laws of Nature are by themselves incomplete and become complete only if one adds the requirement that conditions should allow intelligent life to arise, for only in the presence of intelligent life does science become meaningful. It is apparent that such a formulation places intelligent life or science at the centre of attention as far as the Universe is concerned. From this perspective one cannot help but notice that the anthropic principle becomes reminiscent of the Tolemaic model. Additionally, to quote Weinberg [429]:

"...although science is clearly impossible without scientists, it is not clear that the Universe is impossible without science".

The third and most moderate version of the anthropic principle, known as the "weak anthropic principle" states essentially that observers will only observe conditions which allow for observers. This version is the one mostly discussed from a scientific perspective and even though it might seem tautological, it acquires a meaning if one invokes probability theory.

To be more concrete, as opposed to the second stronger formulation, the weak anthropic principle does not assume some sort of conspiracy of Nature aimed at creating intelligent life. It merely states that, since the existence of intelligent observes requires certain conditions, it is not possible for them in practice to observe any other conditions, something that introduces a bias in any probabilistic analysis. This, of course, requires one extra assumption: that parts of the Universe, either in space or time, might indeed be in alternative conditions. Unfortunately we cannot conclude at this point whether this last statement is true. Assuming that it is, one could put constrains on the value of the cosmological constant by requiring that it should be small enough for galaxies to form as in [427] and arrive at the conclusion that the currently observed value of the cosmological constant is by no means unlikely. Some modern theories do allow such alternative states of the Universe to co-exist (multiverse), and for this reason it has recently been argued that the anthropic principle could even be placed on firm ground by using the ideas of string theory for the "anthropic or string landscape", consisting of a large number of different false vacua [400]. However, admitting that there are limits on our ability to unambiguously and directly explain the observable Universe inevitably comes with a disappointment. It is for this reason that many physicists would refrain from using the anthropic principle or at least they would consider it only as a last resort, when all other possibilities have failed.

Let us now proceed to the indirect ways of solving problems related with the cosmological constant. As already mentioned, the main approach of this kind is to dismiss the cosmological constant completely and assume that there is some form of dynamical dark energy. In this sense, dark energy and vacuum energy are not related and therefore the cosmological constant problem ceases to exist, at least in the strict formulation given above. However, this comes with a cost: as mentioned previously, observational data seem to be in very good agreement with having a cosmological constant, therefore implying that any form of dynamical dark energy should be able to mimic a cosmological constant very precisely at present times. This is not something easy to achieve. In order to be clearer and also to have the possibility to discuss how well dynamical forms of dark energy can address the cosmological constant and coincidence problems, let us use an example.

Given the discussion presented earlier about inflation, it should be clear by now that if a matter field is to account for accelerated expansion, it should have a special characteristic: negative pressure or more precisely $p \leq -\rho/3$. Once again, as in the inflationary paradigm, the obvious candidate is a scalar field. When such a field is used to represent dark energy, it is usually called *quintessence* [26,34,73,125,318,328,339,424,432]. Quintessence is one of the simplest and probably the most common alternative to the cosmological constant.

If the scalar field is taken to be spatially homogeneous, its equation of motion in an Friedmann-Robertson-Walker space-time will be given by Eq. (1.73) and its energy density and pressure will be given by Eqs. (1.71) and (1.72) respectively, just like the inflaton. As dictated by observations through Eq. (1.58), a viable candidate for dark energy should have an effective equation of state with w very close to minus one. In the previous section it was mentioned that this can be achieved for a scalar field if the condition $\dot{\phi}^2 \ll V(\phi)$ holds. This should not be confused with the slow-roll condition for inflation, which just requires that $\dot{\phi}^2 < V(\phi)$ and also places a constraint for $\ddot{\phi}$. However, there is a similarity in the spirit of the two conditions, namely that in both cases the scalar field is required, roughly speaking, to be slowly-varying. It is worth mentioning that the condition $\dot{\phi}^2 \ll V(\phi)$ effectively restricts the form of the potential V.

Let us see how well quintessence can address the cosmological constant problem. One has to bear in mind that the value given in Eq. (1.68) for the energy density of the cosmological constant now becomes the current value of the energy density of the scalar ρ_ϕ. Since we have asked that the potential terms should be very dominant with respect to the kinetic terms, this value for the energy density effectively constrains the current value of the potential. What is more, the equation of motion for the scalar field, Eq. (1.73) is that of a damped oscillator, $3H\dot{\phi}$ being the friction term. This implies that, for ϕ to be rolling slowly enough so that $\dot{\phi}^2 \ll V(\phi)$ could be satisfied, then $H \sim \sqrt{V''(\phi)}$. Consequently, this means that the current value of $V''(\phi)$ should be that of the observed cosmological constant or, taking also into account that $\sqrt{V''(\phi)}$ represents the effective mass of the scalar m_ϕ, that

$$m_\phi \sim 10^{-33}\ \text{eV}. \tag{1.89}$$

Such a small value for the mass of the scalar field raises doubts about whether quintessence really solves the cosmological constant problem or actually just transfers it from the domain of Cosmology to the domain of particle physics. The reason for this is that the scalar fields usually present in quantum field theory have masses many orders of magnitude larger than that given in Eq. (1.89) and, hence, this poses a naturalness question (see [123] for more details). For instance, one of the well-known problems in particle physics, the hierarchy problem, concerns explaining why the Higgs field appears to have a mass of 10^{11} eV which is much smaller that the grand unification/Planck scale, 10^{25}-10^{28} eV. As commented in [124], one can then imagine how hard it could be to explain the existence of a field with a mass equal to 10^{-33} eV. In all fairness to quintessence, however, it should be stated that the current value of the energy density of dark energy (or vacuum, depending on the approach) is an observational fact, and so it does not seem possible to completely dismiss this number in some way. All that is left to do, therefore, is to put the cosmological constant problem on new grounds that will hopefully be more suitable for explaining it.

One should not forget, however, also the coincidence problem. There are attempts to address it within the context of quintessence mainly based on what is referred to as tracker models [144,191,273,353,396,445,446]. These are specific models of quintessence whose special characteristic is that the energy density of the scalar parallels that of matter or radiation for a part of the evolution which is significant enough so as to remove the coincidence problem. What is interesting is that these models do not in general require specific initial conditions, which means that the coincidence problem is not just turned into an initial conditions fine-tuning problem. Of course, the dependence of such approaches on the parameters of the potential remains inevitable.

It is also worth mentioning that ϕ should give rise to some force, which judging from its mass should be long-range, if the scalar couples to ordinary matter. From a particle physics point of view, one could expect that this is indeed the case, even if those interactions would have to be seriously suppressed by powers of the Planck scale [125,215]. However, current limits based on experiments concerning a fifth-force or time dependence of coupling constants, appear to be several orders of magnitude lower than this expectation [125,215]. This implies that, if quintessence really exists, then there should be a mechanism — probably a symmetry — that suppresses these couplings.

Yet another possibility for addressing the cosmological constant problems, or more precisely for dismissing them, comes when one adopts the approach that the accelerated expansion as inferred by observations is not due to some new physics but is actually due to a misinterpretation or an abuse of the underlying model being used. The Big Bang model is based on certain assumptions, such as homogeneity and isotropy, and apparently all calculations made rely on these assumptions. Even though, at present, one cannot claim that there is compelling evidence for this, it could be, for example, that the role of inhomogeneities is underestimated in the standard cosmological model and a more detailed model may provide a natural solution to the problem of dark energy, even by changing our interpretation of current observations (for instance see [70] and references therein).

1.4.6. The Dark Matter Problem

The presence of dark matter is indirectly inferred from observations through its gravitational interaction. Therefore, if one accepts that General Relativity describes gravity correctly, then an explanation for the nature of dark matter as some form of matter yet to be observed in the Universe or in the laboratory should be given. Note that dark matter is used here generically to mean matter that does not emit light. So, to begin with, its nature could be either baryonic and non-baryonic. The candidates for baryonic dark matter are mostly quite conventional astrophysical objects such as brown dwarfs, massive black holes and cold diffuse gas. However, there is precise evidence from observations that only a small fraction of dark matter can be baryonic (see for example [377] and [51,398] for reviews). Therefore, the real puzzle regards the nature of non-baryonic dark matter.

One can separate the candidates into two major categories: hot dark matter, i.e. non-baryonic particles which move (ultra-)relativistically, and cold dark matter i.e. non-baryonic particles which move non-relativistically. The most prominent candidate for hot dark matter is the neutrino. However, studies of the cosmic microwave background, numerical simulations and other astrophysical observations indicate that dark matter has clumped to form some structures on rather small scales and therefore it cannot consist mainly of particles with high velocities, since this clumping would then have been suppressed (see for example [425, 434] and references in [51]). For this reason, and because of its simplicity, cold dark matter currently gives the favoured picture.

There are many cold dark matter candidates and so we will refrain from listing them all or discussing their properties in detail here and refer the reader to the literature [51]. The most commonly considered ones are the axion and a number of weakly interacted massive particles (WIMPs) naturally predicted in supersymmetry theories, such as the neutralino, the sneutrino, the gravitino, the axino etc. There are a number of experiments aiming for direct and indirect detection of dark matter and some of them, such as the DAMA/NaI experiment [49], even claim to have already achieved that (see [222] for a full list of dark matter detection experiments and [398] for a review of experimental searches for dark matter). Great hope is also being placed on the Large Hadron Collider (LHC) [223], which is starting to operate and to constrain the parameter space of particles arising from supersymmetric theories. Finally, the improvement of cosmological and astrophysical observations obviously plays a crucial role. Let us close by saying that the general flavour or expectation seems to be that one of the proposed candidates will soon be detected and that the relevant dark matter scenario will be verified. Of course expectations are not always fulfilled and it is best to be prepared for surprises.

1.5. The Status of Gravity

In this introductory Chapter, an attempt has been made to pose clearly a series of open questions related, in one way or the other, to gravity and to discuss some of the most common approaches currently being pursued for their solution. This brings us to the main question motivating the research presented in this book: could all or at least some of the problems mentioned earlier be somehow related and is the fact that General Relativity is now facing so many challenges indicative of a need for some new gravitational physics, even at a classical level?

Let us be more analytic. In Section 1.1. and 1.2. we presented a brief chronological review of some landmarks in the passage from Newtonian gravity to General Relativity. One could find striking similarities with what has happened in the last decades with General Relativity itself. For instance, the cosmological and astrophysical observations which are interpreted as indicating the existence of dark matter and/or dark energy could be compared with Le Verrier's observation of the excess precession of Mercury's orbit. Remarkably, the first attempt to explain this phenomenon, was exactly the suggestion that an extra unseen — and therefore dark, in a way — planet orbited the Sun inside Mercury's orbit. The basic motivation behind this attempt, much like the contemporary proposals for matter fields to describe dark matter and dark energy, was to solve the problem within the context of an otherwise successful theory, instead of questioning the theory itself. Another example one could give, is the theoretical problems faced by Newtonian gravity once Special Relativity was established. The desire for a unified description of coordinate frames, inertial or not, and the need for a gravitational theory that is in good accordance with the conceptual basis of Special Relativity (e.g. Lorentz invariance) does not seem to be very far from the current desire for a unified description of forces and the need to resolve the conceptual clash between General Relativity and Quantum Field Theory.

The idea of looking for an alternative theory to describe the gravitational interaction is obviously not new. We already mentioned previously that attempts to unify gravity with quantum theory have included such considerations in the form of making quantum corrections to the gravitational field equations (or to the action, from a field theory perspective). Such corrections became effective at small scales or high energies. Additionally, many attempts have been made to modify General Relativity on both small and large scales, in order to address specific problems, such as those discussed earlier. Since we will refer to such modification extensively in the forthcoming Chapters, we will refrain from listing them here to avoid repetition. At present, we will confine ourselves to giving two very early examples of such attempts which were not triggered so much by a theoretical or observational need for a new theory, but by another important issue in our opinion: the desire to test the uniqueness of General Relativity as the only viable gravitational theory and the need to verify its conceptual basis.

Eddington, who performed the deflection of light experiment during the Solar eclipse of 1919 (which was one of the early experimental verifications of General Relativity), was one of fthe irst people to question whether Einstein's theory was the unique theory that could describe gravity [170]. Eddington tried to develop alternative theories sharing the same conceptual basis with General Relativity, most probably for the sake of theoretical completeness, since at the time there was no apparent reason coming from observations. Dicke was also one of the pioneers in exploring the conceptual basis of General Relativity and questioning Einstein's equivalence principle. He reformulated Mach's principle and together with Brans developed an alternative theory, known as Brans–Dicke theory [61, 162, 163]. Part of the value of Dicke's work lies on the fact that it helped people to understand that we do not know as much as we thought about the basic assumptions of General Relativity, a subject that we will discuss shortly.

Even though the idea of an alternative theory for gravitation is not new, a new perspective about it has emerged quite recently. The quantum corrections predicted in the 1960s were expected to appear only at small scales. On the other hand, Eddington's modification or Brans-Dicke theory were initially pursued as a conceptual alternative to General Relativity and had phenomenological effects on large scales as well. Now, due to both the shortcomings of Quantum Gravity and the puzzling cosmological and astrophysical observations, these ideas have stopped being considered unrelated. It seem worthwhile to consider the possibility of developing a gravitation theory that will be in agreement with observations and, at the same time, will be closer to the theories that emerge as a classical limit of our current approaches to Quantum Gravity, especially since it has been understood that quantum corrections might have an effect on large scale phenomenology as well.

Unfortunately, constructing a viable alternative to General Relativity with the above characteristics is far from being an easy task since there are numerous theoretical and observational restrictions. Two main paths have been followed towards achieving this goal: proposing phenomenological models tailored to fit observations, with the hope that they will soon gain some theoretical motivation from high energy physics and current Quantum

Gravity candidates, and developing ideas for Quantum Gravity, with the hope that they will eventually give the answer in the form of an effective gravitational theory through their classical limit which will account for unexplained observations. In this book a different approach will be followed in an attempt to combine and complement these two. At least according to the our opinion, we seem to be still at too early a stage in the development of our ideas about Quantum Gravity to be able to give precise answers about the type and form of the expected quantum corrections to General Relativity. Current observations still leave room for a wide range of different phenomenological models and so it seems a good idea to attempt exploring the limits of classical gravity by combining theory and observations. In a sense, this approach lies somewhere in the middle between the more conventional approaches mentioned before. Instead of starting from something known in order to extrapolate to the unknown, we attempt here to jump directly into the unknown, hoping that we will find an answer.

To this end, we will examine theories of gravity, trying to determine how far one can go from General Relativity. These theories have been chosen in such a way as to present a resemblance with the low energy effective actions of contemporary candidates for Quantum Gravity in a quest to study the phenomenology of the induced corrections. Their choice has also been motivated by a desire to fit recent unexplained observations. However, it should be stressed that both of these criteria have been used in a loose manner, since the main scope of this study is to explore the limits of alternative theories of gravity and hopefully shed some light on the strength and validity of the several assumptions underlying General Relativity. The main motivation comes from the fear that we may not know as much as we think or as much as needed to be known before making the key steps pursued in the last 50 years in gravitational physics; and from the hope that a better understanding of classical gravity might have a lot to offer in this direction.

As a conclusion to this Introduction, it is worth saying the following: it is probably too early to conclude whether General Relativity needs to be modified or replaced by some other gravitational theory or whether other solutions to the problems presented in this Chapter will eventually give the required answers. However, in scientific research, pursuing more than one possible solution to a problem has always been the wisest and most rewarding choice; not only because there is an already explored alternative when one of the proposed solutions fails, but also due to the fact that trial and error is one of the most efficient ways to get a deeper understanding of a physical theory. Exploring alternative theories of gravity, although having some disadvantages such as complexity, also presents a serious advantage: it is bound to be fruitful even if it leads to the conclusion that General Relativity is the only correct theory for gravitation, as it will have helped us both to understand General Relativity better and to secure our faith in it.

Before reviewing some popular alternatives and extensions to General Relativity in Chapter 4, we want to discuss the problem at its very origin: what can generate gravity? Are the Invariance Principles related to gravity? How is gravity related to the other gauge theories? Space-time deformations related to the gravitational field? These are the topics which will be discussed in Chapters 2 and 3.

Chapter 2

Gravity Emerging from Poincaré Gauge Invariance

2.1. General Considerations on the Gauge Theories

Since the perturbative scheme is unsatisfactory because it fails over *one-loop* level and cannot be renormalized, as we have seen in Chapter 1, we can ask if exists an invariance principles that produces the gravitation. Here we discuss how gravity and spin can be obtained as the realization of the local Conformal-Affine group of symmetry transformations. In particular, we show how gravitation is a gauge theory which can be obtained starting from some local invariance as the Poincaré local symmetry. Let us we start by considering some aspects of the gauge theories.

As we have seen, General Relativity and Quantum Mechanics are the two fundamental theories of modern physics and the Standard Model of particles is currently the most successful relativistic quantum field theory. It is a non-Abelian gauge theory (Yang-Mills theory) associated with the internal symmetry group $SU(3) \times SU(2) \times U(1)$, in which the $SU(3)$ color symmetry for the strong force in quantum chromodynamics is treated as exact whereas the $SU(2) \times U(1)$ symmetry, responsible for generating the electro-weak gauge fields, is spontaneously broken. So far as we know, there are four fundamental forces in Nature; namely, electromagnetic force, weak force, strong force and gravitational force. The Standard Model covers the first three, but not the gravitational interaction.

Here we intend to give a short summary of the various attempts to put together gravitation an the other interactions in view of a self-contained unified theory.

In General Relativity, the geometrized gravitational field is described by the metric tensor $g_{\mu\nu}$ of pseudo-Riemannian space-time, and the field equations that the metric tensor satisfies are nonlinear. This nonlinearity, as we have shortly discussed above, is indeed a source of difficulty in quantization of General Relativity. Since the successful Standard Model of particle physics is a gauge theory in which all the fields mediating the interactions are represented by gauge potentials, a question arises to understand why the fields mediating the gravitational interaction are different from those of other fundamental forces. It is reasonable to expect that there may be a gauge theory in which the gravitational fields stand on the same footing as those of other fields. This expectation has prompted a re-examination of General Relativity from the gauge theoretical point of view.

While the gauge groups involved in the Standard Model are all internal symmetry groups (e.g. spin is an internal symmetry), the gauge groups in General Relativity must be associated to external space-time symmetries. Therefore, the gauge theory of gravity would not be a usual Yang-Mills theory. It must be one in which gauge objects are not only the gauge potentials but also tetrads that relate the symmetry group to the external space-time. For this reason, we have to consider a more complex nonlinear gauge theory. In General Relativity, Einstein took the space-time metric as the basic variable representing gravity, whereas Ashtekar employed the tetrad fields and the connection forms as the fundamental variables. We also consider the tetrads and the connection forms as the fundamental fields.

Utiyama was the first to suggest that gravitation may be viewed as a gauge theory [443] in analogy to the Yang-Mills theory [442]. He identified the gauge potential, due to the Lorentz group, with the symmetric connection of Riemann geometry, and constructed Einstein's General Relativity as a gauge theory of the Lorentz group $SO(3, 1)$ with the help of tetrad fields introduced in an *ad hoc* manner. Although the tetrads were necessary components of the theory to relate the Lorentz group, adopted as an internal gauge group to the external space-time, they were not introduced as gauge fields. After, Kibble [248] constructed a gauge theory based on the Poincaré group $P(3, 1) = T(3, 1) \rtimes SO(3, 1)$ ($\rtimes$ represents the semi-direct product) which resulted in the Einstein-Cartan theory characterized by curvature and torsion. The translation group $T(3, 1)$ is considered responsible for generating the tetrads as gauge fields. Cartan [130] generalized the Riemann geometry to include torsion in addition to curvature. The torsion (tensor) arises from an asymmetric connection. Sciama [369], and others (Fikelstein [192], Hehl [210, 211]) pointed out that intrinsic spin may be the source of torsion of the underlying space-time manifold.

Since the form and role of the tetrad fields are very different from those of gauge potentials, it has been thought that even Kibble's attempt is not satisfactory as a full gauge theory. There have been a number of gauge theories of gravitation based on several Lie groups [133, 203, 210, 211, 213, 292, 293]. It was argued that a gauge theory of gravitation, corresponding to General Relativity, can be constructed with the translation group alone in the so-called teleparallel scheme. Inomata *et al.* [233] proposed that Kibble's gauge theory could be obtained in a way closer to the Yang-Mills approach by considering the de Sitter group $SO(4, 1)$ which is reducible to the Poincaré group by group-contraction. Unlike the Poincaré group, the de Sitter group is homogeneous and the associated gauge fields are all of gauge potential type. By the Wigner-Inönu group contraction procedure, one of the five vector potentials reduces to the tetrad.

It is standard to use the fiber-bundle formulation by which gauge theories can be constructed on the basis of any Lie group. A work by Hehl *et al.* [213] on the so-called Metric-Affine Gravity adopted, as a gauge group, the affine group $A(4, \mathbb{R}) = T(4) \rtimes GL(4, \mathbb{R})$ which was realized linearly. The tetrad was identified with the nonlinearly realized translational part of the affine connection on the tangent bundle. In metric-affine gravity, the Lagrangian is quadratic in both curvature and torsion in contrast to the Einstein-Hilbert Lagrangian of General Relativity which is linear in the scalar curvature. The theory has the Einstein limit on one hand and leads to the Newtonian inverse distance potential plus the linear confinement potential in the weak field approximation on the other. This approach has been recently developed also for more general theories as $f(R)$-gravity (see [82, 83] and also [290]). As we have seen above, there are many attempts to formulate gravitation as a gauge theory. Currently no theory has been uniquely accepted as the gauge theory of gravitation.

The nonlinear approach to group realizations was originally introduced by Coleman, Wess and Zumino [75,141] in the context of internal symmetry groups. It was later extended to the case of space-time symmetries by Isham, Salam, and Strathdee [232, 356] considering the nonlinear action of $GL(4, \mathbb{R})$ mod the Lorentz subgroup. In 1974, Borisov, Ivanov and Ogievetsky [60,234] considered the simultaneous nonlinear realization of the affine and conformal groups. They showed that General Relativity can be viewed as a consequence of spontaneous breakdown of the affine symmetry in much the same manner that chiral dynamics in quantum chromodynamics is a result of spontaneous breakdown of chiral symmetry. In their model, gravitons are considered as Goldstone bosons associated with the affine symmetry breaking. In 1978, Chang and Mansouri [134] used the nonlinear realization scheme employing $GL(4, \mathbb{R})$ as the principal group. In 1980, Stelle and West [393] investigated the nonlinear realization induced by the spontaneous breakdown of $SO(3, 2)$. In 1982 Ivanov and Niederle considered nonlinear gauge theories of the Poincaré, de Sitter, conformal and special conformal groups [235,236]. In 1983, Ivanenko and Sardanashvily [237] considered gravity to be a spontaneously broken $GL(4, \mathbb{R})$ gauge theory. The tetrads fields arise in their formulation as a result of the reduction of the structure group of the tangent bundle from the general linear Lorentz group. In 1987, Lord and Goswami [280, 281] developed the nonlinear realization in the fiber bundle formalism based on the bundle structure $G(G/H, H)$ as suggested by Ne'eman and Regge [305]. In this approach the quotient space G/H is identified with physical space-time. Most recently, in a series of papers, Lopez-Pinto, Julve, Tiemblo, Tresguerres and Mielke discussed nonlinear gauge theories of gravity on the basis of the Poincaré, affine and conformal groups [240, 282, 406, 407, 415, 416].

Now, following the prescriptions of General Relativity, the physical space-time is assumed to be a four-dimensional differential manifold. In Special Relativity, this manifold is the Minkwoski flat-space-time M_4 while, in General Relativity, the underlying space-time is assumed to be curved in order to describe the effects of gravitation.

As we said, Utiyama [443] proposed that General Relativity can be seen as a gauge theory based on the local Lorentz group in the same way that the Yang-Mills gauge theory [442] is developed on the basis of the internal iso-spin gauge group. In this formulation the Riemannian connection is the gravitational counterpart of the Yang-Mills gauge fields. While $SU(2)$, in the Yang-Mills theory, is an internal symmetry group, the Lorentz symmetry represents the local nature of space-time rather than internal degrees of freedom. The Einstein Equivalence Principle, asserted for General Relativity, requires that the local space-time structure can be identified with the Minkowski space-time possessing Lorentz symmetry as discussed in Chapter 3.25.

In order to relate local Lorentz symmetry to the external space-time, we need to solder the local space to the external space. The soldering tools can be the tetrad fields. Utiyama regarded the tetrads as objects given *a priori* while they can be dynamically generated [38] and the space-time has necessarily to be endowed with torsion in order to accommodate spinor fields. In other words, the gravitational interaction of spinning particles requires the modification of the Riemann space-time of General Relativity to be a (non-Riemannian) curved space-time with torsion. Although Sciama used the tetrad formalism for his gauge-like handling of gravitation, his theory fell shortcomings in treating tetrad fields as gauge fields. Following the Kibble approach [248], it can be demonstrated how gravitation can be formulated starting from a pure gauge viewpoint.

After this short summary of thirty years long attempts to put General Relativity on the same footing of non-Abelian gauge theories, the aim of this Chapter is to show, in details, how a theory of gravitation is a gauge theory which can be obtained starting from some local invariance, e.g. the local Poincaré symmetry (see [86, 87] and references therein). This dynamical structure give rise to a gauge theory of gravity, based on a nonlinear realization of the local conformal-affine group of symmetry transformations and on conservation principles [39]. In particular, we want to show how invariance properties and conservation laws induce the gravitational field and internal (spin) fields generalizing results in [86, 91].

Specifically, in this Chapter, we are going to consider the general problem of how gravity, as a gauge theory, could be achieved by the nonlinear realization of the conformal-affine group. We give all the mathematical tools for this realization discussing in details the bundle approach to the gauge theories and investigating also how internal symmetries as spin could be achieved under the same standard. The result is the "Invariance Induced Gravity" which can be considered as the prototype of all extensions of General Relativity.

2.2. The Bundle Approach to the Gauge Theories

Let us start by briefly reviewing the standard bundle approach to gauge theories. The bundle formalism, together with the conformal-affine group, will give us the mathematical tools to achieve gravity as an interaction induced from invariance properties. Besides, internal degrees of freedom of conformal-affine algebra will be related to the spin fields.

First of all, one has to verify that the usual gauge potential Ω is the pullback of connection 1-form ω by local sections of the bundle. After, the transformation laws of the ω and Ω under the action of the structure group G can be deduced.

Modern formulations of gauge field theories are expressible geometrically in the language of principal fiber bundles. A fiber bundle is a structure $\langle \mathbb{P}, M, \pi; \mathbb{F} \rangle$ where $\mathbb{P}$ (the total bundle space) and M (the base space) are smooth manifolds, $\mathbb{F}$ is the fiber space and the surjection π (a canonical projection) is a smooth map of $\mathbb{P}$ onto M,

$$\pi : \mathbb{P} \to M. \tag{2.1}$$

The inverse image π^{-1} is diffeomorphic to $\mathbb{F}$

$$\pi^{-1}(x) \equiv \mathbb{F}_x \approx \mathbb{F}, \tag{2.2}$$

and is called the fiber at $x \in M$. The partitioning $\bigcup_x \pi^{-1}(x) = \mathbb{P}$ is referred to as the fibration. Note that a smooth map is one whose coordinatization is C^∞ differentiable; a smooth manifold is a space that can be covered with coordinate patches in such a manner that a change from one patch to any overlapping patch is smooth, see A. S. Schwarz [362]. Fiber bundles that admit decomposition as a direct product, locally looking like $\mathbb{P} \approx M \times \mathbb{F}$, is called trivial. Given a set of open coverings $\{\mathcal{U}_i\}$ of M with $x \in \{\mathcal{U}_i\} \subset M$ satisfying

$\bigcup_\alpha \mathcal{U}_\alpha = M$, the diffeomorphism map is given by

$$\chi_i : \mathcal{U}_i \times_M G \rightarrow \pi^{-1}(\mathcal{U}_i) \in \mathbb{P}, \tag{2.3}$$

($\times_M$ represents the fiber product of elements defined over space M) such that $\pi(\chi_i(x,g)) = x$ and $\chi_i(x,g) = \chi_i(x,(id)_G)g = \chi_i(x)g\ \forall x \in \{\mathcal{U}_i\}$ and $g \in G$. Here, $(id)_G$ represents the identity element of group G. In order to obtain the global bundle structure, the local charts χ_i must be glued together continuously. Consider two patches $\mathcal{U}_n$ and $\mathcal{U}_m$ with a non-empty intersection $\mathcal{U}_n \cap \mathcal{U}_m \neq \emptyset$. Let ρ_{nm} be the restriction of χ_n^{-1} to $\pi^{-1}(\mathcal{U}_n \cap \mathcal{U}_m)$ defined by $\rho_{nm} : \pi^{-1}(\mathcal{U}_n \cap \mathcal{U}_m) \rightarrow (\mathcal{U}_n \cap \mathcal{U}_m) \times_M G_n$. Similarly let $\rho_{mn} : \pi^{-1}(\mathcal{U}_m \cap \mathcal{U}_n) \rightarrow (\mathcal{U}_m \cap \mathcal{U}_n) \times_M G_m$ be the restriction of χ_m^{-1} to $\pi^{-1}(\mathcal{U}_n \cap \mathcal{U}_m)$. The composite diffeomorphism $\Lambda_{nm} \in G$

$$\Lambda_{mn} : (\mathcal{U}_n \cap \mathcal{U}_m) \times G_n \rightarrow (\mathcal{U}_m \cap \mathcal{U}_n) \times_M G_m, \tag{2.4}$$

defined as

$$\Lambda_{ij}(x) \equiv \rho_{ji} \circ \rho_{ij}^{-1} = \chi_{i,x} \circ \chi_{j,x}^{-1} : \mathbb{F} \rightarrow \mathbb{F} \tag{2.5}$$

constitute the transition function between bundle charts ρ_{nm} and ρ_{mn} ($\circ$ represents the group composition operation) where the diffeomorphism $\chi_{i,x} : \mathbb{F} \rightarrow \mathbb{F}_x$ is written as $\chi_{i,x}(g) := \chi_i(x,g)$ and satisfies $\chi_j(x,g) = \chi_i(x, \Lambda_{ij}(x)g)$. The transition functions $\{\Lambda_{ij}\}$ can be interpreted as passive gauge transformations. They satisfy the identity $\Lambda_{ii}(x)$, inverse $\Lambda_{ij}(x) = \Lambda_{ji}^{-1}(x)$ and cocycle $\Lambda_{ij}(x)\Lambda_{jk}(x) = \Lambda_{ik}(x)$ consistency conditions. For trivial bundles, the transition function reduces to

$$\Lambda_{ij}(x) = g_i^{-1} g_j, \tag{2.6}$$

where $g_i : \mathbb{F} \rightarrow \mathbb{F}$ is defined by $g_i := \chi_{i,x}^{-1} \circ \tilde{\chi}_{i,x}$ provided the local trivializations $\{\chi_i\}$ and $\{\tilde{\chi}_i\}$ give rise to the same fiber bundle.

A section is defined as a smooth map

$$s : M \rightarrow \mathbb{P}, \tag{2.7}$$

such that $s(x) \in \pi^{-1}(x) = \mathbb{F}_x\ \forall x \in M$ and satisfies

$$\pi \circ s = (id)_M, \tag{2.8}$$

where $(id)_M$ is the identity element of M. It assigns to each point $x \in M$ a point in the fiber over x. Trivial bundles admit global sections.

A bundle is a principal fiber bundle $\langle \mathbb{P}, \mathbb{P}/G, G, \pi \rangle$ provided the Lie group G acts freely (i.e. if $pg = p$ then $g = (id)_G$) on $\mathbb{P}$ to the right $R_g p = pg$, $p \in \mathbb{P}$, preserves fibers on $\mathbb{P}$ ($R_g : \mathbb{P} \rightarrow \mathbb{P}$), and is transitive on fibers. Furthermore, there must exist local trivializations compatible with the G action. Hence, $\pi^{-1}(\mathcal{U}_i)$ is homeomorphic to $\mathcal{U}_i \times_M G$ and the fibers of $\mathbb{P}$ are diffeomorphic to G. The trivialization or inverse diffeomorphism map is given by

$$\chi_i^{-1} : \pi^{-1}(\mathcal{U}_i) \rightarrow \mathcal{U}_i \times_M G \tag{2.9}$$

such that $\chi^{-1}(p) = (\pi(p), \varphi(p)) \in \mathcal{U}_i \times_M G$, $p \in \pi^{-1}(\mathcal{U}_i) \subset \mathbb{P}$, where we see from the above definition that φ is a local mapping of $\pi^{-1}(\mathcal{U}_i)$ into G satisfying $\varphi(L_g p) = \varphi(p)g$ for any $p \in \pi^{-1}(\mathcal{U})$ and any $g \in G$. Observe that the elements of $\mathbb{P}$ which are projected onto the same $x \in \{\mathcal{U}_i\}$ are transformed into one another by the elements of G. In other words, the fibers of $\mathbb{P}$ are the orbits of G and at the same time, the set of elements which are projected onto the same $x \in \mathcal{U} \subset M$. This observation motivates calling the action of the group vertical and the base manifold horizontal. The diffeomorphism map χ_i is called the local gauge since χ_i^{-1} maps $\pi^{-1}(\mathcal{U}_i)$ onto the direct (Cartesian) product $\mathcal{U}_i \times_M G$. The action L_g of the structure group G on $\mathbb{P}$ defines an isomorphism of the Lie algebra $\mathfrak{g}$ of G onto the Lie algebra of vertical vector fields on $\mathbb{P}$ tangent to the fiber at each $p \in \mathbb{P}$ called fundamental vector fields

$$\lambda_g : T_p(\mathbb{P}) \to T_{gp}(\mathbb{P}) = T_{\pi(p)}(\mathbb{P}), \tag{2.10}$$

where $T_p(\mathbb{P})$ is the space of tangents at p, i.e. $T_p(\mathbb{P}) \in T(\mathbb{P})$. The map λ is a linear isomorphism for every $p \in \mathbb{P}$ and is invariant with respect to the action of G, that is, λ_g : $(\lambda_{g*} T_p(\mathbb{P})) \to T_{gp}(\mathbb{P})$, where λ_{g*} is the differential push forward map induced by λ_g defined by $\lambda_{g*} : T_p(\mathbb{P}) \to T_{gp}(\mathbb{P})$.

Since the principal bundle $\mathbb{P}(M, G)$ is a differentiable manifold, we can define tangent $T(\mathbb{P})$ and cotangent $T^*(\mathbb{P})$ bundles. The tangent space $T_p(\mathbb{P})$ defined at each point $p \in \mathbb{P}$ may be decomposed into a vertical $V_p(\mathbb{P})$ and horizontal $H_p(\mathbb{P})$ subspace as $T_p(\mathbb{P}) := V_p(\mathbb{P}) \oplus H_p(\mathbb{P})$ (where $\oplus$ represents the direct sum). The space $V_p(\mathbb{P})$ is a subspace of $T_p(\mathbb{P})$ consisting of all tangent vectors to the fiber passing through $p \in \mathbb{P}$, and $H_p(\mathbb{P})$ is the subspace complementary to $V_p(\mathbb{P})$ at p. The vertical subspace $V_p(\mathbb{P}) := \{X \in T(\mathbb{P}) \,|\, \pi(X) \in \mathcal{U}_i \subset M\}$ is uniquely determined by the structure of $\mathbb{P}$, whereas the horizontal subspace $H_p(\mathbb{P})$ cannot be uniquely specified. Thus we require the following condition: when p transforms as $p \to p' = pg$, $H_p(\mathbb{P})$ transforms as [304],

$$R_{g*} H_p(\mathbb{P}) \to H_{p'}(\mathbb{P}) = R_g H_p(\mathbb{P}) = H_{pg}(\mathbb{P}). \tag{2.11}$$

Let the local coordinates of $\mathbb{P}(M, G)$ be $p = (x, g)$ where $x \in M$ and $g \in G$. Let $\boldsymbol{G}_A$ denote the generators of the Lie algebra $\mathfrak{g}$ corresponding to group G satisfying the commutators $[\boldsymbol{G}_A, \boldsymbol{G}_B] = f_{AB}{}^C \boldsymbol{G}_C$, where $f_{AB}{}^C$ are the structure constants of G. Let Ω be a connection form defined by $\Omega^A := \Omega_i^A dx^i \in \mathfrak{g}$. Let ω be a connection 1-form defined by

$$\omega := \tilde{g}^{-1} \pi^*_{\mathbb{P}M} \Omega \tilde{g} + \tilde{g}^{-1} d\tilde{g} \tag{2.12}$$

($*$ represents the differential pullback map) belonging to $\mathfrak{g} \otimes T_p^*(\mathbb{P})$ where $T_p^*(\mathbb{P})$ is the space dual to $T_p(\mathbb{P})$. The differential pullback map applied to a test function φ and p-forms α and β satisfy $f^*\varphi = \varphi \circ f$, $(g \circ f)^* = f^* g^*$ and $f^*(\alpha \wedge \beta) = f^*\alpha \wedge f^*\beta$. If G is represented by a d-dimensional $d \times d$ matrix, then $\boldsymbol{G}_A = [\boldsymbol{G}_{\alpha\beta}]$, $\tilde{g} = [\tilde{g}^{\alpha\beta}]$, where $\alpha, \beta = 1, 2, 3, \ldots, d$. Thus, ω assumes the form

$$\omega_\alpha^\beta = (\tilde{g}^{-1})_{\alpha\gamma} d\tilde{g}^{\gamma\beta} + (\tilde{g}^{-1})_{\rho\gamma} \pi^*_{\mathbb{P}M} \Omega^\rho{}_{\sigma i} \boldsymbol{G}_{\alpha}^{\gamma} \tilde{g}^{\sigma\beta} \otimes dx^i. \tag{2.13}$$

If M is n-dimensional, the tangent space $T_p(\mathbb{P})$ is $(n+d)$-dimensional. Since the vertical subspace $V_p(\mathbb{P})$ is tangential to the fiber G, it is d-dimensional. Accordingly, $H_p(\mathbb{P})$

is n-dimensional. The basis of $V_p(\mathbb{P})$ can be taken to be $\partial_{\alpha\beta} := \frac{\partial}{\partial g^{\alpha\beta}}$. Now, let the basis of $H_p(\mathbb{P})$ be denoted by

$$E_i := \partial_i + \Gamma_i^{\alpha\beta}\partial_{\alpha\beta},\ i = 1, 2, 3, \ldots, n \text{ and } \alpha, \beta = 1, 2, 3, \ldots, d \tag{2.14}$$

where $\partial_i = \frac{\partial}{\partial x^i}$. The connection 1-form ω projects $T_p(\mathbb{P})$ onto $V_p(\mathbb{P})$. In order for $X \in T_p(\mathbb{P})$ to belong to $H_p(\mathbb{P})$, that is for $X \in H_p(\mathbb{P})$, $\omega_p(X) = \langle \omega(p) | X \rangle = 0$. In other words,

$$H_p(\mathbb{P}) := \{X \in T_p(\mathbb{P}) | \omega_p(X) = 0\}, \tag{2.15}$$

from which $\Omega_i^{\alpha\beta}$ can be determined. The inner product appearing in $\omega_p(X) = \langle \omega(p) | X \rangle = 0$ is a map $\langle \cdot | \cdot \rangle : T_p^*(\mathbb{P}) \times T_p(\mathbb{P}) \to \mathbb{R}$ defined by $\langle W | V \rangle = W_\mu V^\nu \left\langle dx^\mu | \frac{\partial}{\partial x^\nu} \right\rangle = W_\mu V^\nu \delta^\mu_\nu$, where the 1-form W and vector V are given by $W = W_\mu dx^\mu$ and $V = V^\mu \frac{\partial}{\partial x^\nu}$. Observe also that, $\left\langle dg^{\alpha\beta} | \partial_{\rho\sigma} \right\rangle = \delta^\alpha_\rho \delta^\beta_\sigma$.

We parameterize an arbitrary group element $\tilde{g}_\lambda$ as $\tilde{g}(\lambda) = e^{\lambda^A G_A} = e^{\lambda \cdot \boldsymbol{G}}$, $A = 1, \ldots, dim(\mathfrak{g})$. The right action $R_{\tilde{g}(\lambda)} = R_{\exp(\lambda \cdot G)}$ on $p \in \mathbb{P}$, i.e. $R_{\exp(\lambda \cdot \boldsymbol{G})} p = p \exp(\lambda \cdot \boldsymbol{G})$, defines a curve through p in $\mathbb{P}$. Define a vector $G^\# \in T_p(\mathbb{P})$ by [304]

$$G^\# f(p) := \frac{d}{dt} f(p \exp(\lambda \cdot \boldsymbol{G}))\,|_{\lambda=0} \tag{2.16}$$

where $f : \mathbb{P} \to \mathbb{R}$ is an arbitrary smooth function. Since the vector $G^\#$ is tangent to $\mathbb{P}$ at p, $G^\# \in V_p(\mathbb{P})$, the components of the vector $G^\#$ are the fundamental vector fields at p which constitute $V(\mathbb{P})$. The components of $G^\#$ may also be viewed as a basis element of the Lie algebra $\mathfrak{g}$. Given $G^\# \in V_p(\mathbb{P})$, $\boldsymbol{G} \in \mathfrak{g}$,

$$\begin{aligned} \omega_p(G^\#) &= \langle \omega(p) | G^\# \rangle = \tilde{g}^{-1} d\tilde{g}(G^\#) + \tilde{g}^{-1} \pi^*_{\mathbb{P}M} \Omega \tilde{g}(G^\#) \\ &= \tilde{g}_p^{-1} \tilde{g}_p \frac{d}{d\lambda} (\exp(\lambda \cdot \boldsymbol{G}))\,|_{\lambda=0}, \end{aligned} \tag{2.17}$$

where use was made of $\pi_{\mathbb{P}M*} G^\# = 0$. Hence, $\omega_p(G^\#) = \boldsymbol{G}$. An arbitrary vector $X \in H_p(\mathbb{P})$ may be expanded in a basis spanning $H_p(\mathbb{P})$ as $X := \beta^i E_i$. By direct computation, one can show

$$\left\langle \omega^\beta_\alpha | X \right\rangle = \left(\tilde{g}^{-1}\right)_{\alpha\gamma} \beta^i \Gamma_i^{\gamma\beta} + \left(\tilde{g}^{-1}\right)_{\alpha\gamma} \pi^*_{\mathbb{P}M} \Omega^\rho{}_{\sigma i} \beta^i \boldsymbol{G}^\gamma_\rho \tilde{g}^{\sigma\beta} = 0,\ \forall \beta^i \tag{2.18}$$

Equation (2.18) yields

$$\left(\tilde{g}^{-1}\right)_{\alpha\gamma} \Gamma_i^{\gamma\beta} + \left(\tilde{g}^{-1}\right)_{\alpha\gamma} \pi^*_{\mathbb{P}M} \Omega^\rho{}_{\sigma i} \boldsymbol{G}^\gamma_\rho \tilde{g}^{\sigma\beta} = 0, \tag{2.19}$$

from which we obtain

$$\Gamma_i^{\gamma\beta} = -\pi^*_{\mathbb{P}M} \Omega^\rho{}_{\sigma i} \boldsymbol{G}^\gamma_\rho \tilde{g}^{\sigma\beta}. \tag{2.20}$$

In this manner, the horizontal component is completely determined. An arbitrary tangent vector $\mathfrak{X} \in T_p(\mathbb{P})$ defined at $p \in \mathbb{P}$ takes the form

$$\mathfrak{X} = A^{\alpha\beta} \partial_{\alpha\beta} + B^i \left(\partial_i - \pi^*_{\mathbb{P}M} \Omega^\rho{}_{\sigma i} \boldsymbol{G}^\alpha_\rho \tilde{g}^{\sigma\beta} \partial_{\alpha\beta} \right), \tag{2.21}$$

where $A^{\alpha\beta}$ and B^i are constants. The vector field $\mathfrak{X}$ is comprised of horizontal $\mathfrak{X}^H := B^i\left(\partial_i - \pi^*_{\mathbb{P}M}\Omega^\rho{}_{\sigma i} G^\alpha_\rho \tilde{g}^{\sigma\beta}\partial_{\alpha\beta}\right) \in H(\mathbb{P})$ and vertical $\mathfrak{X}^V := A^{\alpha\beta}\partial_{\alpha\beta} \in V(\mathbb{P})$ components.

Let $\mathfrak{X} \in T_p(\mathbb{P})$ and $g \in \boldsymbol{G}$, then

$$R^*_g\omega(\mathfrak{X}) = \omega(R_{g*}\mathfrak{X}) = \tilde{g}^{-1}_{pg}\Omega(R_{g*}\mathfrak{X})\tilde{g}_{pg} + \tilde{g}^{-1}_{pg}d\tilde{g}_{pg}(R_{g*}\mathfrak{X}), \tag{2.22}$$

Observing that $\tilde{g}_{pg} = \tilde{g}_p g$ and $\tilde{g}^{-1}_{gp} = g^{-1}\tilde{g}^{-1}_p$ the first term on the r.h.s. of (2.22) reduces to $\tilde{g}^{-1}_{pg}\Omega(R_{g*}\mathfrak{X})\tilde{g}_{pg} = g^{-1}\tilde{g}^{-1}_p\Omega(R_{g*}\mathfrak{X})\tilde{g}_p g$ while the second term gives $\tilde{g}^{-1}_{pg}d\tilde{g}_{pg}(R_{g*}\mathfrak{X}) = g^{-1}\tilde{g}^{-1}_p d(R_{g*}\mathfrak{X})\tilde{g}_p g$. We therefore conclude

$$R^*_g\omega_\lambda = ad_{g^{-1}}\omega_\lambda, \tag{2.23}$$

where the adjoint map ad is defined by

$$ad_g Y := L_{g*} \circ R_{g^{-1}*} \circ Y = gYg^{-1}, \; ad_{g^{-1}}Y := g^{-1}Yg. \tag{2.24}$$

The potential Ω^A can be obtained from ω as $\Omega^A = s^*\omega$. To demonstrate this, let $Y \in T_p(M)$ and $\tilde{g}$ be specified by the inverse diffeomorphism or trivialization map (2.9) with $\chi^{-1}_\lambda(p) = (x, \tilde{g}_\lambda)$ for $p(x) = s_\lambda(x)\cdot\tilde{g}_\lambda$. We find $s^*_i\omega(Y) = \tilde{g}^{-1}\Omega(\pi_* s_{i*}Y)\tilde{g} + \tilde{g}^{-1}d\tilde{g}(s_{i*}Y)$, where we [304] have used $s_{i*}Y \in T_{s_i}(\mathbb{P})$, $\pi_* s_{i*} = (id)_{T_p(M)}$ and $\tilde{g} = (id)_G$ at s_i implying $\tilde{g}^{-1}d\tilde{g}(s_{i*}Y) = 0$. Hence,

$$s^*_i\omega(Y) = \Omega(Y). \tag{2.25}$$

To determine the gauge transformation of the connection 1-form ω we use the fact that $R_{\tilde{g}*}X = X\tilde{g}$ for $X \in T_p(M)$ and the transition functions $\tilde{g}_{nm} \in G$ defined between neighboring bundle charts (2.6). By direct computation we get

$$\begin{aligned} c_{j*}X &= \frac{d}{dt}c_j(\lambda(t))\,|_{t=0} = \frac{d}{dt}\left[c_i(\lambda(t))\cdot\tilde{g}_{ij}\right]|_{t=0} \\ &= R_{\tilde{g}_{ij}*}c^*_i(X) + \left(\tilde{g}^{-1}_{ji}(x)\,d\tilde{g}_{ij}(X)\right)^{\#}. \end{aligned} \tag{2.26}$$

where $\lambda(t)$ is a curve in M with boundary values $\lambda(0) = m$ and $\frac{d}{dt}\lambda(t)\,|_{t=0} = X$. Thus, we obtain the useful result

$$c_*X = R_{\tilde{g}*}(c_*X) + \left(\tilde{g}^{-1}d\tilde{g}(X)\right)^{\#}. \tag{2.27}$$

Applying ω to (6.35) we get

$$\omega(c_*X) = c^*\omega(X) = ad_{\tilde{g}^{-1}}c^*\omega(X) + \tilde{g}^{-1}d\tilde{g}(X), \forall X. \tag{2.28}$$

Hence, the gauge transformation of the local gauge potential Ω reads,

$$\Omega \to \Omega' = ad_{\tilde{g}^{-1}}(d + \Omega) = \tilde{g}^{-1}(d + \Omega)\tilde{g}. \tag{2.29}$$

Since $\Omega = c^*\omega$ we obtain from (2.29) the gauge transformation law of ω

$$\omega \to \omega' = \tilde{g}^{-1}(d + \omega)\tilde{g}. \tag{2.30}$$

Now we are ready to undertake the task to construct the bundle structure of gravitation.

2.3. The Bundle Structure

Let us recall the definition of gauge transformations in the context of ordinary fiber bundles. This step will be extremely relevant to induce metric and dynamics from invariance properties. Given a principal fiber bundle $\mathbb{P}(M, G; \pi)$ with base space M and standard G-diffeomorphic fiber, gauge transformations are characterized by bundle isomorphisms [199] $\lambda : \mathbb{P} \to \mathbb{P}$ exhausting all diffeomorphisms λ_M on M. This mapping is called an automorphism of $\mathbb{P}$ provided it is equivariant with respect to the action of G. This amounts to restricting the action λ of G along local fibers leaving the base space unaffected. Indeed, with regard to gauge theories of internal symmetry groups, a gauge transformation is a fiber preserving bundle automorphism, i.e. diffeomorphisms λ with $\lambda_M = (id)_M$. The automorphisms λ form a group called the automorphism group $Aut_{\mathbb{P}}$ of $\mathbb{P}$. The gauge transformations form a subgroup of $Aut_{\mathbb{P}}$ called the gauge group $G\,(Aut_{\mathbb{P}})$ (or G in short) of $\mathbb{P}$.

The map λ is required to satisfy two conditions, namely its commutability with the right action of G [the equivariance condition $\lambda\,(R_g(p)) = \lambda\,(pg) = \lambda\,(p)\,g$]

$$\lambda \circ R_g(p) = R_g(p) \circ \lambda, \;\; p \in \mathbb{P}, g \in G \tag{2.31}$$

according to which fibers are mapped into fibers, and the verticality condition

$$\pi \circ \lambda\,(u) = \pi\,(u), \tag{2.32}$$

where u and $\lambda\,(u)$ belong to the same fiber. The last condition ensures that no diffeomorphisms $\lambda_M : M \to M$ given by

$$\lambda_M \circ \pi\,(u) = \pi \circ \lambda\,(u), \tag{2.33}$$

be allowed on the base space M. In a gauge description of gravitation, one is interested in gauging external transformation groups. That is to say the group action on space-time coordinates cannot be neglected. The spaces of internal fiber and external base must be interlocked in the sense that transformations in one space must induce corresponding transformations in the other. The usual definition of a gauge transformation, i.e. as a displacement along local fibers not affecting the base space, must be generalized to reflect this interlocking. One possible way of framing this interlocking is to employ a nonlinear realization of the gauge group G, provided a closed subgroup $H \subset G$ exist. The interlocking requirement is then transformed into the interplay between groups G and one of its closed subgroups H.

Denote by G a Lie group with elements $\{g\}$. Let H be a closed subgroup of G specified by [37, 67]

$$H := \{h \in G | \Pi\,(R_h g) = \pi\,(g), \forall g \in G\}, \tag{2.34}$$

with elements $\{h\}$ and known linear representations $\rho\,(h)$. Here Π is the first of the two projection maps in (2.37), and R_h is the right group action. Let M be a differentiable manifold with points $\{x\}$ to which G and H may be referred, i.e. $g = g(x)$ and $h = h(x)$. Being that G and H are Lie groups, they are also manifolds. The right action of H on

G induce a complete partition of G into mutually disjoint orbits gH. Since $g = g(x)$, all elements of $gH = \{gh_1, gh_2, gh_3, \cdots, gh_n\}$ are defined over the same x. Thus, each orbit gH constitute an equivalence class of point x, with equivalence relation $g \equiv g'$ where $g' = R_h g = gh$. By projecting each equivalence class onto a single element of the quotient space $\mathcal{M} := G/H$, the group G becomes organized as a fiber bundle in the sense that $G = \bigcup_i \{g_i H\}$. In this manner the manifold G is viewed as a fiber bundle $G(\mathcal{M}, H; \Pi)$ with H-diffeomorphic fibers $\Pi^{-1}(\xi) : G \to \mathcal{M} = gH$ and base space $\mathcal{M}$. A composite principal fiber bundle $\mathbb{P}(M, G; \pi)$ is one whose G-diffeomorphic fibers possess the fibered structure $G(\mathcal{M}, H; \Pi) \simeq \mathcal{M} \times H$ described above. The bundle $\mathbb{P}$ is then locally isomorphic to $M \times G(\mathcal{M}, H)$. Moreover, since an element $g \in G$ is locally homeomorphic to $\mathcal{M} \times H$ the elements of $\mathbb{P}$ are - by transitivity - also locally homeomorphic to $M \times \mathcal{M} \times H \simeq \Sigma \times H$ where (locally) $\Sigma \simeq M \times \mathcal{M}$. Thus, an alternative view [416] of $\mathbb{P}(M, G; \pi)$ is provided by the $\mathbb{P}$-associated H-bundle $\mathbb{P}(\Sigma, H; \tilde{\pi})$. The total space $\mathbb{P}$ may be regarded as $G(\mathcal{M}, H; \Pi)$-bundles over base space M or equivalently as H-fibers attached to manifold $\Sigma \simeq M \times \mathcal{M}$.

The nonlinear realization technique [75, 141] provides a way to determine the transformation properties of fields defined on the quotient space G/H. The nonlinear realization of Diff(4, $\mathbb{R}$) becomes tractable due to a theorem given by V. I. Ogievetsky. According to the Ogievetsky theorem [60], the algebra of the infinite dimensional group Diff(4, $\mathbb{R}$) can be taken as the closure of the finite dimensional algebras of $SO(4, 2)$ and $A(4, \mathbb{R})$. Remind that the Lorentz group generates transformations that preserve the quadratic form on Minkowski space-time built from the metric tensor, while the special conformal group generates infinitesimal angle-preserving transformations on Minkowski space-time. The affine group is a generalization of the Poincaré group where the Lorentz group is replaced by the group of general linear transformations [86]. As such, the affine group generates translations, Lorentz transformations, volume preserving shear and volume changing dilation transformations. As a consequence, the nonlinear realization of Diff(4, $\mathbb{R}$) $/SO(3, 1)$ can be constructed by taking a simultaneous realization of the conformal group $SO(4, 2)$ and the affine group $A(4, \mathbb{R}) := \mathbb{R}^4 \rtimes GL(4, \mathbb{R})$ on the coset spaces $A(4, \mathbb{R})/SO(3, 1)$ and $SO(4, 2)/SO(3, 1)$. One possible interpretation of this theorem is that the conform-affine group (defined below) may be the largest subgroup of Diff(4, $\mathbb{R}$) whose transformations may be put into the form of a generalized coordinate transformation. We remark that a nonlinear realization can be made linear by embedding the representation in a sufficiently higher dimensional space. Alternatively, a linear group realization becomes nonlinear when subject to constraints. One type of relevant constraints may be those responsible for symmetry reduction from Diff(4, $\mathbb{R}$) to $SO(3, 1)$ for instance.

We take the group $CA(3, 1)$ as the basic symmetry group G. The conformal-affine group consists of the groups $SO(4, 2)$ and $A(4, \mathbb{R})$. In particular, conformal-affine is proportional to the union $SO(4, 2) \cup A(4, \mathbb{R})$. We know however (see section *Conform-Affine Lie Algebra*) that the affine and special conformal groups have several group generators in common. These common generators reside in the intersection $SO(4, 2) \cap A(4, \mathbb{R})$ of the two groups, within which there are *two copies* of $\Pi := D \times P(3, 1)$, where D is the group of scale transformations (dilations) and $P(3, 1) := T(3, 1) \rtimes SO(3, 1)$ is the Poincaré group. We define the conformal-affine group as the union of the affine and conformal groups minus *one copy* of the overlap Π, i.e. $CA(3, 1) := SO(4, 2) \cup A(4, \mathbb{R}) - \Pi$. Being defined in this way we recognize that $CA(3, 1)$ is a 24 parameter Lie group representing the action of Lorentz transformations (6), translations (4), special conformal transformations (4), space-time shears (9) and scale transformations (1). In this section, we are obtaining the nonlinear realization of $CA(3,1)$ modulo $SO(3,1)$.

2.4. The Conformal-Affine Lie Algebra

In order to implement the nonlinear realization procedure, we choose the partition Diff(4, $\mathbb{R}$) with respect to the Lorentz group. By Ogievetsky's theorem [60], we identify representations of Diff(4, $\mathbb{R}$)$/SO(3,1)$ with those of $CA(3,1)/SO(3,1)$. The 20 generators of affine transformations can be decomposed into the 4 translational $\boldsymbol{P}_{\mu}^{\mathrm{Aff}}$ and 16 $GL(4,\mathbb{R})$ transformations $\boldsymbol{\Lambda}_{\alpha}^{\beta}$. The 16 generators $\boldsymbol{\Lambda}_{\alpha}^{\beta}$ may be further decomposed into the 6 Lorentz generators $\boldsymbol{L}_{\alpha}^{\beta}$ plus the remaining 10 generators of symmetric linear transformation $\boldsymbol{S}_{\alpha}^{\beta}$, that is, $\boldsymbol{\Lambda}_{\beta}^{\alpha}=\boldsymbol{L}_{\beta}^{\alpha}+\boldsymbol{S}_{\beta}^{\alpha}$. The 10 parameter symmetric linear generators $\boldsymbol{S}_{\alpha}^{\beta}$ can be factored into the 9 parameter shear (the traceless part of $\boldsymbol{S}_{\alpha}^{\beta}$) generator defined by ${}^{\dagger}\boldsymbol{S}_{\alpha}^{\beta}=\boldsymbol{S}_{\alpha}^{\beta}-\frac{1}{4}\delta_{\alpha}^{\beta}\boldsymbol{D}$, and the 1 parameter dilaton generator $\boldsymbol{D}=tr\left(\boldsymbol{S}_{\alpha}^{\beta}\right)$. Shear transformations generated by ${}^{\dagger}\boldsymbol{S}_{\alpha}^{\beta}$ describe shape changing, volume preserving deformations, while the dilaton generator gives rise to volume changing transformations. The four diagonal elements of $\boldsymbol{S}_{\alpha}^{\beta}$ correspond to the generators of projective transformations. The 15 generators of conformal transformations are defined in terms of the set $\{J_{AB}\}$ where $A=0,1,2,\ldots,5$. The elements J_{AB} can be decomposed into translations $\boldsymbol{P}_{\mu}^{\mathrm{Conf}}:=J_{5\mu}+J_{6\mu}$, special conformal generators $\boldsymbol{\Delta}_{\mu}:=J_{5\mu}-J_{6\mu}$, dilatons $\boldsymbol{D}:=J_{56}$ and the Lorentz generators $\boldsymbol{L}_{\alpha\beta}:=J_{\alpha\beta}$. The Lie algebra of $CA(3,1)$ is characterized by the commutation relations

$$\begin{gathered}
\left[\boldsymbol{\Lambda}_{\alpha\beta},\boldsymbol{D}\right]=\left[\boldsymbol{\Delta}_{\alpha},\boldsymbol{\Delta}_{\beta}\right]=0,\ \left[\boldsymbol{P}_{\alpha},\boldsymbol{P}_{\beta}\right]=\left[\boldsymbol{D},\boldsymbol{D}\right]=0,\\
\left[\boldsymbol{L}_{\alpha\beta},\boldsymbol{P}_{\mu}\right]=io_{\mu[\alpha}\boldsymbol{P}_{\beta]},\ \left[\boldsymbol{L}_{\alpha\beta},\boldsymbol{\Delta}_{\gamma}\right]=io_{[\alpha|\gamma}\boldsymbol{\Delta}_{|\beta]},\\
\left[\boldsymbol{\Lambda}_{\beta}^{\alpha},\boldsymbol{P}_{\mu}\right]=i\delta_{\mu}^{\alpha}\boldsymbol{P}_{\beta},\ \left[\boldsymbol{\Lambda}_{\beta}^{\alpha},\boldsymbol{\Delta}_{\mu}\right]=i\delta_{\mu}^{\alpha}\boldsymbol{\Delta}_{\beta},\\
\left[\boldsymbol{S}_{\alpha\beta},\boldsymbol{P}_{\mu}\right]=io_{\mu(\alpha}\boldsymbol{P}_{\beta)},\ \left[\boldsymbol{P}_{\alpha},\boldsymbol{D}\right]=-i\boldsymbol{P}_{\alpha},\\
\left[\boldsymbol{L}_{\alpha\beta},\boldsymbol{L}_{\mu\nu}\right]=-i\left(o_{\alpha[\mu}\boldsymbol{L}_{\nu]\beta}-o_{\beta[\mu}\boldsymbol{L}_{\nu]\alpha}\right),\\
\left[\boldsymbol{S}_{\alpha\beta},\boldsymbol{S}_{\mu\nu}\right]=i\left(o_{\alpha(\mu}\boldsymbol{L}_{\nu)\beta}-o_{\beta(\mu}\boldsymbol{L}_{\nu)\alpha}\right),\\
\left[\boldsymbol{L}_{\alpha\beta},\boldsymbol{S}_{\mu\nu}\right]=i\left(o_{\alpha(\mu}\boldsymbol{S}_{\nu)\beta}-o_{\beta(\mu}\boldsymbol{S}_{\nu)\alpha}\right),\\
\left[\boldsymbol{\Delta}_{\alpha},\boldsymbol{D}\right]=i\boldsymbol{\Delta}_{\alpha},\ \left[\boldsymbol{S}_{\mu\nu},\boldsymbol{\Delta}_{\alpha}\right]=io_{\alpha(\mu}\boldsymbol{\Delta}_{\nu)},\\
\left[\boldsymbol{\Lambda}_{\beta}^{\alpha},\boldsymbol{\Lambda}_{\nu}^{\mu}\right]=i\left(\delta_{\nu}^{\alpha}\boldsymbol{\Lambda}_{\beta}^{\mu}-\delta_{\beta}^{\mu}\boldsymbol{\Lambda}_{\nu}^{\alpha}\right),\\
\left[\boldsymbol{P}_{\alpha},\boldsymbol{\Delta}_{\beta}\right]=2i\left(o_{\alpha\beta}\boldsymbol{D}-\boldsymbol{L}_{\alpha\beta}\right),
\end{gathered}\tag{2.35}$$

where $o_{\alpha\beta}=diag(-1,1,1,1)$ is Lorentz group metric. Here we adopt this signature since it is more suitable to dela with tetrad gravity and gauge theories. The above algebra is the core of the nonlinear realization and, in some sense, of the Invariance Induced Gravity.

2.5. Group Actions and Bundle Morphisms

Let us now introduce the main ingredients required to specify the structure of the fiber bundle, namely the canonical projection, the sections, etc. We follow the prescription in [416] for constructing the composite fiber bundle, but implement the program for the conformal-affine group.

The composite bundle $\mathbb{P}(\Sigma, H; \widetilde{\pi})$ is comprised of H-fibers, base space $\Sigma(M, \mathcal{M})$ and a composite map

$$\widetilde{\pi} \stackrel{\text{def}}{=} \widetilde{\pi}_{\Sigma M} \circ \Pi_{\mathbb{P}\Sigma} : \mathbb{P} \to \Sigma \to M, \tag{2.36}$$

with component projections

$$\Pi_{\mathbb{P}\Sigma} : \mathbb{P} \to \Sigma, \; \widetilde{\pi}_{\Sigma M} : \Sigma \to M. \tag{2.37}$$

The projection $\Pi_{\mathbb{P}\Sigma}$ maps the point $(p \in \mathbb{P}, R_h p \in \mathbb{P})$ into point $(x, \xi) \in \Sigma$. There is a correspondence between sections $s_{M\Sigma} : M \to \Sigma$ and the projection $\Pi_{\mathbb{P}\Sigma} : \mathbb{P} \to \Sigma$ in the sense that both maps project their functional argument onto elements of Σ. This is formalized by the relation, $\Pi_{\mathbb{P}\Sigma}(p) = s_{M\Sigma} \circ \pi_{\mathbb{P}M}(p)$. Hence, the total projection is given by

$$\widetilde{\pi} := \pi_{\mathbb{P}M} = \widetilde{\pi}_{\Sigma M} \circ \Pi_{\mathbb{P}\Sigma}. \tag{2.38}$$

Associated with the projections $\widetilde{\pi}_{\Sigma M}$ and $\Pi_{\mathbb{P}\Sigma}$ are the corresponding local sections

$$s_{M\Sigma} : \mathcal{U} \to \widetilde{\pi}^{-1}_{\Sigma M}(\mathcal{U}) \subset \Sigma, \, s_{\Sigma\mathbb{P}} : \mathcal{V} \to \Pi^{-1}_{\mathbb{P}\Sigma}(\mathcal{V}) \subset \mathbb{P}, \tag{2.39}$$

with neighborhoods $\mathcal{U} \subset M$ and $\mathcal{V} \subset \Sigma$ satisfying

$$\widetilde{\pi}_{\Sigma M} \circ s_{M\Sigma} = (id)_M, \, \Pi_{\mathbb{P}\Sigma} \circ s_{\Sigma\mathbb{P}} = (id)_\Sigma. \tag{2.40}$$

The bundle injection $\widetilde{\pi}^{-1}(\mathcal{U})$ is the inverse image of $\widetilde{\pi}(\mathcal{U})$ and is called the fiber over $\mathcal{U}$. The equivalence class $R_h p = pH \in \widetilde{\pi}^{-1}_{\Sigma M}(\mathcal{U})$ of left cosets is the fiber of $\mathbb{P}(\Sigma, H)$ while each orbit pH through $p \in \mathbb{P}$ projects into a single element $Q \in \Sigma$. In analogy to the total bundle projection (2.37), a total section of $\mathbb{P}$ is given by the total section composition

$$s_{M\mathbb{P}} = s_{\Sigma\mathbb{P}} \circ s_{M\Sigma}. \tag{2.41}$$

Let elements of G/H be labeled by the parameter ξ. Functions on G/H are represented by continuous coset functions $c(\xi)$ parameterized by ξ. These elements are referred to as cosets to the right of H with respect to $g \in G$. Indeed, the orbits of the right action of H on G are the left cosets $R_h g = gH$. For a given section $s_{M\mathbb{P}}(x \in M) \in \pi^{-1}_{\mathbb{P}M}$ with local coordinates (x, g) one can perform decompositions of the partial fibers $s_{M\Sigma}$ and $s_{\Sigma\mathbb{P}}$ as:

$$s_{M\Sigma}(x) = \widetilde{c}_{M\Sigma}(x) \cdot c = R_{c'} \circ \widetilde{c}_{M\Sigma}(x); \, c = c(\xi), \tag{2.42}$$

$$s_{\Sigma\mathbb{P}}(x, \xi) = \widetilde{c}_{\Sigma\mathbb{P}}(x, \xi) \cdot a' = R_{a'} \circ \widetilde{c}_{\Sigma\mathbb{P}}(x, \xi); \, a' \in H, \tag{2.43}$$

with the null sections $\{\widetilde{c}_{M\Sigma}(x)\}$ and $\{\widetilde{c}_{\Sigma\mathbb{P}}(x, \xi)\}$ having coordinates $(x, (id)_M)$ and $(x, \xi, (id)_H)$ respectively. A null or zero section is a map that sends every point $x \in M$

to the origin of the fiber $\pi^{-1}(x)$ over x, i.e. $\chi_i^{-1}(\widetilde{c}(x)) = (x, 0)$ in any trivialization. The trivialization map χ_i^{-1} is defined in (2.9) The identity map appearing in the above trivializations are defined as $(id)_{\mathcal{M}} : \mathcal{M} \to \mathcal{M}$ and $(id)_H : H \to H$. We assume the total null bundle section be given by the composition law

$$\widetilde{c}_{M\mathbb{P}} = \widetilde{c}_{\Sigma\mathbb{P}} \circ \widetilde{c}_{M\Sigma}. \tag{2.44}$$

The images of two sections $s_{\Sigma\mathbb{P}}$ and $s_{M\Sigma}$ over $x \in M$ must coincide, implying $s_{\Sigma\mathbb{P}}(x, \xi) = s_{M\Sigma}(x)$. Using (2.41) with (2.42), (2.43) and (2.44), we arrive at the total bundle section decomposition

$$s_{M\mathbb{P}}(x) = \widetilde{c}_{M\mathbb{P}}(x) \cdot g = R_g \circ \widetilde{c}_{M\mathbb{P}}(x) \tag{2.45}$$

provided $g = c \cdot a$ and

$$\widetilde{c}_{\Sigma\mathbb{P}} = R_{c^{-1}} \circ \widetilde{c}_{\Sigma\mathbb{P}}(x, \xi) \circ R_c. \tag{2.46}$$

The pullback of $\widetilde{c}_{\Sigma\mathbb{P}}$, defined [416] as

$$\widetilde{c}_{\xi}(x) = \left(s^*_{M\Sigma}\widetilde{c}_{\Sigma\mathbb{P}}\right)(x) = \widetilde{c}_{\Sigma\mathbb{P}} \circ s_{M\Sigma} = \widetilde{c}_{\Sigma\mathbb{P}}(x, \xi), \tag{2.47}$$

ensures the coincidence of images of sections $\widetilde{c}_{\xi}(x) : M \to \mathbb{P}$ and $\widetilde{c}_{\Sigma\mathbb{P}}(x, \xi) : \Sigma \to \mathbb{P}$, respectively. With the aid of the above results, we arrive at the equation

$$\widetilde{c}_{\Sigma\mathbb{P}}(x, \xi) = \widetilde{c}_{M\mathbb{P}}(x) \cdot c(\xi), \tag{2.48}$$

which will be extremely useful in the following.

2.6. Nonlinear Realizations and Generalized Gauge Transformations

The generalized gauge transformation law is obtained by comparing bundle elements $p \in \mathbb{P}$ that differ by the left action of elements of the principal group G, $L_{g\in G}$. An arbitrary element $p \in \mathbb{P}$ can be written in terms of the null section with the aid of (2.45), (2.46) and (2.48) as

$$p = s_{M\mathbb{P}}(x) = R_a \circ \widetilde{c}_{\Sigma\mathbb{P}}(x, \xi), \, a \in H. \tag{2.49}$$

Performing a gauge transformation on p, we obtain the orbit $\lambda(p)$ defining a curve through (x, ξ) in Σ

$$\lambda(p) = L_{g(x)} \circ p = R_{a'} \circ \widetilde{c}_{\Sigma\mathbb{P}}(x, \xi'); \; g(x) \in G, \; a' \in H. \tag{2.50}$$

Comparison of (2.49) with (2.50) leads to

$$L_{g(x)} \circ R_a \circ \widetilde{c}_{\Sigma\mathbb{P}}(x, \xi) = R_{a'} \circ \widetilde{c}_{\Sigma\mathbb{P}}(x, \xi'). \tag{2.51}$$

By virtue of the commutability [304] of left and right group translations of elements belonging to G, i.e. $L_g \circ R_h = R_h \circ L_g$, Eq. (2.229) may be recast as

$$L_{g(x)} \circ \widetilde{c}_{\Sigma\mathbb{P}}(x, \xi) = R_h \circ \widetilde{c}_{\Sigma\mathbb{P}}(x, \xi'). \tag{2.52}$$

where $R_{a^{-1}} \circ R_{a'} \equiv R_{a'a^{-1}} := R_h$ and $a'a^{-1} \equiv h \in H$. Equation (2.52) constitute a generalized gauge transformation. Performing the pullback of (2.52) with respect to the section $s_{M\Sigma}$ leads to

$$L_{g(x)} \circ \widetilde{c}_{\xi}(x) = R_{h(\xi,\, g(x))} \circ \widetilde{c}_{\xi'}(x). \tag{2.53}$$

Thus, the left action L_g of G is a map that acts on $\mathbb{P}$ and Σ. In particular, L_g acting on fibers defined as orbits of the right action describes diffeomorphisms that transforming fibers over $\widetilde{c}_{\xi}(x)$ into the fibers $\widetilde{c}_{\xi'}(x)$ of Σ while simultaneously being displaced along H fibers via the action of R_h. Equation (2.53) states that nonlinear realizations of G mod H is determined by the action of an arbitrary element $g \in G$ on the quotient space G/H transforming one coset into another as

$$L_g : G/H \to G/H,\ c(\xi) \to c(\xi') \tag{2.54}$$

inducing a diffeomorphism $\xi \to \xi'$ on G/H. To simplify the action induced by (2.53) for calculation purposes we proceed as follows. Departing from (2.47) and substituting $s_{M\Sigma} = R_c \circ \widetilde{c}_{M\mathbb{P}}$ we get

$$\widetilde{c}_{\xi}(x) = \widetilde{c}_{\Sigma\mathbb{P}} \circ R_c \circ \widetilde{c}_{M\Sigma}. \tag{2.55}$$

Using $\widetilde{c}_{M\mathbb{P}} \circ R_c = R_c \circ \widetilde{c}_{M\mathbb{P}}$, (2.55) becomes $\widetilde{c}_{\xi}(x) = R_c \circ \widetilde{c}_{\Sigma\mathbb{P}} \circ \widetilde{c}_{M\Sigma} = R_c \circ \widetilde{c}_{M\mathbb{P}}$, where the last equality follows from use of $\widetilde{c}_{M\mathbb{P}} = \widetilde{c}_{\Sigma\mathbb{P}} \circ \widetilde{c}_{M\Sigma}$. By way of analogy, we assume $\widetilde{c}_{\xi'}(x) \equiv R_{c'} \circ \widetilde{c}_{M\mathbb{P}}$. Upon substitution of $\widetilde{c}_{\xi'}$ into (2.53) we obtain

$$L_g \circ R_c \circ \widetilde{c}_{M\mathbb{P}} = R_{h(\xi,\, g(x))} \circ R_{c'} \circ \widetilde{c}_{M\mathbb{P}}, \tag{2.56}$$

which after implementing the group actions is equivalent to,

$$g \cdot \widetilde{c}_{M\mathbb{P}} \cdot c = \widetilde{c}_{M\mathbb{P}} \cdot c' \cdot h. \tag{2.57}$$

Operating on (2.57) from the left by $\widetilde{c}^{-1}_{M\mathbb{P}}$ and making use of $g = \widetilde{c}^{-1}_{M\mathbb{P}} g \widetilde{c}_{M\mathbb{P}}$, we get $\left(\widetilde{c}^{-1}_{M\mathbb{P}} \cdot g \cdot \widetilde{c}_{M\mathbb{P}}\right) \cdot c = c' \cdot h$ which leads to $g \cdot c_{\xi} = c_{\xi'} \cdot h$, or

$$c' = g \cdot c \cdot h^{-1} \tag{2.58}$$

in short, where $c \equiv c_{\xi}$ and $c' \equiv c_{\xi'}$. Observe that the element h is a function whose argument is the couple $(\xi, g(x))$. The transformation rule (2.58) is in fact the key equation to determine the nonlinear realizations of G and specifies a unique H-valued field $h(\xi, g(x))$ on G/H.

Consider a family of sections $\{\widehat{c}(x, \xi)\}$ defined [406] on Σ by

$$\widehat{c}(x, \xi) := c \circ \widetilde{c}(x, \xi) = c(\widetilde{c}(x, \xi)). \tag{2.59}$$

Taking $\Pi_{\mathbb{P}\Sigma} \circ R_h \circ \widetilde{c}_{\Sigma\mathbb{P}} = \Pi_{\mathbb{P}\Sigma} \circ \widetilde{c}_{\Sigma\mathbb{P}} = (id)_{\Sigma}$ into account, we can explicitly exhibit the fact that the left action L_g of G on the null sections $\widetilde{c}_{\Sigma\mathbb{P}} : \mathbb{P} \to \Sigma$ induces an equivalence relation between differing elements $\widetilde{c}_{\xi}, \widetilde{c}_{\xi'} \in \Sigma$ given by

$$\Pi_{\mathbb{P}\Sigma} \circ L_g \circ \widehat{c}_{\xi} = \Pi_{\mathbb{P}\Sigma} \circ R_{h(\xi,\, g(x))} \circ \widehat{c}_{\xi'} = R_{h(\xi,\, g(x))} \circ \widehat{c}_{\xi'}, \tag{2.60}$$

so that

$$\widetilde{c}'_{\xi} := R_{h(\xi,\, g(x))} \circ \widetilde{c}_{\xi'} = L_g \circ \widetilde{c}_{\xi}. \quad (2.61)$$

From (2.233) we can write

$$\widetilde{c}_{\xi} \overset{L_g}{\longmapsto} \widetilde{c}'_{\xi} = R_{h(\xi,\, g(x))} \circ \widetilde{c}_{\xi'} \;\forall h \in H. \quad (2.62)$$

Equation (2.247) gives rise to a complete partition of G/H into equivalence classes $\Pi^{-1}_{\mathbb{P}\Sigma}(\xi)$ of left cosets [406, 408]

$$cH = \left\{R_{h(\xi,\, g(x))} \circ c / c \in G/H, \forall h \in H\right\} = \{ch_1, ch_2, \ldots, ch_n\}, \quad (2.63)$$

where $c \in (G-H)$ plays the role of the fibers attached to each point of Σ. The elements ch_i are single representatives of each equivalence class $R_{h(\xi,\, g(x))} \circ c = cH \in \widetilde{\pi}^{-1}_{\Sigma M}(\mathcal{U})$. Thus, any diffeomorphism $L_g \circ \widetilde{c}_{\xi}$ on Σ together with the H-valued function $h(\xi, g(x))$ determine a unique gauge transformation $\widetilde{c}'_{\xi} = R_{h(\xi,\, g(x))} \circ \widetilde{c}_{\xi'}$. This demonstrates that gauge transformations are those diffeomorphisms on Σ that map fibers over $c(\xi)$ into fibers over $c(\xi')$ and simultaneously preserve the action of H.

2.7. The Covariant Coset Field Transformations

We now proceed to determine the transformation behavior of parameters belonging to G/H. The elements of the conformal-affine and Lorentz groups are respectively parameterized about the identity element as

$$g = e^{i\varepsilon^{\alpha} P_{\alpha}} e^{i\alpha^{\mu\nu}\, {}^{\dagger}S_{\mu\nu}} e^{i\beta^{\mu\nu} L_{\mu\nu}} e^{ib^{\alpha}\Delta_{\alpha}} e^{i\varphi D},\; h = e^{iu^{\mu\nu} L_{\mu\nu}}. \quad (2.64)$$

Elements of the coset space G/H are coordinatized by

$$c = e^{-i\xi^{\alpha} P_{\alpha}} e^{ih^{\mu\nu}\, {}^{\dagger}S_{\mu\nu}} e^{i\zeta^{\alpha}\Delta_{\alpha}} e^{i\phi D}. \quad (2.65)$$

We consider transformations with infinitesimal group parameters ε^{α}, $\alpha^{\mu\nu}$, $\beta^{\mu\nu}$, b^{α} and φ. The transformed coset parameters read $\xi'^{\alpha} = \xi^{\alpha} + \delta\xi^{\alpha}$, $h'^{\mu\nu} = h^{\mu\nu} + \delta h^{\mu\nu}$, $\zeta'^{\alpha} = \zeta^{\alpha} + \delta\zeta^{\alpha}$ and $\phi' = \phi + \delta\phi$. Note that $u^{\mu\nu}$ is infinitesimal. The translational coset field variations reads

$$\delta\xi^{\alpha} = -\left(\alpha_{\beta}{}^{\alpha} + \beta_{\beta}{}^{\alpha}\right)\xi^{\beta} - \varepsilon^{\alpha} - \varphi\xi^{\alpha} - \left[|\xi|^2 b^{\alpha} - 2(b\cdot\xi)\xi^{\alpha}\right]. \quad (2.66)$$

For the dilatons we get,

$$\delta\phi = \varphi + 2(b\cdot\xi) - \left\{u^{\alpha}_{\beta}\xi^{\beta} + \varepsilon^{\alpha} + \varphi\xi^{\alpha} + \left[b^{\alpha}|\xi|^2 - 2(b\cdot\xi)\xi^{\alpha}\right]\right\}\partial_{\alpha}\phi. \quad (2.67)$$

Similarly for the special conformal 4-boosts we find,

$$\begin{aligned}\delta\zeta^{\alpha} \;&=\; u^{\alpha}_{\beta}\zeta^{\beta} + b^{\alpha} - \varphi\zeta^{\alpha} + 2\left[(b\cdot\xi)\zeta^{\alpha} - (b\cdot\zeta)\xi^{\alpha}\right] \\ &\quad - \left\{u^{\beta}_{\lambda}\xi^{\lambda} + \varepsilon^{\beta} + \varphi\xi^{\beta} + \left[b^{\beta}|\xi|^2 - 2(b\cdot\xi)\xi^{\beta}\right]\right\}\partial_{\beta}\zeta^{\alpha}. \end{aligned} \quad (2.68)$$

Observe the homogeneous part of the special conformal coset parameter ζ^α has the same structure as that of the translational parameter ξ^α (with the substitutions: $\zeta^\alpha \to -\xi^\alpha$ and $-\varepsilon^\alpha \to b^\alpha$). For the shear parameters we obtain

$$\delta r^{\alpha\beta} = \left(\alpha^{\gamma\alpha} + \beta^{\gamma\alpha}\right) r_\gamma{}^\beta + u^\beta_\gamma r^{\alpha\gamma} + 2b^{[\alpha}\xi^{\rho]} r_\rho{}^\beta, \tag{2.69}$$

where $r^{\alpha\beta} := e^{h^{\alpha\beta}}$. From $\delta r^{\alpha\beta}$ we obtain the nonlinear Lorentz transformation

$$u^{\alpha\beta} = \beta^{\alpha\beta} + 2b^{[\alpha}\xi^{\beta]} - \alpha^{\mu\nu} \tanh\left\{\frac{1}{2}\ln\left[r^\alpha_\mu \left(r^{-1}\right)^\beta_\nu\right]\right\}. \tag{2.70}$$

In the limit of vanishing special conformal 4-boost, this result coincides with that of Pinto *et al.* [282]. For vanishing shear, the result of Julve *et al* [240] is obtained.

In this section, all covariant coset field transformations have been determined directly from the nonlinear transformation law (2.58). We observe that the translational coset parameter transforms as a coordinate under the action of G. From the shear coset variation, the explicit form of the nonlinear Lorentz-like transformation was obtained. From (2.70) it is clear that $u^{\alpha\beta}$ contains the linear Lorentz parameter in addition to conformal and shear contributions via the nonlinear 4-boosts and symmetric GL_4 parameters.

2.8. The Decomposition of Connections in $\pi_{\mathbb{P}M} : \mathbb{P} \to M$ into Components in $\pi_{\mathbb{P}\Sigma} : \mathbb{P} \to \Sigma$ and in $\pi_{\Sigma M} : \Sigma \to M$

At this point, it is useful to classify all possible decompositions of connections in order to achieve all the conformal-affine nonlinear gauge potentials. This step is essential to have all the ingredients to construct the induced metric and dynamics related to the conformal-affine group.

Depending on which bundle is considered, either the total bundle $\mathbb{P} \to M$ or the intermediate bundles $\mathbb{P} \to \Sigma$, $\Sigma \to M$, we may construct corresponding Ehresmann connections for the respective space. With respect to M, we have the connection form

$$\omega = \widetilde{g}^{-1}\left(d + \pi^*_{\mathbb{P}M}\Omega_M\right)\widetilde{g}. \tag{2.71}$$

The gauge potential Ω_M is defined in the standard manner as the pullback of the connection ω by the null section $\widetilde{c}_{M\mathbb{P}}$, $\Omega_M = \widetilde{c}^*_{M\mathbb{P}}\omega \in T^*(M)$. With regard to the space Σ an alternative form of the connection is given by

$$\omega = a^{-1}\left(d + \pi^*_{\mathbb{P}\Sigma}\Gamma_\Sigma\right)a, \tag{2.72}$$

where the connection on Σ reads $\Gamma_\Sigma = \widetilde{c}^*_{\Sigma\mathbb{P}}\omega$. Carrying out a similar analysis and evaluating the tangent vector $X \in T_p(\Sigma)$ at each point ξ along the curve c_ξ on the coset space G/H that coincides with the section $\widetilde{c}^*_{\Sigma\mathbb{P}}$, we find the gauge transformation law

$$\omega \to \omega' = ad_{h^{-1}}\left(d + \omega\right). \tag{2.73}$$

Comparison of (2.71) and 2.72 leads to $\pi^*_{\mathbb{P}\Sigma}\Gamma_\Sigma = c^{-1}\left(d + \pi^*_{\mathbb{P}M}\Omega_M\right)c$. Taking account of

$\widetilde{c}^{*}_{\Sigma\mathbb{P}}\Pi^{*}_{\mathbb{P}\Sigma}=(id)_{T^{*}(\Sigma)}$ which follows from $\Pi_{\mathbb{P}\Sigma}\circ\widetilde{c}_{\Sigma\mathbb{P}}=(id)_{\Sigma}$, we deduce

$$\Gamma_{\Sigma}=\widetilde{c}^{*}_{\Sigma\mathbb{P}}\left[c^{-1}\left(d+\pi^{*}_{\mathbb{P}M}\Omega_{M}\right)c\right]. \tag{2.74}$$

By use of the family of sections pulled back to Σ introduced in (2.59) we find $\widetilde{c}^{*}_{\Sigma\mathbb{P}}\left(c^{-1}dc\right)=\widehat{c}^{\,-1}d\widehat{c}$ and $\widetilde{c}^{*}_{\Sigma\mathbb{P}}R^{*}_{c}=R^{*}_{\widehat{c}}\widetilde{c}^{*}_{\Sigma\mathbb{P}}$. Recalling $\widetilde{\pi}^{*}_{\mathbb{P}M}=\widetilde{\pi}^{*}_{\mathbb{P}\Sigma}\widetilde{\pi}^{*}_{\Sigma M}$, we get $c^{-1}\widetilde{\pi}^{*}_{\mathbb{P}M}\Omega_{M}c=R^{*}_{c}\widetilde{\pi}^{*}_{\mathbb{P}M}\Omega_{M}$. With these results in hand, we obtain the alternative form of the connection Γ_{Σ},

$$\Gamma_{\Sigma}=\widehat{c}^{-1}\left(d+\pi^{*}_{\Sigma M}\Omega_{M}\right)\widehat{c}. \tag{2.75}$$

Completing the pullback of Γ_{Σ} to M by means of $\widetilde{c}_{M\Sigma}$ we obtain, $\Gamma_{M}=\widetilde{c}^{*}_{M\Sigma}\Gamma_{\Sigma}$. By use of $\Gamma_{\Sigma}=\widetilde{c}^{*}_{\Sigma\mathbb{P}}\omega$ and (2.47) we find $\Gamma_{M}=s^{*}_{M\Sigma}\widetilde{c}^{*}_{\Sigma\mathbb{P}}\omega=\widetilde{c}^{*}_{\xi}\omega$. In terms of the substitution $\widehat{c}(x,\xi)\rightarrow\overline{c}(x)$ where $\overline{c}(x)$ is the pullback of $\widehat{c}(x,\xi)$ to M defined as $\overline{c}(x)=s^{*}_{M\Sigma}\widehat{c}=c\left(\widetilde{c}_{\xi}(x)\right)$, we arrive at the desired result

$$\boldsymbol{\Gamma}\equiv\Gamma_{M}=\overline{c}^{-1}\left(d+\Omega_{M}\right)\overline{c}, \tag{2.76}$$

which explicitly relates the connection $\boldsymbol{\Gamma}$ on Σ pulled back to M to its counterpart Ω_{M}.

The gauge transformation behavior of $\boldsymbol{\Gamma}$ may be determined directly by use of (2.29) and the transformation $\widetilde{c}'=g\widetilde{c}h^{-1}$. We calculate

$$\boldsymbol{\Gamma}'=h\widetilde{c}^{-1}g^{-1}d\left(g\widetilde{c}h^{-1}\right)+h\widetilde{c}^{-1}\Omega\widetilde{c}h^{-1}+h\widetilde{c}^{-1}\left(dg^{-1}\right)g\widetilde{c}h^{-1}. \tag{2.77}$$

Observing however, that

$$h\widetilde{c}^{-1}g^{-1}d\left(g\widetilde{c}h^{-1}\right)=h\widetilde{c}^{-1}\left(g^{-1}dg\right)\widetilde{c}h^{-1}+h\widetilde{c}^{-1}d\widetilde{c}h^{-1}+hdh^{-1}, \tag{2.78}$$

we obtain

$$\boldsymbol{\Gamma}'=h\left[\widetilde{c}^{-1}\left(d+\Omega\right)\widetilde{c}\right]h^{-1}+hdh^{-1}+h\widetilde{c}^{-1}d\left(gg^{-1}\right)\widetilde{c}h^{-1}. \tag{2.79}$$

Thus, we arrive at the gauge transformation law

$$\boldsymbol{\Gamma}'=h\boldsymbol{\Gamma}h^{-1}+hdh^{-1}. \tag{2.80}$$

According to the Lie algebra decomposition of $\mathfrak{g}$ into $\mathfrak{h}$ and $\mathfrak{c}$, the connection Γ_{Σ} can be divided into $\boldsymbol{\Gamma}_{H}$ defined on the subgroup H and $\boldsymbol{\Gamma}_{G/H}$ defined on G/H. From the transformation law (2.80) it is clear that $\boldsymbol{\Gamma}_{H}$ transforms inhomogeneously

$$\boldsymbol{\Gamma}'_{H}=h\boldsymbol{\Gamma}_{H}h^{-1}+hdh^{-1}, \tag{2.81}$$

while $\Gamma_{G/H}$ transforms as a tensor

$$\boldsymbol{\Gamma}'_{G/H}=h\boldsymbol{\Gamma}_{G/H}h^{-1}. \tag{2.82}$$

In this regard, only Γ_{H} transforms as a true connection. We use the gauge potential $\boldsymbol{\Gamma}$ to define the gauge covariant derivative

$$\boldsymbol{\nabla}:=\left(d+\rho\left(\boldsymbol{\Gamma}\right)\right) \tag{2.83}$$

acting on ψ as $\nabla\psi = (d+\rho(\Gamma))\psi$ with the desired transformation property

$$(\nabla\psi(c(\xi)))' = \rho(h(\xi,g))\nabla\psi(c(\xi)) \simeq (1+iu(\xi,g)\rho(H))\nabla\psi(c(\xi)) \tag{2.84}$$

leading to

$$\delta(\nabla\psi(c(\xi))) = iu(\xi,g)\rho(H)\nabla\psi(c(\xi)). \tag{2.85}$$

Let us now classify the conformal-affine gauge potentials considering the various components of the decomposition.

2.8.1. Conformal-Affine Nonlinear Gauge Potential in $\pi_{\mathbb{P}M} : \mathbb{P} \to M$

The ordinary gauge potential defined on the total base space M reads

$$\Omega = -i\left(\overset{\mathrm{T}}{\Gamma}{}^{\alpha}\boldsymbol{P}_\alpha + \overset{\mathrm{C}}{\Gamma}{}^{\alpha}\boldsymbol{\Delta}_\alpha + \overset{\mathrm{D}}{\Gamma}\boldsymbol{D} + \overset{\mathrm{GL}}{\Gamma}{}^{\alpha\beta}\,{}^{\dagger}\boldsymbol{\Lambda}_{\alpha\beta}\right). \tag{2.86}$$

The horizontal basis vectors that span the horizontal tangent space $\mathbb{H}(\mathbb{P})$ of $\pi_{\mathbb{P}M} : \mathbb{P} \to M$ are given by

$$E_i = \widetilde{c}_{M\mathbb{P}*}\partial_i - \Omega_i. \tag{2.87}$$

The explicit form of the connections (2.86) are given by

$$\omega = -i\Big[V^{\mu}_{M}\widetilde{\chi}_{\mu}{}^{\nu}\boldsymbol{P}_\nu - i\left(i\overline{\Theta}{}^{\alpha\beta}_{(\dagger\Lambda)} + \widetilde{\pi}^{*}_{\mathbb{P}M}\overset{\mathrm{GL}}{\Gamma}{}^{\alpha\beta}\right)\widetilde{\chi}_{\alpha}{}^{\nu}\widetilde{\chi}_{\beta}{}^{\nu}\,{}^{\dagger}\boldsymbol{\Lambda}_{\mu\nu}$$
$$+\vartheta^{\mu}_{M}\widetilde{\beta}_{\mu}{}^{\nu}\boldsymbol{\Delta}_\nu - i\widetilde{\pi}^{*}_{\mathbb{P}M}\Phi_M\boldsymbol{D}\Big] \tag{2.88}$$

where $\overline{\Theta}{}^{\alpha\beta}_{(\dagger\Lambda)} = \overline{\Theta}{}^{\alpha\beta}_{(L)} + \overline{\Theta}{}^{\alpha\beta}_{(SY)}$, with right invariant Maurer-Cartan forms

$$\overline{\Theta}{}^{\mu\nu}_{(L)} = i\widetilde{\beta}^{[\nu|}{}_{\gamma}d\widetilde{\beta}^{|\mu]\gamma} - 2idb^{\mu}\varepsilon^{\nu} \text{ and } \overline{\Theta}{}^{\mu\nu}_{(SY)} = i\widetilde{\alpha}^{(\nu|}{}_{\gamma}d\widetilde{\alpha}^{|\mu)\gamma}. \tag{2.89}$$

The linear connection Ω_M varies under the action of G as

$$\delta\Omega = \Omega' - \Omega = \delta\overset{\mathrm{T}}{\Gamma}{}^{\mu}\boldsymbol{P}_\mu + \delta\overset{\mathrm{C}}{\Gamma}{}^{\mu}\boldsymbol{\Delta}_\mu + \delta\overset{\mathrm{D}}{\Gamma}\boldsymbol{D} + \delta\overset{\mathrm{GL}}{\Gamma}{}^{\beta\nu}\,{}^{\dagger}\boldsymbol{\Lambda}_{\beta\nu} \tag{2.90}$$

where

$$\begin{aligned}
\delta\overset{\mathrm{T}}{\Gamma}{}^{\mu} &= {}^{\dagger}\overset{\mathrm{GL}}{D}\varepsilon^{\mu} - \overset{\mathrm{T}}{\Gamma}{}^{\alpha}\left(\alpha_{\alpha}^{\mu} + \beta_{\alpha}^{\mu} + \varphi\delta_{\alpha}^{\mu}\right) - \overset{\mathrm{D}}{\Gamma}\varepsilon^{\mu},\\
\delta\overset{\mathrm{C}}{\Gamma}{}^{\mu} &= {}^{\dagger}\overset{\mathrm{GL}}{D}b^{\mu} - \overset{\mathrm{C}}{\Gamma}{}^{\alpha}\left(\alpha_{\alpha}^{\mu} + \beta_{\alpha}^{\mu} - \varphi\delta_{\alpha}^{\mu}\right) + \overset{\mathrm{D}}{\Gamma}b^{\mu},\\
\delta\overset{\mathrm{GL}}{\Gamma}{}^{\alpha\beta} &= {}^{\dagger}\overset{\mathrm{GL}}{D}\left(\alpha^{\alpha\beta} + \beta^{\alpha\beta}\right) + \left(\overset{\mathrm{T}}{\Gamma}{}^{[\alpha}b^{\beta]} + \overset{\mathrm{C}}{\Gamma}{}^{[\alpha}\varepsilon^{\beta]}\right),\\
\delta\overset{\mathrm{D}}{\Gamma} &= d\varphi + 2\left(\overset{\mathrm{C}}{\Gamma}{}^{\alpha}\varepsilon_{\alpha} - \overset{\mathrm{T}}{\Gamma}{}^{\alpha}b_{\alpha}\right).
\end{aligned} \tag{2.91}$$

The components of $\overline{\omega}$ on M are identified as space-time quantities and are determined from the pullback of the corresponding (quotient space) quantities defined on Σ:

$$V^{\mu}_{M}=s^{*}_{M\Sigma}V^{\mu}_{\Sigma},\ \vartheta^{\mu}_{M}=s^{*}_{M\Sigma}\vartheta^{\mu}_{\Sigma},\ \Phi_{M}=s^{*}_{M\Sigma}\Phi_{\Sigma}\ \text{and}\ \Gamma^{\mu\nu}_{M}=s^{*}_{M\Sigma}\Gamma^{\mu\nu}_{\Sigma}. \tag{2.92}$$

In the following, we depart from the alternative form of the connection $\omega=a^{-1}(d+\Pi^{*}_{\mathbb{P}\Sigma}\Gamma_{\Sigma})a,\ \forall\, a\in H$ on Σ.

2.8.2. Conformal-Affine Nonlinear Gauge Potential in $\pi_{\mathbb{P}\Sigma}:\mathbb{P}\to\Sigma$

The components of ω in $\mathbb{P}\to\Sigma$ are oriented along the Lie algebra basis of H

$$\overset{L}{\omega}=a^{-1}\left(d+i\widetilde{\pi}^{*}_{\mathbb{P}\Sigma}\overset{\circ}{\Gamma}{}^{\alpha\beta}\boldsymbol{L}_{\alpha\beta}\right)a=-i\overset{L}{\omega}{}^{\alpha\beta}\boldsymbol{L}_{\alpha\beta}, \tag{2.93}$$

where

$$\overset{L}{\omega}{}^{\alpha\beta}:=\left(i\widetilde{\Theta}^{\rho\sigma}_{(L)}+\widetilde{\pi}^{*}_{\mathbb{P}\Sigma}\Gamma^{\rho\sigma}_{[L]}\right)\widetilde{\beta}^{\alpha}_{[\rho}\widetilde{\beta}^{\beta}_{\sigma]}. \tag{2.94}$$

2.8.3. Conformal-Affine Nonlinear Gauge Potential on $\Pi_{\Sigma M}:\Sigma\to M$

The components of ω in $\Pi_{\Sigma M}:\Sigma\to M$ are oriented [416] along the Lie algebra basis of the quotient space G/H belonging to Σ

$$\overset{P}{\omega}=-ia^{-1}\left(\widetilde{\pi}^{*}_{\Sigma M}V^{\nu}_{\Sigma}\boldsymbol{P}_{\nu}\right)a=-i\overset{P}{\omega}{}^{\mu}\boldsymbol{P}_{\mu}, \tag{2.95}$$

$$\overset{\Delta}{\omega}=-ia^{-1}\left(\widetilde{\pi}^{*}_{\Sigma M}\vartheta^{\nu}_{\Sigma}\boldsymbol{\Delta}_{\nu}\right)a=-i\overset{\Delta}{\omega}{}^{\mu}\boldsymbol{\Delta}_{\mu}, \tag{2.96}$$

$$\overset{D}{\omega}=-ia^{-1}\left(\widetilde{\pi}^{*}_{\Sigma M}\Phi_{\Sigma}\boldsymbol{D}\right)a=-i\omega_{[D]}\boldsymbol{D}, \tag{2.97}$$

$$\overset{SY}{\omega}=-ia^{-1}\left(\widetilde{\pi}^{*}_{\Sigma M}\Upsilon^{\alpha\beta}\boldsymbol{S}_{\alpha\beta}\right)a=-i\overset{SY}{\omega}{}^{\alpha\beta}\boldsymbol{S}_{\alpha\beta}, \tag{2.98}$$

where

$$\overset{P}{\omega}{}^{\mu}:=\widetilde{\pi}^{*}_{\Sigma M}V^{\nu}_{\Sigma}\widetilde{\beta}^{\mu}_{\nu},\ \overset{\Delta}{\omega}{}^{\mu}:=\widetilde{\pi}^{*}_{\Sigma M}\vartheta^{\nu}_{\Sigma}\widetilde{\beta}^{\mu}_{\nu}, \tag{2.99}$$

$$\omega_{[D]}:=\widetilde{\pi}^{*}_{\Sigma M}\Phi_{\Sigma},\ \overset{SY}{\omega}{}^{\alpha\beta}:=\widetilde{\pi}^{*}_{\mathbb{P}\Sigma}\Upsilon^{\rho\sigma}\widetilde{\alpha}^{\alpha}_{(\rho}\widetilde{\alpha}^{\beta}_{\sigma)}. \tag{2.100}$$

By direct computation we obtain

$$\boldsymbol{\Gamma}^{CA}_{\Sigma}=-i\left(V^{\mu}_{\Sigma}\boldsymbol{P}_{\mu}+i\vartheta^{\mu}_{\Sigma}\boldsymbol{\Delta}_{\mu}+\Phi_{\Sigma}\boldsymbol{D}+\Gamma^{\alpha\beta}_{\Sigma}\boldsymbol{\Lambda}_{\alpha\beta}\right). \tag{2.101}$$

The nonlinear translational and special conformal connection coefficients V_Σ^ν and ϑ_Σ^ν read

$$V_\Sigma^\beta = \widetilde{\pi}_{\Sigma M}^{*}\left[e^{\phi}\left(\upsilon^\beta(\xi) + r^\alpha_\sigma \overset{C}{\Gamma}{}^\sigma \mathfrak{B}_\alpha^\beta(\xi)\right)\right], \tag{2.102}$$

$$\vartheta_\Sigma^\beta = \widetilde{\pi}_{\Sigma M}^{*}\left[e^{-\phi}\left(\upsilon^\beta(\zeta) + \upsilon^\sigma(\xi)\,\mathfrak{B}_\sigma^\beta(\zeta)\right)\right], \tag{2.103}$$

with

$$\upsilon_i^\beta(\xi) := r_\sigma^\beta\left({}^{\dagger}\overset{GL}{D}_i\xi^\sigma + \overset{D}{\Gamma}_i\xi^\sigma + \overset{T}{\Gamma}{}_i^\sigma\right),\ \mathfrak{B}_\alpha^\rho(\xi) := \left(|\xi|^2\delta_\alpha^\rho - 2\xi_\alpha\xi^\rho\right). \tag{2.104}$$

The nonlinear GL_4 and dilaton connections are given by

$$\Gamma_\Sigma^{\mu\nu} = \widehat{\Gamma}{}^{\mu\nu} + 2\zeta^{[\mu}\varpi^{\nu]}, \tag{2.105}$$

$$\Phi = \widetilde{\pi}_{\Sigma M}^{*}\left(\zeta_\beta\varpi^\beta\right) - \frac{1}{2}d\phi, \tag{2.106}$$

with

$$\widehat{\Gamma}{}^{\mu\nu} := \widetilde{\pi}_{\Sigma M}^{*}\left[\left(r^{-1}\right)^\mu_\sigma \overset{GL}{\Gamma}{}^{\sigma\beta} r_\beta^\nu - \left(r^{-1}\right)^\mu_\sigma dr^{\sigma\nu}\right] \tag{2.107}$$

and

$$\varpi^\nu := \upsilon^\nu + r_\alpha^\nu \overset{C}{\Gamma}{}^\alpha. \tag{2.108}$$

The nonlinear GL_4 connection can be expanded in the GL_4 Lie algebra according to $\Gamma^{\alpha\beta}\ {}^{\dagger}\boldsymbol{\Lambda}_{\alpha\beta} = \overset{\circ}{\Gamma}{}^{\alpha\beta}\boldsymbol{L}_{\alpha\beta} + \Upsilon^{\alpha\beta}\ {}^{\dagger}\boldsymbol{S}_{\alpha\beta}$, where

$$\overset{\circ}{\Gamma}{}_\Sigma^{\alpha\beta} := \widehat{\Gamma}{}^{[\alpha\beta]} + 2\zeta^{[\alpha}\varpi^{\beta]},\ \Upsilon_\Sigma^{\alpha\beta} := \widehat{\Gamma}{}^{(\alpha\beta)}. \tag{2.109}$$

The symmetric GL_4 (shear) gauge fields Υ are distortion fields describing the difference between the general linear connection and the Levi-Civita connection.

We define the (group) algebra bases e_ν and h_ν dual to the translational and special conformal 1-forms V^μ and ϑ^μ as

$$e_\mu \ : \ = e_\mu^i s_{M\Sigma *}\partial_i = \partial_{\xi^\mu} - e_\mu^i \widetilde{e}_i, \tag{2.110}$$

$$h_\mu \ : \ = h_\mu^i s_{M\Sigma *}\partial_i = \partial_{\zeta^\mu} - h_\mu^i \widetilde{h}_i, \tag{2.111}$$

with corresponding tetrad-like components

$$e_i^\mu(\xi) \ = \ e^{\phi}\left(\upsilon_i^\mu(\xi) + r^\alpha_\sigma \overset{C}{\Gamma}{}_i^\sigma \mathfrak{B}_\alpha^\mu(\xi)\right), \tag{2.112}$$

$$h_i^\mu(\xi,\zeta) \ = \ e^{-\phi}\left(\upsilon_\rho^\mu(\zeta) + \upsilon_i^\sigma(\xi)\,\mathfrak{B}_\sigma^\mu(\zeta)\right), \tag{2.113}$$

and basis vectors (on M)

$$\widetilde{e}_j(\xi) = \widetilde{c}_{M\Sigma*}\partial_j - e^{\phi}\left[r_\mu{}^\nu\left(\overset{GL}{\Gamma}{}_{j\alpha}{}^{\mu}\xi^\alpha + \overset{D}{\Gamma}_j\xi^\mu + \overset{T}{\Gamma}{}_j^{\ \mu}\right) + \overset{C}{\Gamma}{}_j^{\ \sigma} r^\mu{}_\sigma \mathfrak{B}_\mu^{\ \nu}(\xi)\right]\partial_{\xi^\nu} \tag{2.114}$$

and

$$\widetilde{h}_j(\xi,\zeta) = \widetilde{c}_{M\Sigma*}\partial_j + e^{-\phi}\Bigg[r^\mu{}_\rho\left(\overset{GL}{\Gamma}{}_{j\alpha}{}^{\rho}\zeta^\alpha + \overset{C}{\Gamma}{}_j^{\ \rho}\right)$$
$$+ r^\gamma{}_\sigma\left(\overset{GL}{\Gamma}{}_{j\alpha}{}^{\sigma}\xi^\alpha + \overset{D}{\Gamma}_j\xi^\sigma + \overset{T}{\Gamma}{}_j^{\ \sigma}\right)\mathfrak{B}_\gamma^\mu(\zeta)\Bigg]\partial_{\zeta^\mu}. \tag{2.115}$$

Here $\upsilon^\beta(\zeta) = \upsilon^\beta(\xi\to\zeta)$, $\mathfrak{B}_\alpha^\beta(\zeta) = \mathfrak{B}_\alpha^\rho(\xi\to\zeta)$. By definition, the basis vectors satisfy the orthogonality relations

$$\langle V_\Sigma^\mu|\widetilde{e}_j\rangle = 0,\ \left\langle \vartheta_\Sigma^\mu|\widetilde{h}_j\right\rangle = 0,\ \langle V^\mu|e_\nu\rangle = \delta_\nu^\mu,\ \langle \vartheta^\mu|h_\nu\rangle = \delta_\nu^\mu. \tag{2.116}$$

We introduce the dilatonic and symmetric GL_4 *algebra* bases

$$\flat := \partial_\phi - d^i\widetilde{d}_i,\ f_{\mu\nu} := \partial_{\alpha^{\mu\nu}} - f^{\,i}_{\mu\nu}\widetilde{f}_i \tag{2.117}$$

with *auxiliary soldering* components d_i and $f_i^{\,\mu\nu}$,

$$d_i = \zeta_\sigma r^\sigma{}_\rho\left({}^{\dagger}\overset{GL}{D}_i\xi^\rho + \overset{D}{\Gamma}_i\xi^\rho + \overset{T}{\Gamma}{}_i^{\ \rho} + \overset{C}{\Gamma}{}_i^{\ \rho}\right) - \frac{1}{2}\partial_i\phi, \tag{2.118}$$

$$f_i^{\,\mu\nu} = \left(r^{-1}\right)^\mu_{\ \sigma}\overset{GL}{\Gamma}{}_i^{\ \sigma\beta} r_\beta^{\ \nu} - \left(r^{-1}\right)^\mu_{\ \sigma}\partial_i r^{\sigma\nu}. \tag{2.119}$$

The *coordinate* bases $\widetilde{d}_j$ and $\widetilde{f}_j$ read

$$\widetilde{d}_j(\xi,\zeta,\phi,h) := \widetilde{c}_{M\Sigma*}\partial_j - \zeta_\sigma r^\sigma{}_\rho\left({}^{\dagger}\overset{GL}{\Gamma}{}^{\rho}_{j\gamma}\xi^\gamma + \overset{D}{\Gamma}_j\xi^\rho + \overset{T}{\Gamma}{}_j^{\ \rho} + \overset{C}{\Gamma}{}_j^{\ \rho}\right)\partial_\phi, \tag{2.120}$$

and

$$\widetilde{f}_j(\xi,h) := \widetilde{c}_{M\Sigma*}\partial_j - \left(\left(r^{-1}\right)^{(\mu|}{}_\sigma\overset{GL}{\Gamma}{}_j^{\ \sigma\beta} r_\beta^{\ |\nu)} - \left(r^{-1}\right)^{(\mu|}{}_\sigma\partial_j r^{\sigma|\nu)}\right)\partial_{h^{\mu\nu}}. \tag{2.121}$$

The bases satisfy

$$\left\langle\Phi|\widetilde{d}_i\right\rangle = 0,\ \left\langle\Upsilon^{\alpha\beta}|\widetilde{f}_i\right\rangle = 0,\ \langle\Phi|\flat\rangle = I,\ \left\langle\Upsilon^{\alpha\beta}|f_{\mu\nu}\right\rangle = \delta_\mu^\alpha\delta_\nu^\beta. \tag{2.122}$$

With the basis vectors and tetrad components in hand, we observe

$$V_M^\mu := dx^i\otimes e_i^\mu,\ \vartheta_M^\mu := dx^i\otimes h_i^\mu,$$
$$\Phi_M := dx^i\otimes e_i^\alpha\langle\Phi|e_\alpha\rangle = dx^i\otimes d_i. \tag{2.123}$$

The symmetric and antisymmetric GL_4 connection pulled back to M is given by

$$\Upsilon^{\mu\nu}_{M} = dx^i \otimes e^{\alpha}_{i} \left\langle \Upsilon^{\mu\nu}_{\Sigma} | e_\alpha \right\rangle := dx^i \otimes f_i^{\;\mu\nu},$$
$$\overset{\circ}{\Gamma}{}^{\mu\nu}_{M} = dx^i \otimes e^{\alpha}_{i} \left\langle \overset{\circ}{\Gamma}{}^{\mu\nu}_{\Sigma} | e_\alpha \right\rangle := dx^i \otimes \overset{\circ}{\Gamma}{}^{\mu\nu}_{i}. \quad (2.124)$$

With the aid of (2.259) and (2.286), we determine

$$V^{\beta}_{i} := e_i^{\;\alpha} \left\langle V^{\beta}_{\Sigma} | e_\alpha \right\rangle = e_i^{\;\alpha} \delta^{\beta}_{\alpha} = e_i^{\;\beta},\ \vartheta^{\beta}_{i} \equiv h^{\beta}_{i},\ \Upsilon^{\mu\nu}_{i} \equiv f_i^{\;\mu\nu},\ \Phi_i \equiv d_i. \quad (2.125)$$

The horizontal tangent subspace vectors in $\widetilde{\pi}_{\mathbb{P}\Sigma} : \mathbb{P} \to \Sigma$ are given by

$$\widehat{E}_i = \widetilde{c}_{M\mathbb{P}*} \widetilde{e}_i + i \widetilde{c}_{M\Sigma*} \left\langle \overset{\circ}{\Gamma}{}^{\alpha\beta} | \widetilde{e}_i \right\rangle \overset{\text{Int}}{\widehat{\mathfrak{R}}}{}^{(L)}_{\alpha\beta}, \quad (2.126)$$

$$\widehat{E}_\mu = \widetilde{c}_{\Sigma\mathbb{P}*} \widetilde{e}_\mu + i \left\langle \overset{\circ}{\Gamma}{}^{\alpha\beta} | \widetilde{e}_\mu \right\rangle \overset{\text{Int}}{\widehat{\mathfrak{R}}}{}^{(L)}_{\alpha\beta}, \quad (2.127)$$

and satisfy

$$\left\langle \overset{L}{\omega} | \widehat{E}_j \right\rangle = 0 = \left\langle \overset{L}{\omega} | \widehat{E}_\mu \right\rangle. \quad (2.128)$$

The right invariant fundamental vector operator appearing in (2.126) or (2.127) is given by

$$\widehat{\mathfrak{R}}{}^{(L)}_{\mu\nu} = i \left(\widetilde{\beta}_{[\mu|}{}^{\gamma} \frac{\partial}{\partial \widetilde{\beta}^{|\nu]\gamma}} + \varepsilon_{[\mu} \frac{\partial}{\partial \varepsilon^{\nu]}} \right). \quad (2.129)$$

On the other hand, the vertical tangent subspace vector in $\widetilde{\pi}_{\mathbb{P}\Sigma} : \mathbb{P} \to \Sigma$ satisfies

$$\left\langle \overset{L}{\omega} | \widehat{\mathfrak{L}}{}^{(L)}_{\mu\nu} \right\rangle = \boldsymbol{L}_{\mu\nu} = \left\langle \overset{L}{\omega} | \widehat{\mathfrak{R}}{}^{(L)}_{\mu\nu} \right\rangle, \quad (2.130)$$

where

$$\widehat{\mathfrak{L}}{}^{(L)}_{\mu\nu} = i \widetilde{\beta}_{\gamma[\mu|} \frac{\partial}{\partial \widetilde{\beta}_{\gamma}{}^{|\nu]}},\ \widehat{\mathfrak{R}}{}^{(L)}_{\mu\nu} = i \left(\widetilde{\beta}_{[\mu|}{}^{\gamma} \frac{\partial}{\partial \widetilde{\beta}^{|\nu]\gamma}} + \varepsilon_{[\mu} \frac{\partial}{\partial \varepsilon^{\nu]}} \right). \quad (2.131)$$

and $\widetilde{\beta}_{\mu}^{\;\nu} := e^{\beta_{\mu}^{\;\nu}} = \delta_{\mu}^{\;\nu} + \beta_{\mu}^{\;\nu} + \frac{1}{2!} \beta_{\mu}^{\;\gamma} \beta_{\gamma}^{\;\nu} + \cdots$. The horizontal tangent subspace vectors in $\Pi_{\Sigma M} : \Sigma \to M$ are given by

$$\widetilde{E}_j = \widetilde{c}_{\Sigma\mathbb{P}*} \widetilde{e}_j,\ \widetilde{H}_j = \widetilde{c}_{\Sigma\mathbb{P}*} \widetilde{h}_j,\ \widehat{E}{}^{(D)}_{i} = \widetilde{c}_{\Sigma\mathbb{P}*} \widetilde{d}_j,\ \breve{E}_j = \widetilde{c}_{\Sigma\mathbb{P}*} \widetilde{f}_j, \quad (2.132)$$

and satisfy

$$\left\langle \overset{P}{\omega} | \widetilde{E}_j \right\rangle = 0,\ \left\langle \overset{\Delta}{\omega} | \widetilde{H}_j \right\rangle = 0,\ \left\langle \overset{\text{SY}}{\omega} | \breve{E}_j \right\rangle = 0,\ \left\langle \overset{D}{\omega} | \widehat{E}{}^{(D)}_{i} \right\rangle = 0. \quad (2.133)$$

The vertical tangent subspace vectors in $\Pi_{\Sigma M} : \Sigma \to M$ are given by

$$\widetilde{E}_\mu = \widetilde{c}_{\Sigma\mathbb{P}*}\widehat{\mathfrak{L}}^{(P)}_{\mu},\ \breve{E}_{\alpha\beta} = \widetilde{c}_{\Sigma\mathbb{P}*}\widehat{\mathfrak{L}}^{(SY)}_{\alpha\beta},\ \widetilde{H}_\mu = \widetilde{c}_{\Sigma\mathbb{P}*}\widehat{\mathfrak{L}}^{(\Delta)}_{\mu},\ \widehat{E}^{(D)} = \widetilde{c}_{\Sigma\mathbb{P}*}\widehat{\mathfrak{L}}^{(D)}, \tag{2.134}$$

and satisfy

$$\left\langle \overset{P}{\omega} | \widetilde{E}_\mu \right\rangle = \boldsymbol{P}_\mu,\ \left\langle \overset{\Delta}{\omega} | \widetilde{H}_\mu \right\rangle = \boldsymbol{\Delta}_\mu,\ \left\langle \overset{SY}{\omega} | \breve{E}_{\alpha\beta} \right\rangle = {}^{\dagger}\boldsymbol{S}_{\alpha\beta},\ \left\langle \overset{D}{\omega} | \widehat{E}^{(D)} \right\rangle = \boldsymbol{D}. \tag{2.135}$$

The left invariant fundamental vector operators appearing in (2.134) are readily computed, the result being

$$\widehat{\mathfrak{L}}^{(P)}_{\mu} = i\widetilde{Q}^{\nu}{}_{\mu}\frac{\partial}{\partial\varepsilon^{\nu}},\ \widehat{\mathfrak{L}}^{(\Delta)}_{\mu} = i\widetilde{W}^{\nu}{}_{\mu}\frac{\partial}{\partial b^{\nu}},$$

$$\widehat{\mathfrak{L}}^{(SY)}_{\alpha\beta} = i\widetilde{\alpha}_{\gamma(\mu|}\frac{\partial}{\partial\widetilde{\alpha}_{\gamma}{}^{|\nu)}},\ \widehat{\mathfrak{L}}^{(D)} = -i\varepsilon^{\beta}\frac{\partial}{\partial\varepsilon^{\beta}}, \tag{2.136}$$

where $\widetilde{\alpha}_{\mu}^{\ \nu} := e^{\alpha_{\mu}^{\ \nu}} = \alpha_{\mu}^{\ \nu} + \alpha_{\mu}^{\ \nu} + \frac{1}{2!}\alpha_{\mu}^{\ \gamma}\alpha_{\gamma}^{\ \nu} + \cdots$, $\widetilde{Q}_{\sigma}^{\alpha} := (\widetilde{\chi}_{\sigma}^{\alpha} + \delta_{\sigma}^{\alpha}e^{\varphi})$, $\widetilde{W}_{\sigma}^{\alpha} := (\widetilde{\chi}_{\sigma}^{\alpha} + \delta_{\sigma}^{\alpha}e^{-\varphi})$ satisfying $\left(\widetilde{Q}^{-1}\right)_{\sigma}^{\alpha} = \widetilde{Q}_{\sigma}^{\alpha}$ and $\left(\widetilde{W}^{-1}\right)_{\sigma}^{\alpha} = \widetilde{W}_{\sigma}^{\alpha}$. Making use of the transformation law of the nonlinear connection (2.80) we obtain

$$\delta\Gamma = \delta V^{\alpha}\boldsymbol{P}_{\alpha} + \delta\vartheta^{\alpha}\boldsymbol{\Delta}_{\alpha} + 2\delta\Phi\boldsymbol{D} + \delta\boldsymbol{\Gamma}^{\alpha\beta}\,{}^{\dagger}\boldsymbol{\Lambda}_{\alpha\beta} \tag{2.137}$$

where

$$\delta V^{\nu} = u_{\alpha}{}^{\nu}V^{\alpha},\ \delta\vartheta^{\nu} = u_{\alpha}{}^{\nu}\vartheta^{\alpha},\ \delta\Phi = 0,\ \delta\boldsymbol{\Gamma}^{\alpha\beta} = {}^{\dagger}\overset{GL}{\nabla}u^{\alpha\beta}. \tag{2.138}$$

From $\delta\boldsymbol{\Gamma}^{\alpha\beta} = {}^{\dagger}\overset{GL}{\nabla}u^{\alpha\beta}$ we observe that

$$\delta\Gamma^{[\alpha\beta]} = \overset{\circ}{\nabla}u^{\alpha\beta},\ \delta\Upsilon_{\alpha\beta} = 2u^{\rho}{}_{(\alpha|}\Upsilon_{\rho|\beta)}. \tag{2.139}$$

According to (2.138), the nonlinear translational and special conformal gauge fields transform as contravariant vector valued 1-forms under H, the antisymmetric part of $\boldsymbol{\Gamma}^{\alpha\beta}$ transforms inhomogeneously as a gauge potential and the nonlinear dilaton gauge field $\boldsymbol{\Phi}$ transforms as a scalar valued 1-form. From (2.139) it is clear that the symmetric part of $\boldsymbol{\Gamma}^{\alpha\beta}$ is a tensor valued 1-form. Being 4-covectors we identify V^{ν} as coframe fields. The connection coefficient $\overset{\circ}{\Gamma}{}^{\alpha\beta}$ serves as the gravitational gauge potential. The remaining components of $\boldsymbol{\Gamma}$, namely ϑ, Υ and Φ are dynamical fields of the theory. As will be seen in the following Section, the tetrad components of the coframe are used in conjunction with the H-metric to induce a space-time metric on M.

At this point, we have discussed all the mathematical tools that we will use to realize the Invariance Induced Gravity. In next Sections, we will proceed with the program of constructing the induced metric, the action functional and the field equations.

2.9. The Induced Metric

The bundle structure of gravitation, together with the conformal-affine algebra and the nonlinear realizations of gauge transformations (in particular the classification of gauge potentials), provide us all the tools to realize the Invariance Induced Gravity. In the following part of the section, we will derive the gravitational field and internal symmetry (spin) quantities showing that they are nothing else but realizations of the local conformal-affine transformations. In other words, the deformations of the Poincaré group give rise to gravity and internal symmetries (see also [86, 91].

Since the Lorentz group H is a subgroup of G, we inherit the invariant ($\delta o_{\alpha\beta} = \delta o^{\alpha\beta} = 0$) (constant) metric of H, where $o^{\alpha\beta} = o_{\alpha\beta} = diag\,(-, +, +, +)$. With the aid of $o_{\alpha\beta}$ and the tetrad components $e_i^{\ \alpha}$, we define the space-time metric

$$g_{ij} = e_i^{\ \alpha} e_j^{\ \beta} o_{\alpha\beta}. \tag{2.140}$$

Observing $\overset{\text{GL}}{{}^\dagger\nabla} o_{\alpha\beta} = -2\Upsilon_{\alpha\beta}$ (where we used $do_{\alpha\beta} = 0$) and taking account of the (second) transformation property (2.139), we interpret $\Upsilon_{\alpha\beta}$ as a sort of nonmetricity, i.e. a deformation (or distortion) gauge field that describes the difference between the general linear connection and the Levi-Civita connection of Riemannian geometry [91]. In the limit of vanishing gravitational interactions, we have $\overset{\text{T}}{\Gamma}{}^{\sigma} \sim \overset{\text{C}}{\Gamma}{}^{\sigma} \sim \overset{\circ}{\Gamma}{}^{\alpha}_{\ \beta} \sim \Upsilon^{\alpha}_{\ \beta} \sim \Phi \to 0$, $r^{\beta}_{\sigma} \to \delta^{\beta}_{\sigma}$ (to first order) and $\overset{\text{GL}}{{}^\dagger D}\xi^{\sigma} \to d\xi^{\sigma}$. Under these conditions, the coframe reduces to $V^{\beta} \to e^{\phi}\delta^{\beta}_{\alpha} d\xi^{\alpha}$ leading to the space-time metric

$$g_{ij} \to e^{2\phi}\delta^{\rho}_{\alpha}\delta^{\sigma}_{\beta}\left(\partial_i\xi^{\alpha}\right)\left(\partial_j\xi^{\beta}\right)o_{\rho\sigma} = e^{2\phi}\left(\partial_i\xi^{\alpha}\right)\left(\partial_j\xi^{\beta}\right)o_{\alpha\beta} \tag{2.141}$$

characteristic of a Weyl geometry. In this sense, the invariance properties induce the gravitational field and generalize results in [91]. It is worth noting that conformal transformations of the metric tensor constitute only a part of the whole deformation field as we discuss in the next Chapter.

2.10. The Cartan Structure Equations

Our task is now to deduce the dynamics. Using the nonlinear gauge potentials derived in Eqs. (2.103), (2.105), (2.106), the covariant derivative defined on Σ pulled back to M has the form

$$\nabla := d - iV^{\alpha}\boldsymbol{P}_{\alpha} - i\vartheta^{\alpha}\boldsymbol{\Delta}_{\alpha} - 2i\Phi\boldsymbol{D} - i\Gamma^{\alpha\beta}\,{}^{\dagger}\boldsymbol{\Lambda}_{\alpha\beta}. \tag{2.142}$$

By using (2.142) together with the relevant Lie algebra commutators, we obtain the the bundle curvature

$$\mathbb{F} := \nabla \wedge \nabla = -i\mathcal{T}^{\alpha}\boldsymbol{P}_{\alpha} - i\mathcal{K}^{\alpha}\boldsymbol{\Delta}_{\alpha} - i\mathcal{Z}\boldsymbol{D} - i\mathbb{R}_{\alpha}^{\ \beta}\,{}^{\dagger}\boldsymbol{\Lambda}^{\alpha}_{\ \beta}. \tag{2.143}$$

The field strength components of $\mathbb{F}$ are given by the first Cartan structure equations. They are respectively, the projectively deformed, Υ-distorted translational field strength

$$\mathcal{T}^{\alpha} := {}^{\dagger}\overset{\mathrm{GL}}{\nabla} V^{\alpha} + 2\Phi \wedge V^{\alpha}, \tag{2.144}$$

the projectively deformed, Υ-distorted special conformal field strength

$$\mathcal{K}^{\alpha} := {}^{\dagger}\overset{\mathrm{GL}}{\nabla} \vartheta^{\alpha} - 2\Phi \wedge \vartheta^{\alpha}, \tag{2.145}$$

the Ψ-deformed Weyl homothetic curvature 2-form (dilaton field strength)

$$\mathcal{Z} := d\Phi + \Psi, \, \Psi = V \cdot \vartheta - \vartheta \cdot V \tag{2.146}$$

and the general conformal-affine curvature

$$\mathbb{R}^{\alpha\beta} := \widehat{R}^{\alpha\beta} + \Psi^{\alpha\beta}, \tag{2.147}$$

with

$$\widehat{R}^{\alpha\beta} := \mathfrak{R}^{\alpha\beta} + \mathcal{R}^{\alpha\beta}, \quad \Psi^{\alpha\beta} := V^{[\alpha} \wedge \vartheta^{\beta]}. \tag{2.148}$$

Operator ${}^{\dagger}\overset{\mathrm{GL}}{\nabla}$ denotes the nonlinear covariant derivative built from volume preserving (VP) connection (i.e. excluding Φ) forms respectively. The Υ and $\overset{\circ}{\Gamma}$-affine curvatures in (2.148) read

$$\mathfrak{R}^{\alpha\beta} \; : = \overset{\circ}{\nabla}\Upsilon^{\alpha\beta} + \Upsilon^{\alpha}_{\gamma} \wedge \Upsilon^{\gamma\beta}, \tag{2.149}$$

$$\mathcal{R}^{\alpha\beta} \; : = d\overset{\circ}{\Gamma}{}^{\alpha\beta} + \overset{\circ}{\Gamma}{}_{\gamma}^{\alpha} \wedge \overset{\circ}{\Gamma}{}^{\gamma\beta}, \tag{2.150}$$

respectively. Operator $\overset{\circ}{\nabla}$ is defined with respect to the restricted connection $\overset{\circ}{\Gamma}{}^{\alpha\beta}$ given in (2.109).

The field strength components of the bundle curvature have the following group variations

$$\delta\mathbb{R}^{\beta}_{\alpha} = u^{\gamma}_{\alpha}\mathbb{R}^{\beta}{}_{\gamma} - u^{\beta}_{\gamma}\mathbb{R}^{\gamma}_{\alpha}, \, \delta\mathcal{Z} = 0, \, \delta\mathcal{T}^{\alpha} = -u^{\alpha}_{\beta}\mathcal{T}^{\beta}, \, \delta\mathcal{K}^{\alpha} = -u^{\alpha}_{\beta}\mathcal{K}^{\beta}. \tag{2.151}$$

A gauge field Lagrangian is built from polynomial combinations of the strength $\mathbb{F}$ defined as

$$\mathbb{F}(\Gamma(\Omega, D\xi), d\Gamma) := \nabla \wedge \nabla = d\Gamma + \Gamma \wedge \Gamma. \tag{2.152}$$

Now we have all the ingredients to derive the conservation laws.

2.11. The Bianchi Identities

In what follows, the Bianchi identities play a central role being the conservation laws of the theory. We therefore derive them presently.

1a) The 1st translational Bianchi identity reads,

$$\overset{GL}{\nabla}\mathcal{T}^{a}=\widehat{R}_{\beta}^{\ \alpha}\wedge V^{\beta}+\Phi\wedge T^{a}+2d(\Phi\wedge V^{\alpha}). \tag{2.153}$$

1b) Similarly to the case in (1a), the 1st conformal Bianchi identities are respectively given by

$$\overset{GL}{\nabla}\mathcal{K}^{a}=\widehat{R}_{\beta}^{\ \alpha}\wedge\vartheta^{\beta}-\Phi\wedge\mathcal{K}^{a}-2d(\Phi\wedge\vartheta^{\alpha}). \tag{2.154}$$

2a) The Υ and $\overset{\circ}{\Gamma}$-affine component of the 2nd Bianchi identity is given by

$$^{\dagger}\overset{GL}{\nabla}\mathfrak{R}^{\alpha\beta}=2\mathfrak{R}^{(\alpha|}{}_{\gamma}\Upsilon^{\gamma|\beta)},\ ^{\dagger}\overset{GL}{\nabla}\mathcal{R}^{\alpha\beta}=0, \tag{2.155}$$

respectively. Hence, the generalized 2nd Bianchi identity is given by

$$^{\dagger}\overset{GL}{\nabla}\widehat{R}_{\beta}^{\ \alpha}=2\mathfrak{R}^{(\alpha|}{}_{\gamma}\Upsilon^{\gamma|\rho)}o_{\rho\beta}. \tag{2.156}$$

Since the full curvature $\mathbb{R}^{\alpha\beta}$ is proportional to $\Psi^{\alpha\beta}$, it is necessary to consider

$$^{\dagger}\overset{GL}{\nabla}\Psi^{\alpha\beta}=\ ^{\dagger}\mathcal{T}^{\alpha}\wedge\vartheta^{\beta}+V^{\alpha}\wedge\ ^{\dagger}\mathcal{K}^{\beta}, \tag{2.157}$$

from which we conclude

$$^{\dagger}\overset{GL}{\nabla}\mathbb{R}^{\alpha\beta}=2\mathfrak{R}^{(\alpha|}{}_{\gamma}\Upsilon^{\gamma|\beta)}+\ ^{\dagger}\mathcal{T}^{\alpha}\wedge\vartheta^{\beta}+V^{\alpha}\wedge\ ^{\dagger}\mathcal{K}^{\beta}. \tag{2.158}$$

2c) The dilatonic component of the 2nd Bianchi identity is given by

$$\overset{GL}{\nabla}\mathcal{Z}=dZ+\overset{GL}{\nabla}(V\wedge\vartheta)=\overset{GL}{\nabla}\Psi+\Phi\wedge\Psi, \tag{2.159}$$

From the definition of Ψ, we obtain

$$\nabla\Psi=\mathcal{T}^{\alpha}\wedge\vartheta_{\alpha}+V_{\alpha}\wedge\mathcal{K}^{\alpha}+\Phi\wedge(V_{\alpha}\wedge\vartheta^{\alpha}). \tag{2.160}$$

Defining

$$\Sigma^{\mu\nu}:=\boldsymbol{B}^{\mu\nu}+\Psi^{\mu\nu},\ \boldsymbol{B}^{\mu\nu}:=B^{\mu\nu}+\mathcal{B}^{\mu\nu},\ B^{\mu\nu}:=V^{\mu}\wedge V^{\nu},\ \mathcal{B}^{\mu\nu}:=\vartheta^{\mu}\wedge\vartheta^{\nu}, \tag{2.161}$$

and asserting $V^{\alpha}\wedge\vartheta_{\alpha}=0$, we find $\Sigma_{\mu\nu}\wedge\Sigma^{\mu\nu}=0$. Using this result,we obtain

$$\nabla\Psi=\mathcal{T}^{\alpha}\wedge\vartheta_{\alpha}+V_{\alpha}\wedge\mathcal{K}^{\alpha}. \tag{2.162}$$

The last step is now to derive the field equations.

2.12. The Action Functional and the Gauge Field Equations

We seek an action for a local gauge theory based on the $CA(3,1)$ symmetry group. We consider the $3D$ topological invariants $\mathbb{Y}$ of the non-Riemannian manifold of conformal-affine connections. Our objective is the $4D$ boundary terms $\mathbb{B}$ obtained by means of exterior differentiation of these $3D$ invariants, i.e. $\mathbb{B} = d\mathbb{Y}$. The Lagrangian density of conformal-affine gravity is modelled after $\mathbb{B}$, with appropriate distribution of Lie star operators so as to re-introduce the dual frame fields. The generalized conformal-affine surface topological invariant reads

$$\mathbb{Y} = -\frac{1}{2l^2}\left[\begin{array}{c} \theta_{\mathcal{A}}\left(\mathcal{A}_a{}^b \wedge \widehat{R}_b{}^a + \frac{1}{3}\mathcal{A}_a{}^b \wedge \mathcal{A}_b{}^c \wedge \mathcal{A}_c{}^a\right) \\ -\theta_{\mathcal{V}} \mathcal{V}_a \wedge \boldsymbol{T}^\alpha + \theta_\Phi \Phi \wedge \mathcal{Z} \end{array}\right], \tag{2.163}$$

where $\boldsymbol{T}^\alpha := \mathcal{T}^\alpha + \mathcal{K}^\alpha$. The associated total conformal-affine boundary term is given by,

$$\mathbb{B} = \frac{1}{2l^2}\left[\begin{array}{c} \widehat{R}_{\beta\alpha} \wedge \boldsymbol{B}^{\beta\alpha} + \Sigma^{[\beta\alpha]} \wedge \Sigma_{[\beta\alpha]} - \widehat{R}^{\alpha\beta} \wedge \widehat{R}_{\alpha\beta} - \mathcal{Z} \wedge \mathcal{Z} \\ +\mathcal{K}_\alpha \wedge \mathcal{K}^\alpha + \mathcal{T}_\alpha \wedge \mathcal{T}^\alpha - \Phi \wedge \left(V_\alpha \wedge \mathcal{T}^\alpha + \vartheta_\alpha \wedge \mathcal{K}^\alpha\right) \\ -\Upsilon_{\alpha\beta} \wedge \left(V^\alpha \wedge \mathcal{T}^\beta + \vartheta^\alpha \wedge \mathcal{K}^\beta\right). \end{array}\right] \tag{2.164}$$

Using the boundary term (2.164) as a guide, we choose [48, 51, 54, 56, 66] an action of form

$$I = \int_{\mathcal{M}} \left\{\begin{array}{c} d\left(\mathcal{V}^\alpha \wedge \boldsymbol{T}_\alpha\right) + \widehat{R}^{\alpha\beta} \wedge \Sigma_{\star\alpha\beta} + \mathcal{B}_{\star\alpha\beta} \wedge \mathcal{B}^{\alpha\beta} + \Psi_{\star\alpha\beta} \wedge \Psi^{\alpha\beta} + \eta_{\star\alpha\beta} \wedge \eta^{\alpha\beta} \\ -\frac{1}{2}\left(\mathcal{R}_{\star\mu\nu} \wedge \mathcal{R}^{\mu\nu} + \mathcal{Z} \wedge \star\mathcal{Z}\right) + \mathcal{T}_{\star\alpha} \wedge \mathcal{T}^\alpha + \mathcal{K}_{\star\alpha} \wedge \mathcal{K}^\alpha \\ -\Phi \wedge \left(\mathcal{T}^{\star\alpha} \wedge V_\alpha + \mathcal{K}^{\star\alpha} \wedge \vartheta_\alpha\right) - \Upsilon_{\alpha\beta} \wedge \left(V^\alpha \wedge \mathcal{T}^{\star\beta} + \vartheta^\alpha \wedge \mathcal{K}^{\star\beta}\right). \end{array}\right\} \tag{2.165}$$

Note that the action integral (2.165) is invariant under Lorentz rather than conformal-affine transformations. The Lie star $\star$ operator is defined as $\star V_\alpha = \frac{1}{3!}\eta_{\alpha\beta\mu\nu} V^\beta \wedge V^\mu \wedge V^\nu$.

The field equations are obtained from the variation of I with respect to the independant gauge potentials. It is convenient to define the functional derivatives

$$\begin{aligned} \frac{\delta \mathcal{L}_{\text{gauge}}}{\delta V^\alpha} &:= -\overset{\text{GL}}{\nabla} N_\alpha + \overset{V}{\mathfrak{T}}_\alpha, \\ \frac{\delta \mathcal{L}_{\text{gauge}}}{\delta \vartheta^\alpha} &:= -\overset{\text{GL}}{\nabla} M_\alpha + \overset{\vartheta}{\mathfrak{T}}_\alpha, \\ 3_\alpha^\beta := \frac{\delta \mathcal{L}_{\text{gauge}}}{\delta \widehat{\Gamma}_\beta{}^\alpha} &= -{}^\dagger\overset{\text{GL}}{\nabla} \widehat{M}_\alpha{}^\beta + \widehat{E}_\alpha{}^\beta. \end{aligned} \tag{2.166}$$

where

$$\widehat{M}^{\ \alpha}_{\beta} := -\frac{\partial \mathcal{L}_{\text{gauge}}}{\partial \widehat{R}^{\ \beta}_{\alpha}}, \quad \widehat{E}^{\ \beta}_{\alpha} := \frac{\partial \mathcal{L}_{\text{gauge}}}{\partial \widehat{\Gamma}^{\ \alpha}_{\beta}}, \quad \overset{\mathrm{V}}{\mathfrak{T}}_{\alpha} := \frac{\partial \mathcal{L}_{\text{gauge}}}{\partial V^{\alpha}},$$
$$\overset{\vartheta}{\mathfrak{T}}_{\alpha} := \frac{\partial \mathcal{L}_{\text{gauge}}}{\partial \vartheta^{\alpha}}, \quad \Theta := \frac{\partial \mathcal{L}_{\text{gauge}}}{\partial \Phi}. \tag{2.167}$$

The gauge field momenta are defined by

$$N_{\alpha} := -\frac{\partial \mathcal{L}_{\text{gauge}}}{\partial \mathcal{T}^{\alpha}}, \ M_{\alpha} := -\frac{\partial \mathcal{L}_{\text{gauge}}}{\partial \mathcal{K}^{\alpha}}, \ \Xi := -\frac{\partial \mathcal{L}_{\text{gauge}}}{\partial \mathcal{Z}},$$
$$\widehat{M}_{[\alpha\beta]} := N_{\alpha\beta} = -o_{[\alpha|\gamma}\frac{\partial \mathcal{L}_{\text{gauge}}}{\partial \mathcal{R}_{\gamma}^{\ |\beta]}}, \ \widehat{M}_{(\alpha\beta)} := M_{\alpha\beta} = -2o_{(\alpha|\gamma}\frac{\partial \mathcal{L}_{\text{gauge}}}{\partial \mathfrak{R}_{\gamma}^{\ |\beta)}}. \tag{2.168}$$

Furthermore, the shear (gauge field deformation) and hypermomentum current forms are given by

$$\widehat{E}_{(\alpha\beta)} := U_{\alpha\beta} = -V_{(\alpha} \wedge \left(M_{\beta)} + N_{\beta)}\right) - M_{\alpha\beta},$$
$$\widehat{E}_{[\alpha\beta]} := E_{\alpha\beta} = -V_{[\alpha} \wedge \left(M_{\beta]} + N_{\beta]}\right), \tag{2.169}$$

The analogue of the Einstein equations read

$$G_{\alpha} + \Lambda \widehat{\eta}_{\alpha} + {}^{\dagger}\overset{\mathrm{GL}}{\nabla} \mathcal{T}_{\star\alpha} + \overset{\mathrm{V}}{\mathfrak{T}}_{\alpha} = 0, \tag{2.170}$$

with Einstein-like three-form

$$G_{\alpha} = \left(\mathcal{R}^{\ \beta\gamma} + \Upsilon^{[\beta|}{}_{\rho} \wedge \Upsilon^{|\gamma]\rho}\right) \wedge \left(\eta_{\beta\gamma\alpha} + \star\left[B_{\beta\gamma} \wedge \vartheta_{\alpha}\right]\right), \tag{2.171}$$

coupling (cosmological) constant Λ and mixed three-form $\widehat{\eta}_{\alpha} = \eta_{\alpha} + \star\left(\vartheta_{\alpha} \wedge V_{\beta}\right) \wedge V^{\beta}$. Let us observe that G_{α} includes symmetric GL_4 (Υ) as well as special conformal (ϑ) contributions. The gauge field 3-form $\overset{\mathrm{V}}{\mathfrak{T}}_{\alpha}$ is given by

$$\begin{aligned}\overset{\mathrm{V}}{\mathfrak{T}}_{\alpha} &= \left\langle \mathcal{L}_{\text{gauge}} | e_{\alpha} \right\rangle + \left\langle \mathcal{Z} | e_{\alpha} \right\rangle \wedge \Xi + \left\langle \mathcal{T}^{\ \beta} | e_{\alpha} \right\rangle \wedge N_{\beta} \\ &+ \left\langle \mathcal{K}^{\ \beta} | e_{\alpha} \right\rangle \wedge M_{\beta} + \left\langle \mathcal{R}_{\gamma}^{\ \beta} | e_{\alpha} \right\rangle \wedge N^{\gamma}_{\ \beta} + \frac{1}{2}\left\langle \mathfrak{R}_{\gamma}^{\ \beta} | e_{\alpha} \right\rangle M^{\gamma}_{\ \beta},\end{aligned} \tag{2.172}$$

We remark that to interpret Eqs. (2.171) as the gravitational field equation analogous to the Einstein equations, we must transform from the Lie algebra index α to the space-time basis index k by contracting over the former (α) with the conformal-affine tetrads e^{α}_{k}. This fact is relevant to read gravity in holonomic and anholonomic frames respectively. It is

$$\begin{aligned}\overset{V}{\mathfrak{T}}_\alpha &= \mathfrak{T}_\alpha[\mathcal{T}]+\mathfrak{T}_\alpha[\mathcal{K}]+\mathfrak{T}_\alpha[\mathcal{R}]+\mathfrak{T}_\alpha[Z]-\left\langle \mathcal{T}^{\beta}|e_\alpha\right\rangle\wedge N_\beta-\left\langle \mathcal{K}^{\beta}|e_\alpha\right\rangle\wedge M_\beta\\ &\quad-\left\langle \mathcal{R}_\gamma{}^{\beta}|e_\alpha\right\rangle\wedge N^{\gamma}_{\beta}-\langle z|e_\alpha\rangle\wedge\Xi+\Psi_{\star\alpha\beta}\wedge\vartheta^\beta+\langle\Sigma_{\star\gamma\beta}|e_\alpha\rangle\wedge\widehat{R}^{\alpha\beta}\\ &\quad+\left\langle \Upsilon^{\gamma\beta}\wedge\left(V_\gamma\wedge\mathcal{T}_{\star\beta}+\vartheta_\gamma\wedge\mathcal{K}_{\star\beta}\right)|e_\alpha\right\rangle+\Sigma_{\star\gamma\beta}\wedge\left\langle\widehat{R}^{\gamma\beta}|e_\alpha\right\rangle\\ &\quad+\mathcal{B}_{\star\gamma\beta}\wedge\left\langle\mathcal{B}^{\gamma\beta}|e_\alpha\right\rangle+\langle\mathcal{B}_{\star\gamma\beta}|e_\alpha\rangle\wedge\mathcal{B}^{\gamma\beta}+\langle\Psi_{\star\gamma\beta}|e_\alpha\rangle\wedge\Psi^{\gamma\beta}\end{aligned} \tag{2.173}$$

respectively, with

$$\begin{aligned}\mathfrak{T}_\alpha[\mathcal{R}]&=\tfrac{1}{2}a_1\left(\mathcal{R}_{\rho\gamma}\wedge\langle\mathcal{R}^{\star\rho\gamma}|e_\alpha\rangle-\langle\mathcal{R}_{\rho\gamma}|e_\alpha\rangle\wedge\mathcal{R}^{\star\rho\gamma}\right),\\ \mathfrak{T}_\alpha[\mathcal{T}]&=\tfrac{1}{2}a_2\left(\mathcal{T}_{\gamma}\wedge\langle\mathcal{T}^{\star\gamma}|e_\alpha\rangle-\langle\mathcal{T}_{\gamma}|e_\alpha\rangle\wedge\mathcal{T}^{\star\gamma}\right),\\ \mathfrak{T}_\alpha[\mathcal{K}]&=\tfrac{1}{2}a_3\left(\mathcal{K}_{\gamma}\wedge\langle\mathcal{K}^{\star\gamma}|e_\alpha\rangle-\langle\mathcal{K}_{\gamma}|e_\alpha\rangle\wedge\mathcal{K}^{\star\gamma}\right),\\ \mathfrak{T}_\alpha[Z]&=\tfrac{1}{2}a_4\left(d\Phi\wedge\langle\star d\Phi|e_\alpha\rangle-\langle d\Phi|e_\alpha\rangle\wedge\star d\Phi\right).\end{aligned} \tag{2.174}$$

From the variation of I with respect to ϑ^α we get

$$\mathfrak{G}_\alpha+\Lambda\widehat{\omega}_\alpha+{}^{\dagger}\overset{GL}{\nabla}\mathcal{K}_{\star\alpha}+\overset{\vartheta}{\mathfrak{T}}_\alpha=0, \tag{2.175}$$

where, in analogy to Eqs. (2.171), we have

$$\mathfrak{G}_\alpha=h^\alpha_i\left(\mathcal{R}^{\beta\gamma}+\Upsilon^{[\beta|}{}_{\rho}\wedge\Upsilon^{|\gamma]\rho}\right)\wedge\left(\omega_{\beta\gamma\alpha}+\star\left[\mathcal{B}_{\beta\gamma}\wedge V_\alpha\right]\right), \tag{2.176}$$

where $\widehat{\omega}_\alpha=\omega_\alpha+\star\left(\vartheta_\alpha\wedge V_\beta\right)\wedge\vartheta^\beta$. The quantity $\overset{\vartheta}{\mathfrak{T}}_i=h^\alpha_i\overset{\vartheta}{\mathfrak{T}}_\alpha$ is similar to (2.172) but with the algebra basis e_α replaced by h_α and the conformal-affine tetrad components e^α_i replaced by h^α_i. The two gravitational field equations (2.171) and (2.176) are $P-\Delta$ symmetric. We may say that they exhibit $P-\Delta$ duality symmetry invariance.

From the variational equation for $\overset{\circ}{\Gamma}{}^{\beta}_{\alpha}$, we obtain the conformal-affine gravitational analogue of the Yang-Mills-torsion type field equation,

$$\overset{\circ}{\nabla}\star\mathcal{R}_\alpha{}^{\beta}+\overset{\circ}{\nabla}\star\Sigma^{\beta}_{\alpha}+\left(V^\beta\wedge\mathcal{T}_{\star\alpha}+\vartheta^\beta\wedge\mathcal{K}_{\star\alpha}\right)=0. \tag{2.177}$$

Variation of I with respect to $\Upsilon^{\beta}_{\alpha}$ leads to

$$\overset{\circ}{\nabla}\star\Sigma_{\alpha\beta}-\Upsilon_{(\alpha|}{}^{\gamma}\wedge\Sigma_{\star\gamma|\beta)}+V_{(\alpha}\wedge\mathcal{T}_{\star\beta)}+\vartheta_{(\alpha}\wedge\mathcal{K}_{\star\beta)}=0. \tag{2.178}$$

Finally, from the variational equation for Φ, the gravi-scalar field equation is given by

$$d\star d\Phi + V_\alpha \wedge \mathcal{T}^{\star\alpha} + \vartheta_\alpha \wedge \mathcal{K}^{\star\alpha} = 0. \tag{2.179}$$

In conclusions, the field equations of conformal-affine gravity have been obtained. The analogue of the Einstein equation, obtained from variation of I with respect to the coframe V, is characterized by an Einstein-like 3-form that includes symmetric GL_4 as well as special conformal contributions. Moreover, the field equation in (2.171) contains a non-trivial torsion contribution. Performing a $P-\Delta$ transformation (i.e. $V \to \vartheta$, $\mathcal{T} \to \mathcal{K}$, $D \to -D$) on (2.171) we obtain (2.176). This result may also be obtained directly by varying I with respect ϑ. A mixed conformal-affine cosmological constant term arises in (2.171), (2.176)) as a consequence of the structure of the 2-form $\mathbb{R}^\alpha_\beta$. This result can be extremely interesting from a physical viewpoint in order to envisage a mechanism capable of producing the "observed" cosmological constant (see also [82, 83]).

The field equation (2.177) is a Yang-Mills-like equation that represents the generalization of the Gauss torsion-free equation $\nabla \star B^{\alpha\beta} = 0$. In our case, we considered a mixed volume form involving both V and ϑ leading to the substitution $B^{\alpha\beta} \to \Sigma^{\alpha\beta}$. Additionally, even in the case of vanishing $T^\rho = \overset{\circ}{\nabla} V^\rho$, the conformal-affine torsion depends on the dilaton potential Φ which in general is non-vanishing. A similar argument holds for the special conformal quantity $\mathcal{K}^\rho$. Admitting the quadratic curvature term $\mathcal{R}_\alpha{}^\beta \wedge \star \mathcal{R}_\beta{}^\alpha$ in the gauge Lagrangian it becomes clear how we draw the analogy between (2.177) and the Gauss equation. Equation (2.178) follow from similar considerations as (2.177), the significant differences being the lack of a $\overset{\circ}{\nabla} \star \mathfrak{R}_\alpha^\beta$ counterpart to $\overset{\circ}{\nabla} \star \mathcal{R}_\alpha{}^\beta$ since $\star \mathfrak{R}_\alpha^\beta = 0$. Finally, (2.179) involves both $\mathcal{T}^\rho$ and $\mathcal{K}^\rho$ in conjunction with a term that resembles the source-free Maxwell equations with the dilaton potential playing a similar role to the electromagnetic vector potential.

2.13. Invariance Principles

The above results have a straightfoward application starting from Poincaré invariance. Gravity can be realized as a local Poincaré gauge theory, in fact as it is well-known, the field equations and conservation laws can be obtained from a least action principle. The same principle is the basis of any gauge theory so we start from it to develop our considerations. Let us start from a least action principle and the Noether theorem.

Let $\chi(x)$ be a multiplet field defined at a space-time point x and $\mathcal{L}\{\chi(x), \partial_j\chi(x); x\}$ be the Lagrangian density of the system. The action integral of the system over a given space-time volume Ω is defined by

$$I(\Omega) = \int_\Omega \mathcal{L}\{\chi(x), \partial_j\chi(x); x\}\, d^4x. \tag{2.180}$$

Now let us consider the infinitesimal variations of the coordinates

$$x^i \to x'^i = x^i + \delta x^i, \tag{2.181}$$

and the field variables

$$\chi(x) \to \chi'(x') = \chi(x) + \delta\chi(x). \tag{2.182}$$

Correspondingly, the variation of the action is given by

$$\delta I = \int_{\Omega'} \mathcal{L}'(x')\, d^4x' - \int_{\Omega} \mathcal{L}(x)\, d^4x = \int_{\Omega} \left[\mathcal{L}'(x')||\partial_j x'^j|| - \mathcal{L}(x)\right] d^4x. \tag{2.183}$$

Since the Jacobian for the infinitesimal variation of coordinates becomes

$$||\partial_j x'^j|| = 1 + \partial_j(\delta x^j), \tag{2.184}$$

the variation of the action takes the form,

$$\delta I = \int_{\Omega} \left[\delta\mathcal{L}(x) + \mathcal{L}(x)\partial_j(\delta x^j)\right] d^4x \tag{2.185}$$

where

$$\delta\mathcal{L}(x) = \mathcal{L}'(x') - \mathcal{L}(x). \tag{2.186}$$

For any function $\Phi(x)$ of x, it is convenient to define the fixed point variation δ_0 by,

$$\delta_0\Phi(x) := \Phi'(x) - \Phi(x) = \Phi'(x') - \Phi(x'). \tag{2.187}$$

Expanding the function to first order in δx^j as

$$\Phi(x') = \Phi(x) + \delta x^j \partial_j \Phi(x), \tag{2.188}$$

we obtain

$$\delta\Phi(x) = \Phi'(x') - \Phi(x) = \Phi'(x') - \Phi(x') + \Phi(x') - \Phi(x) = \delta_0\Phi(x) + \delta x^j \partial_j \Phi(x), \tag{2.189}$$

or

$$\delta_0\Phi(x) = \delta\Phi(x) - \delta x^j \partial_j \Phi(x). \tag{2.190}$$

The advantage to have the fixed point variation is that δ_0 commutes with ∂_j:

$$\delta_0 \partial_j \Phi(x) = \partial_j \delta_0 \Phi(x). \tag{2.191}$$

For $\Phi(x) = \chi(x)$, we have

$$\delta\chi = \delta_0\chi + \delta x^i \partial_i \chi, \tag{2.192}$$

and

$$\delta\partial_i\chi = \partial_i(\delta_0\chi) - \partial(\delta x^j)\partial_i\chi. \tag{2.193}$$

Using the fixed point variation in the integrand of (2.185) gives

$$\delta I = \int_{\Omega} \left[\delta_0\mathcal{L}(x) + \partial_j(\delta x^j \mathcal{L}(x))\right] d^4x. \tag{2.194}$$

If we require the action integral defined over any arbitrary region Ω be invariant, that is,

$\delta I = 0$, then we must have

$$\delta\mathcal{L} + \mathcal{L}\partial_j(\delta x^j) = \delta_0\mathcal{L} + \partial_j(\mathcal{L}\delta x^j) = 0. \tag{2.195}$$

If $\partial_j(\delta x^j) = 0$, then $\delta\mathcal{L} = 0$, that is, the Lagrangian density $\mathcal{L}$ is invariant. In general, however, $\partial_j(\delta x^j) \neq 0$, and $\mathcal{L}$ transforms like a scalar density. In other words, $\mathcal{L}$ is a Lagrangian density unless $\partial_j(\delta x^j) = 0$.

For convenience, let us introduce a function $h(x)$ that behaves like a scalar density, namely

$$\delta h + h\partial_j(\delta x^j) = 0. \tag{2.196}$$

We further assume $\mathcal{L}(\chi,\partial_j\chi;x) = h(x)L(\chi,\partial_j\chi;x)$. Then we see that

$$\delta\mathcal{L} + \mathcal{L}\partial_j(\delta x^j) = h\delta L. \tag{2.197}$$

Hence the action integral remains invariant if

$$\delta L = 0. \tag{2.198}$$

The newly introduced function $L(\chi,\partial_j\chi;x)$ is the scalar Lagrangian of the system.

Let us calculate the integrand of (2.194) explicitly. The fixed point variation of $\mathcal{L}(x)$ is a consequence of a fixed point variation of the field $\chi(x)$,

$$\delta_0\mathcal{L} = \frac{\partial\mathcal{L}}{\partial\chi}\delta_0\chi + \frac{\partial\mathcal{L}}{\partial(\partial_j\chi)}\delta_0(\partial_j\chi) \tag{2.199}$$

which can be cast into the form,

$$\delta_0\mathcal{L} = [\mathcal{L}]_\chi\delta_0\chi + \partial_j\left(\frac{\partial\mathcal{L}}{\partial(\partial_j\chi)}\delta_0\chi\right) \tag{2.200}$$

where

$$[\mathcal{L}]_\chi \equiv \frac{\partial\mathcal{L}}{\partial\chi} - \partial_j\left(\frac{\partial\mathcal{L}}{\partial(\partial_j\chi)}\right). \tag{2.201}$$

Consequently, we have the action integral in the form

$$\delta I = \int_\Omega \left\{[\mathcal{L}]_\chi\delta_0\chi + \partial_j\left(\frac{\partial\mathcal{L}}{\partial(\partial_j\chi)}\delta\chi - T^j_k\,\delta x^k\right)\right\}d^4x, \tag{2.202}$$

where

$$T^j{}_k := \frac{\partial\mathcal{L}}{\partial(\partial_j\chi)}\partial_k\chi - \delta^j_k\mathcal{L} \tag{2.203}$$

is the canonical energy-momentum tensor density. If the variations are chosen in such a way that $\delta x^j = 0$ over Ω and $\delta_0\chi$ vanishes on the boundary of Ω, then $\delta I = 0$ gives us the Euler-Lagrange equation,

$$[\mathcal{L}]_\chi = \frac{\partial\mathcal{L}}{\partial\chi} - \partial_j\left(\frac{\partial\mathcal{L}}{\partial(\partial_j\chi)}\right) = 0. \tag{2.204}$$

On the other hand, if the field variables obey the Euler-Lagrange equation, $[\mathcal{L}]_\chi = 0$, then we have

$$\partial_j \left(\frac{\partial \mathcal{L}}{\partial(\partial_j \chi)} \delta\chi - T^j{}_k \delta x^k \right) = 0, \tag{2.205}$$

which gives rise, considering also the Noether theorem, to conservation laws. These very straightforward considerations are at the basis of our following discussion.

2.14. Global Poincaré Invariance

As standard, we assert that our space-time in the absence of gravitation is a Minkowski space M_4. The isometry group of M_4 is the group of Poincaré transformations (PT) which consists of the Lorentz group $SO(3,\,1)$ and the translation group $T(3,\,1)$. The Poincaré transformations of coordinates are

$$x^i \overset{PT}{\rightarrow} x'^i = a^i{}_j x^j + b^i, \tag{2.206}$$

where a^i_j and b^i are real constants, and a^i_j satisfy the orthogonality conditions $a^i_k a^k_j = \delta^i_j$. For infinitesimal variations,

$$\delta x'^i = \chi'(x') - \chi(x) = \varepsilon^i{}_j x^j + \varepsilon^i \tag{2.207}$$

where $\varepsilon_{ij} + \varepsilon_{ji} = 0$. While the Lorentz transformation forms a six parameter group, the Poincaré group has ten parameters. The Lie algebra for the ten generators of the Poincaré group is

$$\begin{aligned} [\Xi_{ij},\,\Xi_{kl}] &= \eta_{ik}\Xi_{jl} + \eta_{jl}\Xi_{ik} - \eta_{jk}\Xi_{il} - \eta_{il}\Xi_{jk}, \\ [\Xi_{ij},\,T_k] &= \eta_{jk}T_i - \eta_{ik}T_j,\ \ [T_i,\,T_j] = 0, \end{aligned} \tag{2.208}$$

where Ξ_{ij} are the generators of Lorentz transformations, and T_i are the generators of four-dimensional translations. Obviously, $\partial_i(\delta x^i) = 0$ for the Poincaré transformations (2.206). Therefore, our Lagrangian density $\mathcal{L}$, which is the same as L with $h(x) = 1$ in this case, is invariant; namely, $\delta\mathcal{L} = \delta L = 0$ for $\delta I = 0$.

Suppose that the field $\chi(x)$ transforms under the infinitesimal Poincaré transformations as

$$\delta\chi = \frac{1}{2}\varepsilon^{ij} S_{ij}\chi, \tag{2.209}$$

where the tensors S_{ij} are the generators of the Lorentz group, satisfying

$$S_{ij} = -S_{ji},\ \ [S_{ij},\,S_{kl}] = \eta_{ik}S_{jl} + \eta_{jl}S_{ik} - \eta_{jk}S_{il} - \eta_{il}S_{jk}. \tag{2.210}$$

Correspondingly, the derivative of χ transforms as

$$\delta(\partial_k\chi) = \frac{1}{2}\varepsilon^{ij} S_{ij}\partial_k\chi - \varepsilon^i{}_k \partial_i\chi. \tag{2.211}$$

Since the choice of infinitesimal parameters ε^i and ε^{ij} is arbitrary, the vanishing variation of the Lagrangian density $\delta\mathcal{L} = 0$ leads to the identities,

$$\frac{\partial\mathcal{L}}{\partial\chi}S_{ij}\chi + \frac{\partial\mathcal{L}}{\partial(\partial_k\chi)}(S_{ij}\partial_k\chi + \eta_{ki}\partial_j\chi - \eta_{kj}\partial_i\chi) = 0. \tag{2.212}$$

We also obtain the following conservation laws

$$\partial_j T^j_k = 0,\ \partial_k\left(S^k{}_{ij} - x_i T^k{}_j + x_j T^k{}_i\right) = 0, \tag{2.213}$$

where

$$S^k{}_{ij} := -\frac{\partial\mathcal{L}}{\partial(\partial_k\chi)}S_{ij}\chi. \tag{2.214}$$

These conservation laws imply that the energy-momentum and angular momentum, respectively

$$P_l = \int T^0_l\, d^3x,\ J_{ij} = \int \left[S^0{}_{ij} - \left(x_i T^0{}_j - x_j T^0{}_i\right)\right] d^3x, \tag{2.215}$$

are conserved. This means that the system invariant under the ten parameter symmetry group has ten conserved quantities. This is an example of Noether symmetry. The first term of the angular momentum integral corresponds to the spin angular momentum while the second term gives the orbital angular momentum. The global Poincaré invariance of a system means that, for the system, the space-time is homogeneous (all space-time points are equivalent) as dictated by the translational invariance and is isotropic (all directions about a space-time point are equivalent) as indicated by the Lorentz invariance. It is interesting to observe that the fixed point variation of the field variables takes the form

$$\delta_0\chi = \frac{1}{2}\varepsilon^j{}_k\Xi_j{}^k\chi + \varepsilon^j T_j\chi, \tag{2.216}$$

where

$$\Xi_j{}^k = S_j{}^k + \left(x^j\partial_k - x^k\partial_j\right),\ \ T_j = -\partial_j. \tag{2.217}$$

We remark that $\Xi_j{}^k$ are the generators of the Lorentz transformation and T_j are those of the translations.

2.15. Local Poincaré Invariance

As next step, let us consider a modification of the infinitesimal Poincaré transformations (2.207) by assuming that the parameters ε^j_k and ε^j are functions of the coordinates and by writing them altogether as

$$\delta x^\mu = \varepsilon^\mu{}_\nu(x)x^\nu + \varepsilon^\mu(x) = \xi^\mu, \tag{2.218}$$

which we call the local Poincaré transformations (or the general coordinate transformations). In order to make a distinction between the global transformation and the local transformation, we use here the Latin indices $(j,\ k = 0,\ 1,\ 2,\ 3)$ for the former and the Greek indices $(\mu,\ \nu = 0, 1, 2, 3)$ for the latter. The variation of the field variables $\chi(x)$ defined at a point x is still the same as that of the global Poincaré transformations,

$$\delta\chi = \frac{1}{2}\varepsilon_{ij}S^{ij}\chi. \tag{2.219}$$

The corresponding fixed point variation of χ takes the form,

$$\delta_0\chi = \frac{1}{2}\varepsilon_{ij}S^{ij}\chi - \xi^\nu\partial_\nu\chi. \tag{2.220}$$

Differentiating both sides of (2.220) with respect to x^μ, we have

$$\delta_0\partial_\mu\chi = \frac{1}{2}\varepsilon^{ij}S_{ij}\partial_\mu\chi + \frac{1}{2}(\partial_\mu\varepsilon^{ij})\,S^{ij}\chi - \partial_\mu(\xi^\nu\partial_\nu\chi). \tag{2.221}$$

By using these variations, we obtain the variation of the Lagrangian L,

$$\delta\mathcal{L} + \partial_\mu(\delta x^\mu)\mathcal{L} = h\delta L = \delta_0\mathcal{L} + \partial_\nu(\mathcal{L}\delta x^\nu) = -\frac{1}{2}(\partial_\mu\varepsilon^{ij})\,S^\mu{}_{ij} - \partial_\mu\xi^\nu\,T^\mu_\nu, \tag{2.222}$$

which is no longer zero unless the parameters ε^{ij} and ξ^ν become constants. Accordingly, the action integral for the given Lagrangian density $\mathcal{L}$ is not invariant under the local Poincaré transformations. We notice that while $\partial_j(\delta x^j) = 0$ for the local Poincaré transformations, $\partial_\mu\xi^\mu$ does not vanish under local Poincaré transformations. Hence, as expected $\mathcal{L}$ is not a Lagrangian scalar but a Lagrangian density. As mentioned earlier, in order to define the Lagrangian L, we have to select an appropriate non-trivial scalar function $h(x)$ satisfying

$$\delta h + h\partial_\mu\xi^\mu = 0. \tag{2.223}$$

Now we consider a minimal modification of the Lagrangian so as to make the action integral invariant under the local Poincaré transformations. It is rather obvious that if there is a covariant derivative $\nabla_k\chi$ which transforms as

$$\delta(\nabla_k\chi) = \frac{1}{2}\varepsilon^{ij}S_{ij}\nabla_k\chi - \varepsilon^i{}_k\nabla_i\chi, \tag{2.224}$$

then a modified Lagrangian $L'(\chi, \partial_k\chi, x) = L(\chi, \nabla_k\chi, x)$, obtained by replacing $\partial_k\chi$ of $L(\chi, \partial_k\chi, x)$ by $\nabla_k\chi$, remains invariant under the local Poincaré transformations, that is

$$\delta L' = \frac{\partial L'}{\partial\chi}\delta\chi + \frac{\partial L'}{\partial(\nabla_k\chi)}\delta(\nabla_k\chi) = 0. \tag{2.225}$$

To find such a k-covariant derivative, we introduce the gauge fields $A^{ij}{}_\mu = -A^{ji}{}_\mu$ and define the μ-covariant derivative

$$\nabla_\mu\chi := \partial_\mu\chi + \frac{1}{2}A^{ij}{}_\mu S_{ij}\chi, \tag{2.226}$$

in such a way that the covariant derivative transforms as

$$\delta_0\nabla_\mu\chi = \frac{1}{2}S_{ij}\nabla_\mu\chi - \partial_\mu(\xi^\nu\nabla_\nu\chi). \tag{2.227}$$

The transformation properties of $A^{ab}{}_{\mu}$ are determined by $\nabla_{\mu}\chi$ and $\delta\nabla_{\mu}\chi$. Making use of

$$\delta\nabla_{\mu}\chi = \frac{1}{2}\varepsilon^{ij}{}_{,\mu}S_{ij}\chi + \frac{1}{2}\varepsilon^{ij}S_{ij}\partial_{\mu}\chi - (\partial_{\mu}\xi^{\nu})\partial_{\nu}\psi + \frac{1}{2}\delta A^{ij}{}_{\mu}S_{ij}\chi + \frac{1}{4}A^{ij}{}_{\mu}S_{ij}\varepsilon^{kl}S_{kl}\chi \tag{2.228}$$

and comparing with (2.226) we obtain,

$$\delta A^{ij}{}_{\mu}S_{ij}\chi + \varepsilon^{ij}{}_{,\mu}S_{ij}\chi + \frac{1}{2}\left(A^{ij}{}_{\mu}\varepsilon^{kl} - \varepsilon^{ij}A^{kl}{}_{\mu}\right)S_{ij}S_{kl}\chi + (\partial_{\mu}\xi^{\nu})A^{ij}{}_{\nu}S_{ij}\chi = 0. \tag{2.229}$$

Using the antisymmetry in ij and kl to rewrite the term in parentheses on the r.h.s. of (2.229) as $[S_{ij}, S_{kl}]A^{ij}{}_{\mu}\varepsilon^{kl}\chi$, we see the explicit appearance of the commutator $[S_{ij}, S_{kl}]$. Using the expression for the commutator of Lie algebra generators

$$[S_{ij},\, S_{kl}] = \frac{1}{2}c^{[ef]}{}_{[ij][kl]}S_{ef}, \tag{2.230}$$

where $c^{[ef]}{}_{[ij][kl]}$ (the square brackets denote anti-symmetrization) is the structure constants of the Lorentz group (deduced below), we have

$$[S_{ij},\, S_{kl}]\,A^{ij}{}_{\mu}\varepsilon^{kl} = \frac{1}{2}\left(A^{ic}_{\mu}\varepsilon^{j}_{c} - A^{cj}_{\mu}\varepsilon^{i}_{c}\right)S_{ij}. \tag{2.231}$$

The substitution of this equation and the consideration of the antisymmetry of $\varepsilon_{c}^{b} = -\varepsilon^{b}_{c}$ enables us to write

$$\delta A^{ij}{}_{\mu} = \varepsilon^{i}{}_{k}A^{kj}{}_{\mu} + \varepsilon^{j}{}_{k}A^{ik}{}_{\mu} - (\partial_{\mu}\xi^{\nu})A^{ij}{}_{\nu} - \partial_{\mu}\varepsilon^{ij}. \tag{2.232}$$

We require the k-derivative and μ-derivative of χ to be linearly related as

$$\nabla_{k}\chi = V_{k}{}^{\mu}(x)\nabla_{\mu}\chi, \tag{2.233}$$

where the coefficients $V_{k}{}^{\mu}(x)$ are position-dependent and behave like a new set of field variables. From (2.233) it is evident that $\nabla_{k}\chi$ varies as

$$\delta\nabla_{k}\chi = \delta V_{k}^{\mu}\nabla_{\mu}\chi + V_{k}^{\mu}\delta\nabla_{\mu}\chi. \tag{2.234}$$

Comparing with $\delta\nabla_{k}\chi = \frac{1}{2}\varepsilon^{ab}S_{ab}\nabla_{k}\chi - \varepsilon^{j}_{k}\nabla_{j}\chi$ we obtain,

$$V^{k}_{\alpha}\delta V^{\mu}_{k}\nabla_{\mu}\chi - \xi^{\nu}{}_{,\alpha}\nabla_{\nu}\chi + V^{k}_{\alpha}\varepsilon^{j}_{k}\nabla_{j}\chi = 0. \tag{2.235}$$

Exploiting $\delta\left(V^{k}_{\alpha}V^{\mu}_{k}\right) = 0$ we find the quantity $V_{k}{}^{\mu}$ transforms according to

$$\delta V_{k}{}^{\mu} = V_{k}{}^{\nu}\partial_{\nu}\xi^{\mu} - V_{i}{}^{\mu}\varepsilon^{i}{}_{k}. \tag{2.236}$$

It is also important to recognize that the inverse of $\det(V_{k}{}^{\mu})$ transforms like a scalar density as $h(x)$ does. For our minimal modification of the Lagrangian density, we utilize this available quantity for the scalar density h; namely, we let

$$h(x) = [\det(V_k{}^{\mu})]^{-1}. \tag{2.237}$$

In the limiting case, when we consider Poincaré transformations, that are not space-time dependent, $V_k{}^{\mu} \to \delta_k^{\mu}$ so that $h(x) \to 1$. This is a desirable property. Then we replace the Lagrangian density $\mathcal{L}(\chi, \partial_k\chi, x)$, invariant under the global Poincaré transformations, by a Lagrangian density

$$\mathcal{L}(\chi, \partial_\mu\chi; x) \to h(x)L(\chi, \nabla_k\chi). \tag{2.238}$$

The action integral with this modified Lagrangian density remains invariant under the local Poincaré transformations. Since the local Poincaré transformations $\delta x^{\mu} = \xi^{\mu}(x)$ are nothing else but generalized coordinate transformations, the newly introduced gauge fields V_i^{λ} and $A^{ij}{}_{\mu}$ can be interpreted, respectively, as the general tetrad (*vierbein*) fields which set the local coordinate frame and as a local affine connection with respect to the tetrad frame (see also [40]). We are adopting there simbols here since our considerations are devoted to general gauge fields, not only to gravity.

2.16. Spinors, Vectors and Tetrads

Let us consider first the case where the multiplet field χ is the Dirac field $\psi(x)$ which behaves like a four-component spinor under the Lorentz transformations and transforms as

$$\psi(x) \to \psi'(x') = S(\Lambda)\psi(x), \tag{2.239}$$

where $S(\Lambda)$ is an irreducible unitary representation of the Lorentz group. Since the bilinear form $v^k = i\overline{\psi}\gamma^k\psi$ is a vector, it transforms according to

$$v^j = \Lambda^j{}_k v^k, \tag{2.240}$$

where $\Lambda^j{}_i$ is a Lorentz transformation matrix satisfying

$$\Lambda_{ij} + \Lambda_{ji} = 0. \tag{2.241}$$

The invariance of v^i (or the covariance of the Dirac equation) under the transformation $\psi(x) \to \psi'(x')$ leads to

$$S^{-1}(\Lambda)\gamma^{\mu}S(\Lambda) = \Lambda^{\mu}_{\nu}\gamma^{\nu}, \tag{2.242}$$

where the γs are the Dirac γ-matrices satisfying the anticommutator,

$$\gamma_i\gamma_j + \gamma_j\gamma_i = \eta_{ij}\mathbf{1}. \tag{2.243}$$

Furthermore, we notice that the γ-matrices have the following properties:

$$\begin{cases} (\gamma_0)^\dagger = -\gamma_0,\ (\gamma^0)^2 = (\gamma_0)^2 = -1,\ \gamma_0 = -\gamma^0 \text{ and } \gamma_0\gamma^0 = 1 \\ (\gamma_k)^\dagger = \gamma_k\,,\ (\gamma^k)^2 = (\gamma_k)^2 = 1;\ (k = 1,2,3) \text{ and } \gamma_k = \gamma^k \\ (\gamma_5)^\dagger = -\gamma_5,\ (\gamma_5)^2 = -1 \text{ and } \gamma^5 = \gamma_5. \end{cases} \tag{2.244}$$

We assume the transformation $S(\Lambda)$ can be put into the form $S(\Lambda) = e^{\Lambda_{\mu\nu}\gamma^{\mu\nu}}$. Expanding $S(\Lambda)$ about the identity and only retaining terms up to the first order in the infinitesimals and expanding $\Lambda_{\mu\nu}$ to the first order in $\varepsilon_{\mu\nu}$

$$\Lambda_{\mu\nu} = \delta_{\mu\nu} + \varepsilon_{\mu\nu},\ \varepsilon_{ij} + \varepsilon_{ji} = 0, \tag{2.245}$$

we get

$$S(\Lambda) = 1 + \frac{1}{2}\varepsilon^{ij}\gamma_{ij}. \tag{2.246}$$

In order to determine the form of γ_{ij}, we substitute (2.245) and (2.246) into (2.242) to obtain

$$\frac{1}{2}\varepsilon_{ij}\left[\gamma^{ij}, \gamma^k\right] = \eta^{ki}\varepsilon_{ji}\gamma^j. \tag{2.247}$$

Rewriting the r.h.s. of (2.247) using the antisymmetry of ε_{ij} as

$$\eta^{ki}\varepsilon_{ji}\gamma^j = \frac{1}{2}\varepsilon_{ij}\left(\eta^{ki}\gamma^j - \eta^{kj}\gamma^i\right), \tag{2.248}$$

yields

$$\left[\gamma^k, \gamma^{ij}\right] = \eta^{ki}\gamma^j - \eta^{kj}\gamma^i. \tag{2.249}$$

Assuming the solution to have the form of an antisymmetric product of two matrices, we obtain the solution

$$\gamma^{ij} := \frac{1}{2}\left[\gamma^i, \gamma^j\right]. \tag{2.250}$$

If $\chi = \psi$, the group generator S_{ij} appearing in (2.210) is identified with

$$S_{ij} \equiv \gamma_{ij} = \frac{1}{2}(\gamma_i\gamma_j - \gamma_j\gamma_i). \tag{2.251}$$

To be explicit, the Dirac field transforms under Lorentz transformations as

$$\delta\psi(x) = \frac{1}{2}\varepsilon^{ij}\gamma_{ij}\psi(x). \tag{2.252}$$

The Pauli conjugate of the Dirac field is denoted $\overline{\psi}$ and defined by

$$\overline{\psi}(x) := i\psi^\dagger(x)\gamma_0,\ i \in \mathbb{C}. \tag{2.253}$$

The conjugate field $\overline{\psi}$ transforms under Lorentz transformations as,

$$\delta\overline{\psi} = -\overline{\psi}\frac{1}{2}\varepsilon^{ij}\overline{\psi}\gamma_{ij}. \tag{2.254}$$

Under local Lorentz transformations, $\varepsilon_{ab}(x)$ becomes a function of space-time. Now, unlike $\partial_\mu\psi(x)$, the derivative of $\psi'(x')$ is no longer homogenous due to the occurrence of the term $\gamma^{ab}\left[\partial_\mu\varepsilon_{ab}(x)\right]\psi(x)$ in $\partial_\mu\psi'(x')$, which is non-vanishing unless ε_{ab} is constant. When going from locally flat to curved space-time, we must generalize ∂_μ to the covariant derivative ∇_μ to compensate for this extra term, allowing to gauge the group of Lorentz transformations. Thus, by using ∇_μ, we can preserve the invariance of the Lagrangian for arbitrary local Lorentz transformations at each space-time point

$$\nabla_\mu\psi'(x') = S(\Lambda(x))\nabla_\mu\psi(x). \tag{2.255}$$

To determine the explicit form of the connection belonging to ∇_μ, we study the derivative of $S(\Lambda(x))$. The transformation $S(\Lambda(x))$ is given by

$$S(\Lambda(x)) = 1 + \frac{1}{2}\varepsilon_{ab}(x)\gamma^{ab}. \tag{2.256}$$

Since $\varepsilon_{ab}(x)$ is only a function of space-time for local Lorentz coordinates, we express this infinitesimal Lorentz transformations in terms of general coordinates only by shifting all space-time dependence of the local coordinates into tetrad fields as

$$\varepsilon_{ab}(x) = V_a{}^{\lambda}(x)V^{\nu}{}_b(x)\varepsilon_{\lambda\nu}. \tag{2.257}$$

Substituting this expression for $\varepsilon_{ab}(x)$, we obtain

$$\partial_\mu\varepsilon_{ab}(x) = \partial_\mu\left[V_a{}^{\lambda}(x)V^{\nu}{}_b(x)\varepsilon_{\lambda\nu}\right]. \tag{2.258}$$

However, since $\varepsilon_{\lambda\nu}$ has no space-time dependence, this reduces to

$$\partial_\mu\varepsilon_{ab}(x) = V_a{}^{\lambda}(x)\partial_\mu V_{b\lambda}(x) - V_b{}^{\nu}(x)\partial_\mu V_{a\nu}(x). \tag{2.259}$$

Letting

$$\omega_{\mu ba} := V_b{}^{\nu}(x)\partial_\mu V_{a\nu}(x), \tag{2.260}$$

the first and second terms in Eq. (2.259) become $V_a^{\lambda}(x)\partial_\mu V_{b\lambda}(x) = \frac{1}{2}\omega_{\mu ab}$ and $V_b^{\nu}(x)\partial_\mu V_{a\nu}(x) = \frac{1}{2}\omega_{\mu ba}$ respectively. Using the identification

$$\partial_\mu\varepsilon_{ab}(x) = \omega_{\mu ab}, \tag{2.261}$$

we write

$$\partial_\mu S(\Lambda(x)) = -\frac{1}{2}\gamma^{ab}\omega_{\mu ab}. \tag{2.262}$$

According to (2.226), the covariant derivative of the Dirac spinor is

$$\nabla_\mu \psi = \partial_\mu \psi + \frac{1}{2} A^{ij}{}_\mu \gamma_{ij} \psi. \tag{2.263}$$

Correspondingly, the covariant derivative of $\bar{\psi}$ is given by

$$\nabla_\mu \bar{\psi} = \partial_\mu \bar{\psi} - \frac{1}{2} A^{ij}{}_\mu \bar{\psi} \gamma_{ij}. \tag{2.264}$$

Using the covariant derivatives of ψ and $\bar{\psi}$, we can show that

$$\nabla_\mu v_j = \partial_\mu v_j - A^i{}_{j\mu} v_i. \tag{2.265}$$

The same covariant derivative should be used for any covariant vector v_k under the Lorentz transformation. Since $\nabla_\mu (v_i v^i) = \partial_\mu (v_i v^i)$, the covariant derivative for a contravariant vector v^i must be

$$\nabla_\mu v^i = \partial_\mu v^i + A^i{}_{j\mu} v^j. \tag{2.266}$$

Since the tetrad $V_i{}^\mu$ is a covariant vector under Lorentz transformations, its covariant derivative must transform according to the same rule. Using $\nabla_a = V_a^\mu(x)\nabla_\mu$, the covariant derivatives of a tetrad in local Lorentz coordinates read

$$\nabla_\nu V_i{}^\mu = \partial_\nu V_i{}^\mu - A^k{}_{i\nu} V_k{}^\mu, \; \nabla_\nu V^i{}_\mu = \partial_\nu V^i{}_\mu + A^i{}_{k\nu} V^k{}_\mu. \tag{2.267}$$

The inverse of $V_i{}^\mu$ is denoted by $V^i{}_\mu$ and satisfies

$$V^i{}_\mu V_i{}^\nu = \delta_\mu{}^\nu, \quad V^i{}_\mu V_j{}^\mu = \delta^i{}_j. \tag{2.268}$$

To allow the transition to curved space-time, we take account of the general coordinates of objects that are covariant under local Poincaré transformations. Here we define the covariant derivative of a quantity v^λ which behaves like a contravariant vector under the local Poincaré transformation. Namely

$$D_\nu v^\lambda \equiv V_i{}^\lambda \nabla_\nu v^i = \partial_\nu v^\lambda + \Gamma^\lambda{}_{\mu\nu} v^\mu, \; D_\nu v_\mu \equiv V^i{}_\mu \nabla_\nu v_i = \partial_\nu v_\mu - \Gamma^\lambda{}_{\mu\nu} v_\lambda, \tag{2.269}$$

where

$$\Gamma^\lambda{}_{\mu\nu} := V_i{}^\lambda \nabla_\nu V^i{}_\mu \equiv -V^i{}_\mu \nabla_\nu V_i{}^\lambda. \tag{2.270}$$

The definition of $\Gamma^\lambda{}_{\mu\nu}$ implies

$$\begin{aligned} D_\nu V_i{}^\lambda &= \nabla_\nu V_i{}^\lambda + \Gamma^\lambda{}_{\mu\nu} V_i{}^\mu = \partial_\nu V_i{}^\lambda - A^k{}_{i\nu} V_k{}^\lambda + \Gamma^\lambda{}_{\mu\nu} V_i{}^\mu = 0, \\ D_\nu V^i{}_\mu &= \nabla_\nu V^i{}_\mu - \Gamma^\lambda{}_{\mu\nu} V^i{}_\lambda = \partial_\nu V^i{}_\mu + A^i{}_{k\nu} V^k{}_\mu - \Gamma^\lambda{}_{\mu\nu} V^i{}_\lambda = 0. \end{aligned} \tag{2.271}$$

From (2.271) we find,

$$A^i{}_{k\nu} = V^i{}_\lambda \partial_\nu V_k{}^\lambda + \Gamma^\lambda{}_{\mu\nu} V^i{}_\lambda V_k{}^\mu = -V_k{}^\lambda \partial_\nu V^i{}_\lambda + \Gamma^\lambda{}_{\mu\nu} V^i{}_\lambda V_k{}^\mu. \tag{2.272}$$

or, equivalently, in terms of ω defined in (2.260),

$$A^{i}{}_{k\nu} = \omega^{i}{}_{\nu k} + \Gamma^{\lambda}{}_{\mu\nu} V^{i}{}_{\lambda} V_{k}{}^{\mu} = -\omega_{k\nu}{}^{i} + \Gamma^{\lambda}{}_{\mu\nu} V^{i}{}_{\lambda} V_{k}{}^{\mu}. \tag{2.273}$$

Using this in (2.263), we may write

$$\nabla_{\mu}\psi = (\partial_{\mu} - \Gamma_{\mu})\psi, \tag{2.274}$$

where

$$\Gamma_{\mu} = \frac{1}{4}\left(\omega^{i}{}_{j\mu} - \Gamma^{\lambda}{}_{\mu\nu} V^{i}{}_{\lambda} V_{j}{}^{\nu}\right)\gamma_{i}{}^{j}, \tag{2.275}$$

which is known as the Fock-Ivanenko connection.

We now study the transformation properties of $A_{\mu ab}$. Recall $\omega_{\mu ab} = V_{a}{}^{\lambda}(x)\partial_{\mu}V_{\beta\lambda}(x)$ and since $\partial_{\mu}\eta_{ab} = 0$, we write

$$\Lambda^{a}{}_{\bar{a}}\eta_{ab}\partial_{\mu}\Lambda_{\bar{b}}{}^{b} = \Lambda^{a}{}_{\bar{a}}\partial_{\mu}\Lambda_{a\bar{b}}. \tag{2.276}$$

Note that barred indices are equivalent to the primed indices used above. Hence, the spin connection transforms as

$$A_{\bar{a}\bar{b}\bar{c}} = \Lambda_{\bar{a}}{}^{a}\Lambda_{\bar{b}}{}^{b}\Lambda_{\bar{c}}{}^{c}A_{abc} + \Lambda_{\bar{a}}{}^{a}\Lambda_{\bar{c}}{}^{c}V^{\mu}{}_{a}(x)\partial_{\mu}\Lambda_{\bar{b}c}. \tag{2.277}$$

To determine the transformation properties of

$$\Gamma_{abc} = A_{abc} - \left[V^{\mu}{}_{a}(x)\partial_{\mu}V^{\nu}{}_{b}(x)\right]V_{\nu c}(x), \tag{2.278}$$

we consider the local Lorentz transformations of $\left[V_{a}{}^{\mu}(x)\partial_{\mu}V^{\nu}{}_{b}(x)\right]V_{\nu c}(x)$ which is,

$$\left[V^{\mu}{}_{\bar{a}}(x)\partial_{\mu}V^{\nu}{}_{\bar{b}}\right]V_{\nu\bar{c}}(x) = \Lambda_{\bar{a}}{}^{a}\Lambda_{\bar{b}}{}^{b}\Lambda_{\bar{c}}{}^{c}\left[A^{\nu}{}_{ab}V_{\nu c}(x)\right] + \Lambda_{\bar{a}}{}^{a}\Lambda_{\bar{c}}{}^{c}V^{\mu}{}_{a}(x)\partial_{\mu}\Lambda_{c\bar{b}}. \tag{2.279}$$

From this result, we obtain the following transformation law,

$$\Gamma_{\bar{a}\bar{b}\bar{c}} = \Lambda_{\bar{a}}{}^{a}\Lambda_{\bar{b}}{}^{b}\Lambda_{\bar{c}}{}^{c}\Gamma_{abc}. \tag{2.280}$$

We now explore the consequence of the antisymmetry of ω_{abc} in bc. Recalling the equation for Γ_{abc}, exchanging b and c and adding the two equations, we obtain

$$\Gamma_{abc} + \Gamma_{acb} = -V^{\mu}{}_{a}(x)\left[(\partial_{\mu}V^{\nu}{}_{b}(x))V_{\nu c}(x) + (\partial_{\mu}V^{\nu}{}_{c}(x))V_{\nu b}(x)\right]. \tag{2.281}$$

We know however, that

$$\partial_{\mu}\left[V^{\nu}{}_{b}(x)V_{\nu c}(x)\right] = V_{\nu c}(x)\partial_{\mu}V^{\nu}{}_{b}(x) + V_{\lambda b}(x)\partial_{\mu}V^{\lambda}{}_{c}(x) + V_{b}{}^{\nu}(x)V^{\lambda}{}_{c}(x)\partial_{\mu}g_{\lambda\nu}. \tag{2.282}$$

Letting $\lambda \to \nu$ and exchanging b and c, we obtain

$$\partial_{\mu}\left[V^{\nu}{}_{b}(x)V_{\nu c}(x)\right] = -V_{b}{}^{\lambda}(x)V^{\nu}{}_{c}(x)\partial_{\mu}g_{\nu\lambda}, \tag{2.283}$$

so that, finally,

$$\Gamma_{abc} + \Gamma_{acb} = V_{a}{}^{\mu}(x)V_{b}{}^{\lambda}(x)V_{c}{}^{\nu}(x)\partial_{\mu}g_{\nu\lambda}. \tag{2.284}$$

This, however, is equivalent to

$$\Gamma_{\bar{a}\bar{b}\bar{c}} + \Gamma_{\bar{a}\bar{c}\bar{b}} = V^{\bar{\mu}}_{\bar{a}}(x)V^{\bar{\lambda}}_{\bar{b}}(x)V^{\bar{\nu}}_{\bar{c}}(x)\partial_{\bar{\mu}}g_{\bar{\nu}\bar{\lambda}}, \tag{2.285}$$

and then

$$\Gamma_{\mu\lambda\nu} + \Gamma_{\mu\nu\lambda} = \partial_\mu g_{\nu\lambda}, \tag{2.286}$$

which we recognize as the general coordinate connection. It is known that the covariant derivative for general coordinates is

$$\nabla_\mu A_\nu{}^\lambda = \partial_\mu A_\nu{}^\lambda + \Gamma^\lambda{}_{\mu\sigma}A_\nu{}^\sigma - \Gamma^\sigma{}_{\mu\nu}A_\sigma{}^\lambda. \tag{2.287}$$

In a Riemannian manifold, the connection is symmetric under the exchange of $\mu\nu$, that is, $\Gamma^\lambda{}_{\mu\nu} = \Gamma^\lambda{}_{\nu\mu}$. Using the fact that the metric is a symmetric tensor we can now determine the form of the Christoffel connection by cyclically permuting the indices of the general coordinate connection equation (2.286) yielding

$$\Gamma_{\mu\nu\lambda} = \frac{1}{2}\left(\partial_\mu g_{\nu\lambda} + \partial_\nu g_{\lambda\mu} - \partial_\lambda g_{\mu\nu}\right). \tag{2.288}$$

Since $\Gamma_{\mu\nu\lambda} = \Gamma_{\nu\mu\lambda}$ is valid for general coordinate systems, it follows that a similar constraint must hold for local Lorentz transforming coordinates as well, so we expect $\Gamma_{abc} = \Gamma_{bac}$. Recalling the equation for Γ_{abc} and exchanging a and b, we obtain

$$\omega_{abc} - \omega_{bac} = V_{\nu c}(x)\left[V^\mu{}_a(x)\partial_\mu V^\nu{}_b(x) - V^\mu{}_b(x)\partial_\mu V^\nu{}_a(x)\right]. \tag{2.289}$$

We now define the *objects of anholonomicity* as

$$\Omega_{cab} := V_{\nu c}(x)\left[V^\mu{}_a(x)\partial_\mu V^\nu{}_b(x) - V^\mu{}_b(x)\partial_\mu V^\nu{}_a(x)\right]. \tag{2.290}$$

Using $\Omega_{cab} = -\Omega_{cba}$, we permute indices in a similar manner as was done for the derivation of the Christoffel connection above yielding,

$$\omega_{ab\mu} = \frac{1}{2}\left[\Omega_{cab} + \Omega_{bca} - \Omega_{abc}\right]V^c{}_\mu \equiv \Delta_{ab\mu}. \tag{2.291}$$

For completeness, we determine the transformation law of the Christoffel connection. Making use of $\Gamma^\lambda_{\mu\nu}e_\lambda = \partial_\mu e_\nu$ where

$$\partial_{\bar{\mu}}e_{\bar{\nu}} = X^\mu{}_{\bar{\mu}}X^\nu{}_{\bar{\nu}}\partial_\mu e_\nu + X^\mu{}_{\bar{\mu}}\left(\partial_\mu X^\nu{}_{\bar{\nu}}\right)e_\nu, \tag{2.292}$$

we can show

$$\Gamma^{\bar{\lambda}}{}_{\bar{\mu}\bar{\nu}} = X^\mu{}_{\bar{\mu}}X^\nu{}_{\bar{\nu}}X_\lambda{}^{\bar{\lambda}}\Gamma^\lambda{}_{\mu\nu} + X^\mu{}_{\bar{\mu}}X_\nu{}^{\bar{\lambda}}X^\nu{}_{\mu\bar{\nu}}, \tag{2.293}$$

where

$$X^\nu{}_{\mu\bar{\nu}} \equiv \partial_\mu\partial_{\bar{\nu}}x^\nu. \tag{2.294}$$

In the light of the above considerations, we may regard infinitesimal local gauge transformations as local rotations of basis vectors belonging to the tangent space [133, 293] of the manifold. For this reason, given a local frame on a tangent plane to the point x on the base manifold, we can obtain all other frames on the same tangent plane by means of local rotations of the original basis vectors. Reversing this argument, we observe that by knowing all frames residing in the horizontal tangent space to a point x on the base manifold enables us to deduce the corresponding gauge group of symmetry transformations.

2.17. Curvature, Torsion and Metric

From the definition of the Fock-Ivanenko covariant derivative, we can find the second order covariant derivative

$$\begin{aligned} D_\nu D_\mu \psi &= \partial_\nu \partial_\mu \psi + \frac{1}{2} S_{cd} \left(\psi \partial_\nu A_\mu{}^{cd} + A_\mu{}^{cd} \partial_\nu \psi \right) + \Gamma^\rho{}_{\mu\nu} D_\rho \psi + \frac{1}{2} S_{ef} A_\nu{}^{ef} \partial_\mu \psi \\ &+ \frac{1}{4} S_{ef} S_{cd} A_\nu{}^{ef} A_\mu{}^{cd} \psi. \end{aligned} \tag{2.295}$$

Recalling $D_\nu V^{c\mu} = 0$, we can solve for the spin connection in terms of the Christoffel connection

$$A_\mu{}^{cd} = -V^d_\lambda \partial_\mu V^{c\lambda} - \Gamma_\mu{}^{cd}. \tag{2.296}$$

The derivative of the spin connection is then

$$\partial_\mu A^{cd}_{\ \ \nu} = -V^d_\lambda \partial_\mu \partial_\nu V^{c\lambda} - \left(\partial_\nu V^{c\lambda} \right) \partial_\mu V_\lambda{}^d - \partial_\mu \Gamma^{cd}_{\ \ \nu}. \tag{2.297}$$

Noting that the Christoffel connection is symmetric and partial derivatives commute, we find

$$\begin{aligned} [D_\mu, D_\nu] \psi &= \frac{1}{2} S_{cd} \left[\left(\partial_\nu A^{cd}_{\ \ \mu} - \partial_\mu A^{cd}_{\ \ \nu} \right) \psi \right] \\ &+ \frac{1}{4} S_{ef} S_{cd} \left[\left(A^{ef}_{\ \ \nu} A^{cd}_{\ \ \mu} - A^{ef}_{\ \ \mu} A^{cd}_{\ \ \nu} \right) \psi \right], \end{aligned} \tag{2.298}$$

where

$$\partial_\nu A^{cd}_{\ \ \mu} - \partial_\mu A^{cd}_{\ \ \nu} = \partial_\mu \Gamma^{cd}_{\ \ \nu} - \partial_\nu \Gamma^{cd}_{\ \ \mu}. \tag{2.299}$$

Relabeling running indices, we can write

$$\frac{1}{4} S_{ef} S_{cd} \left(A^{ef}_{\ \ \nu} A^{cd}_{\ \ \mu} - A^{ef}_{\ \ \mu} A^{cd}_{\ \ \nu} \right) \psi = \frac{1}{4} [S_{cd}, S_{ef}] A^{ef}_{\ \ \mu} A^{cd}_{\ \ \nu} \psi. \tag{2.300}$$

Using $\{\gamma_a, \gamma_b\} = 2\eta_{ab}$ to deduce

$$\{\gamma_a, \gamma_b\} \gamma_c \gamma_d = 2\eta_{ab} \gamma_c \gamma_d, \tag{2.301}$$

we find that the commutator of bi-spinors is given by

$$[S_{cd}, S_{ef}] = \frac{1}{2} \left[\eta_{ce} \delta^a_d \delta^b_f - \eta_{de} \delta^a_c \delta^b_f + \eta_{cf} \delta^a_e \delta^b_d - \eta_{df} \delta^a_e \delta^b_c \right] S_{ab}. \tag{2.302}$$

Clearly the terms in brackets on the r.h.s. of (2.302) are antisymmetric in cd and ef and also antisymmetric under the exchange of pairs of indices cd and ef. Since the alternating spinor is antisymmetric in ab, so it must be the terms in brackets: this means that the commutator does not vanish. Hence, the term in brackets is totally antisymmetric under interchange of indices ab, cd and ef and exchange of these pairs of indices. We identify this as the structure constant of the Lorentz group [160]

$$\left[\eta_{ce}\delta^a_d\delta^b_f - \eta_{de}\delta^a_c\delta^b_f + \eta_{cf}\delta^a_e\delta^b_d - \eta_{df}\delta^a_e\delta^b_c\right] = c_{[cd][ef]}{}^{[ab]} = c^{[ab]}{}_{[cd][ef]}, \tag{2.303}$$

with the aid of which we can write

$$\frac{1}{4}\left[S_{cd}, S_{ef}\right]A^{ef}{}_{\mu}A^{cd}{}_{\nu}\psi = \frac{1}{2}S_{ab}\left[A^a{}_{e\nu}A^{eb}{}_{\mu} - A^b{}_{e\nu}A^{ae}{}_{\mu}\right]\psi, \tag{2.304}$$

where

$$A^a{}_{e\nu}A^{eb}{}_{\mu} - A^b{}_{e\nu}A^{ae}{}_{\mu} = \Gamma^a_{\nu e}\Gamma^{eb}{}_{\mu} - \Gamma^b_{\nu e}\Gamma^{ea}{}_{\mu}. \tag{2.305}$$

Combining these results, the commutator of two μ-covariant differentiations gives

$$[\nabla_\mu, \nabla_\nu]\chi = -\frac{1}{2}R^{ij}{}_{\mu\nu}S_{ij}\chi, \tag{2.306}$$

where

$$R^i{}_{j\mu\nu} = \partial_\nu A^i{}_{j\mu} - \partial_\mu A^i{}_{j\nu} + A^i{}_{k\nu}A^k{}_{j\mu} - A^i{}_{k\mu}A^k{}_{j\nu}. \tag{2.307}$$

Using the Jacobi identities for the commutator of covariant derivatives, it follows that the field strength $R^i{}_{j\mu\nu}$ satisfies the Bianchi identity

$$\nabla_\lambda R^i{}_{j\mu\nu} + \nabla_\mu R^i{}_{j\nu\lambda} + \nabla_\nu R^i{}_{j\lambda\mu} = 0. \tag{2.308}$$

Permuting indices, this can be put into the cyclic form

$$\varepsilon^{\alpha\beta\rho\sigma}\nabla_\beta R^{ij}{}_{\rho\sigma} = 0, \tag{2.309}$$

where $\varepsilon^{\alpha\beta\rho\sigma}$ is the Levi-Civita alternating symbol. Furthermore, $R^{ij}{}_{\mu\nu} = \eta^{jk}R^i{}_{k\mu\nu}$ is antisymmetric with respect to both pairs of indices,

$$R^{ij}{}_{\mu\nu} = -R^{ji}{}_{\mu\nu} = R^{ji}{}_{\nu\mu} = -R^{ij}{}_{\nu\mu}. \tag{2.310}$$

This condition is known as the first curvature tensor identity.

To determine the analogue of $[\nabla_\mu, \nabla_\nu]\chi$ in local coordinates, we start from $\nabla_k\psi = V^\mu_k\nabla_\mu\psi$. From $\nabla_k\psi$ we obtain,

$$\nabla_l\nabla_k\psi = V^\nu_l\left(\nabla_\nu V^\mu_k\right)\nabla_\mu\psi + V^\nu_l V^\mu_k\nabla_\nu\nabla_\mu\psi. \tag{2.311}$$

Permuting indices and recognizing

$$V_\mu{}^a\nabla_\nu V^\mu_k = -V_k{}^\mu\nabla_\nu V^a{}_\mu, \tag{2.312}$$

(which follows from $\nabla_\nu\left(V^a_\mu V^\mu_k\right) = 0$), we arrive at

$$V^\nu_l\left(\nabla_\nu V^\mu_k\right)\nabla_\mu\psi - V^\mu_k\left(\nabla_\mu V^\nu_l\right)\nabla_\nu\psi = \left(V^\mu_l V^\nu_k - V^\mu_k V^\nu_l\right)\left(\nabla_\nu V_\mu{}^a\right)\nabla_a\psi. \tag{2.313}$$

Defining

$$C^a{}_{kl} := \left(V^\mu_k V^\nu_l - V^\mu_l V^\nu_k\right) \nabla_\nu V_\mu{}^a, \tag{2.314}$$

the commutator of the k-covariant differentiations takes the final form [248]

$$[\nabla_k, \nabla_l]\chi = -\frac{1}{2} R^{ij}{}_{kl} S_{ij} \chi + C^i{}_{kl} \nabla_i \chi, \tag{2.315}$$

where

$$R^{ij}{}_{kl} = V_k{}^\mu V_l{}^\nu R^{ij}{}_{\mu\nu}. \tag{2.316}$$

As done for $R^i{}_{j\mu\nu}$ using the Jacobi identities for the commutator of covariant derivatives, we find the Bianchi identity in Einstein-Cartan space-time [57, 211]

$$\varepsilon^{\alpha\beta\rho\sigma} \nabla_\beta R^{ij}_{\rho\sigma} = \varepsilon^{\alpha\beta\rho\sigma} C_{\beta\rho}{}^\lambda R^{ij}_{\sigma\lambda}. \tag{2.317}$$

The second curvature identity

$$R^k{}_{[\rho\sigma\lambda]} = 2\nabla_{[\rho} C_{\sigma\lambda]}{}^k - 4C_{[\rho\sigma}{}^b C_{\lambda]b}{}^k \tag{2.318}$$

leads to,

$$\varepsilon^{\alpha\beta\rho\sigma} \nabla_\beta C_{\rho\sigma}{}^k = \varepsilon^{\alpha\beta\rho\sigma} R^k{}_{j\rho\sigma} V^j_\beta. \tag{2.319}$$

Notice that if

$$\Gamma^\lambda{}_{\mu\nu} = V_i{}^\lambda \nabla_\nu V^i{}_\mu = -V_\mu{}^i \nabla_\nu V^\lambda_i, \tag{2.320}$$

then

$$\Gamma^\lambda{}_{\mu\nu} - \Gamma^\lambda{}_{\nu\mu} = V^\lambda_i \left(\nabla_\nu V^i{}_\mu - \nabla_\mu V^i{}_\nu\right). \tag{2.321}$$

Contracting by $V^\mu_k V^\nu_l$, we obtain [248],

$$C^a{}_{kl} = V_k{}^\mu V_l{}^\nu V_\lambda{}^a \left(\Gamma^\lambda{}_{\mu\nu} - \Gamma^\lambda{}_{\nu\mu}\right). \tag{2.322}$$

We therefore conclude that C^a_{kl} is related to the antisymmetric part of the affine connection

$$\Gamma^\lambda{}_{[\mu\nu]} = V_\mu{}^k V_\nu{}^l V_a{}^\lambda C^a{}_{kl} \equiv T^\lambda_{\mu\nu}, \tag{2.323}$$

which is usually interpreted as space-time torsion $T^\lambda_{\mu\nu}$. Considering $\Delta_{ab\mu}$ defined in (2.291), we see that the most general connection in the Poincaré gauge approach to gravitation is

$$A_{ab\mu} = \Delta_{ab\mu} - K_{ab\mu} + \Gamma^\lambda{}_{\nu\mu} V_{a\lambda} V_b{}^\nu, \tag{2.324}$$

where

$$K_{abc} = -\left(T^\lambda{}_{\nu\mu} - T_{\nu\mu}{}^\lambda + T_\mu{}^\lambda{}_\nu\right) V_{a\lambda} V_b{}^\nu V_c{}^\mu, \tag{2.325}$$

is the contorsion tensor. Now, the quantity $R^\rho_{\sigma\mu\nu} = V_i{}^\rho R^i{}_{\sigma\mu\nu}$ may be expressed as

$$R^\rho{}_{\sigma\mu\nu} = \partial_\nu \Gamma^\rho{}_{\sigma\mu} - \partial_\mu \Gamma^\rho{}_{\sigma\nu} + \Gamma^\rho{}_{\lambda\nu} \Gamma^\lambda{}_{\sigma\mu} - \Gamma^\rho{}_{\lambda\mu} \Gamma^\lambda{}_{\sigma\nu}. \tag{2.326}$$

Therefore, we can regard $R^\rho{}_{\sigma\mu\nu}$ as the curvature tensor with respect the affine connection

$\Gamma^{\lambda}{}_{\mu\nu}$. By using the inverse of the tetrad, we define the metric of the space-time manifold by

$$g_{\mu\nu} = V^{i}{}_{\mu} V^{j}{}_{\nu} \eta_{ij}. \tag{2.327}$$

From (2.271) and the fact that the Minkowski metric is constant, it is obvious that the metric so defined is covariantly constant, that is,

$$D_{\lambda} g_{\mu\nu} = 0. \tag{2.328}$$

The space-time thus specified by the local Poincaré transformation is said to be metric. It is not difficult to show that

$$\sqrt{-g} = [\det V^{i}{}_{\mu}] = [\det V_{i}{}^{\mu}]^{-1}, \tag{2.329}$$

where $g = \det g_{\mu\nu}$. Hence we may take $\sqrt{-g}$ for the density function $h(x)$.

2.18. The Field Equations of Gravity

Finally, we are able to deduce the field equations for the gravitational field. From the curvature tensor $R^{\rho}{}_{\sigma\mu\nu}$, given in (2.326), the Ricci tensor follows

$$R_{\sigma\nu} = R^{\mu}{}_{\sigma\mu\nu}. \tag{2.330}$$

and the scalar curvature

$$R = R^{\nu}{}_{\nu} = \overset{\mathrm{L}}{R} + \partial_i K_{a}{}^{ia} - T_{a}{}^{bc} K_{bc}{}^{a} \tag{2.331}$$

where $\overset{\mathrm{L}}{R}$ denotes the usual Ricci scalar of General Relativity. Using this scalar curvature R, we choose the Lagrangian density for free Einstein-Cartan gravity

$$\mathcal{L}_G = \frac{1}{2\kappa}\sqrt{-g}\left(\overset{\mathrm{L}}{R} + \partial_i K_{a}{}^{ia} - T_{a}{}^{bc} K_{bc}{}^{a} - 2\Lambda\right), \tag{2.332}$$

where κ is a gravitational coupling constant, and Λ is the cosmological constant. These considerations can be easily extended to any function of $\overset{\mathrm{L}}{R}$ as in [83]. Observe that the second term is a divergence and may be ignored. The field equation can be obtained from the total action,

$$S = \int \left\{\mathcal{L}_{\text{field}}(\chi, \partial_{\mu}\chi, V_{i}{}^{\mu}, A^{ij}{}_{\mu}) + \mathcal{L}_G\right\} d^4x, \tag{2.333}$$

where the matter Lagrangian density is taken to be

$$\mathcal{L}_{\text{field}} = \frac{1}{2}\left[\overline{\psi}\gamma^{a} D_{a}\psi - (D_{a}\overline{\psi})\gamma^{a}\psi\right]. \tag{2.334}$$

Modifying the connection to include Christoffel, spin connection and contorsion contributions so as to operate on general, spinoral arguments, we have

$$\Gamma_\mu = \frac{1}{4} g_{\lambda\sigma} \left(\Delta^{\sigma}{}_{\mu\rho} - \overset{L}{\Gamma}{}^{\sigma}{}_{\rho\mu} - K^{\sigma}{}_{\rho\mu} \right) \gamma^{\lambda\rho}. \tag{2.335}$$

It is important to keep in mind that $\Delta^{\sigma}{}_{\mu\rho}$ act only on multi-component spinor fields, while $\overset{L}{\Gamma}{}^{\sigma}{}_{\rho\mu}$ act on vectors and arbitrary tensors. The gauge covariant derivative for a spinor and adjoint spinor is then given by

$$D_\mu\psi = (\partial_\mu - \Gamma_\mu)\psi, \; D_\mu\overline{\psi} = \partial_\mu\overline{\psi} - \overline{\psi}\Gamma_\mu. \tag{2.336}$$

The variation of the field Lagrangian is

$$\delta\mathcal{L}_{\text{field}} = \overline{\psi}(\delta\gamma^\mu D_\mu + \gamma^\mu \delta\Gamma_\mu)\psi. \tag{2.337}$$

We know that the Dirac gamma matrices are covariantly vanishing, so

$$D_\kappa\gamma_\iota = \partial_\kappa\gamma_\iota - \Gamma^{\mu}_{\iota\kappa}\gamma_\mu + \left[\gamma_\iota, \widehat{\Gamma}_\kappa\right] = 0. \tag{2.338}$$

The 4×4 matrices $\widehat{\Gamma}_\kappa$ are real matrices used to induce similarity transformations on quantities with spinor transformation [63] properties according to

$$\gamma_i' = \widehat{\Gamma}^{-1}\gamma_i\widehat{\Gamma}. \tag{2.339}$$

Solving for $\widehat{\Gamma}_\kappa$ leads to,

$$\widehat{\Gamma}_\kappa = \frac{1}{8}\left[(\partial_\kappa\gamma_\iota)\gamma^\iota - \Gamma^{\mu}{}_{\iota\kappa}\gamma_\mu\gamma^\iota\right]. \tag{2.340}$$

Taking the variation of $\widehat{\Gamma}_\kappa$,

$$\begin{aligned}\delta\widehat{\Gamma}_\kappa &= \frac{1}{8}\left[\begin{array}{c}(\partial_\kappa\delta\gamma_\iota)\gamma^\iota + (\partial_\kappa\gamma_\iota)\delta\gamma^\iota - (\delta\Gamma^{\mu}{}_{\iota\kappa})\gamma_\mu\gamma^\iota \\ -\Gamma^{\mu}{}_{\iota\kappa}((\delta\gamma_\mu)\gamma^\iota + \gamma_\mu\delta\gamma^\iota)\end{array}\right] \\ &= \frac{1}{8}\left[(\partial_\kappa\delta\gamma_\iota)\gamma^\iota - (\delta\Gamma^{\mu}{}_{\iota\kappa})\gamma_\mu\gamma^\iota\right]. \end{aligned} \tag{2.341}$$

Since we require the anticommutator condition on the gamma matrices $\{\gamma^\mu, \gamma^\nu\} = 2g^{\mu\nu}$ to hold, the variation of the metric gives

$$2\delta g^{\mu\nu} = \{\delta\gamma^\mu, \gamma^\nu\} + \{\gamma^\mu\delta\gamma^\nu\}. \tag{2.342}$$

One solution to this equation is,

$$\delta\gamma^\nu = \frac{1}{2}\gamma_\sigma\delta\gamma^{\sigma\nu}. \tag{2.343}$$

With the aid of this result, we can write

$$(\partial_\kappa\delta\gamma_\iota)\gamma^\iota = \frac{1}{2}\partial_\kappa(\gamma^\nu\delta g_{\nu\iota})\gamma^\iota. \tag{2.344}$$

Finally, exploiting the anti-symmetry in $\gamma_{\mu\nu}$ we obtain

$$\delta\widehat{\Gamma}_\kappa = \frac{1}{8}\left[g_{\nu\sigma}\delta\Gamma_{\mu\kappa}^{\ \ \sigma} - g_{\mu\sigma}\delta\Gamma_{\nu\kappa}^{\ \ \sigma}\right]\gamma^{\mu\nu}. \tag{2.345}$$

The field Lagrangian defined in the Einstein-Cartan space-time can be written [57,127,210, 211,364] explicitly in terms of its Lorentzian and contorsion components as

$$\mathcal{L}_{\text{field}} = \frac{1}{2}\left[\left(\overset{\text{L}}{D}_\mu\overline{\psi}\right)\gamma^\mu\psi - \overline{\psi}\gamma^\mu\overset{\text{L}}{D}_\mu\psi\right] - \frac{\hbar c}{8}K_{\mu\alpha\beta}\overline{\psi}\left\{\gamma^\mu,\gamma^{\alpha\beta}\right\}\psi. \tag{2.346}$$

Using the following relations

$$\begin{cases} -\frac{1}{4}K_{\mu\alpha\beta}\overline{\psi}\left\{\gamma^\mu,\gamma^{\alpha\beta}\right\}\psi = \frac{1}{4}K_{\mu\alpha\beta}\overline{\psi}\gamma^{\beta\alpha}\gamma^\mu\psi - \frac{1}{4}K_{\mu\alpha\beta}\overline{\psi}\gamma^\mu\gamma^{\alpha\beta}\psi, \\ \gamma^\mu\gamma^\nu\gamma^\lambda\varepsilon_{\mu\nu\lambda\sigma} = \left\{\gamma^\mu,\gamma^{\nu\lambda}\right\}\varepsilon_{\mu\nu\lambda\sigma} = 3!\gamma_\sigma\gamma_5, \\ \left\{\gamma^\mu,\gamma^{\nu\lambda}\right\} = \gamma^{[\mu}\gamma^\nu\gamma^{\lambda]}, \end{cases} \tag{2.347}$$

we obtain

$$K_{\mu\alpha\beta}\overline{\psi}\left\{\gamma^\mu,\gamma^{\alpha\beta}\right\}\psi = \frac{1}{2i}K_{\mu\alpha\beta}\varepsilon^{\alpha\beta\mu\nu}\left(\overline{\psi}\gamma_5\gamma_\nu\psi\right). \tag{2.348}$$

Here we define the contorsion axial vector

$$K_\nu := \frac{1}{3!}\varepsilon^{\alpha\beta\mu\nu}K_{\alpha\beta\mu}. \tag{2.349}$$

Multiplying through by the axial current $j_\nu^5 = \overline{\psi}\gamma_5\gamma_\nu\psi$, we obtain

$$\left(\overline{\psi}\gamma_5\gamma_\nu\psi\right)\varepsilon^{\alpha\beta\mu\nu}K_{\mu\alpha\beta} = -6ij_\nu^5K^\nu. \tag{2.350}$$

Thus, the field Lagrangian density becomes

$$\mathcal{L}_{\text{field}} = \frac{1}{2}\left[\left(\overset{\text{L}}{D}_\mu\overline{\psi}\right)\gamma^\mu\psi - \overline{\psi}\gamma^\mu\overset{\text{L}}{D}_\mu\psi\right] + \frac{3i\hbar c}{8}K_\mu j_5^\mu. \tag{2.351}$$

The total action reads

$$\begin{aligned} \delta I &= \delta\int\mathcal{L}_G\sqrt{-g}d^4x + \delta\int\mathcal{L}_{\text{field}}\sqrt{-g}d^4x \\ &= \int\left(\delta\mathcal{L}_G + \delta\mathcal{L}_{\text{field}}\right)\sqrt{-g}d^4x. \end{aligned} \tag{2.352}$$

Writing the metric in terms of the tetrads $g^{\mu\nu} = V_i^\mu V^{\nu i}$, we observe

$$\delta\sqrt{-g} = -\frac{1}{2}\sqrt{-g}\left(\delta V_i^\mu V_\mu^{\ i} + V_{\nu i}\delta V^{\nu i}\right). \tag{2.353}$$

By using

$$\delta V^{\nu i} = \delta\left(\eta^{ij}V_j^\nu\right) = \eta^{ij}\delta V_j^\nu, \tag{2.354}$$

we are able to deduce

$$\delta\sqrt{-g} = -\sqrt{-g}V_\mu{}^i\delta V_i{}^\mu. \tag{2.355}$$

For the variation of the Ricci tensor $R_{i\nu} = V_i{}^\mu R_{\mu\nu}$ we have

$$\delta\overset{\mathrm{L}}{R}_{i\nu} = \delta V_i{}^\mu\overset{\mathrm{L}}{R}_{\mu\nu} + V_i{}^\mu\delta\overset{\mathrm{L}}{R}_{\mu\nu}. \tag{2.356}$$

In an inertial frame, the Ricci tensor reduces to

$$\overset{\mathrm{L}}{R}_{\mu\nu} = \partial_\nu\overset{\mathrm{L}}{\Gamma}{}^\beta_{\beta\mu} - \partial_\beta\overset{\mathrm{L}}{\Gamma}{}^\beta_{\nu\mu}, \tag{2.357}$$

so that

$$\delta\overset{\mathrm{L}}{R}_{i\nu} = \delta V_i{}^\mu\overset{\mathrm{L}}{R}_{\mu\nu} + V_i{}^\mu\left(\partial_\nu\delta\overset{\mathrm{L}}{\Gamma}{}^\beta_{\beta\mu} - \partial_\beta\delta\overset{\mathrm{L}}{\Gamma}{}^\beta_{\nu\mu}\right). \tag{2.358}$$

The second term can be converted into a surface term, so it may be ignored. Collecting our results, we have

$$\begin{cases} \delta g^{\mu\nu} = -g^{\mu\rho}g^{\nu\sigma}\delta g_{\rho\sigma}, \\ \delta\sqrt{-g} = -\frac{1}{2}\sqrt{-g}g_{\mu\nu}\delta g^{\mu\nu} = -\sqrt{-g}V_\mu{}^i\delta V_i{}^\mu, \\ \delta R_{\mu\nu} = g_{\rho\mu}\left(\nabla_\lambda\delta\Gamma^{\lambda\rho}{}_\nu - \nabla_\nu\delta\Gamma^{\lambda\rho}{}_\lambda\right) + T_{\lambda\mu}{}^\rho\delta\Gamma^\lambda{}_{\rho\nu},\ \delta\overset{\mathrm{L}}{R}_{i\nu} = \delta V_i{}^\mu\overset{\mathrm{L}}{R}_{\mu\nu} \\ \delta R = \overset{\mathrm{L}}{R}{}^{\mu\nu}\delta g_{\mu\nu} + g^{\mu\nu}\left(\nabla_\lambda\delta\overset{\mathrm{L}}{\Gamma}{}^\lambda{}_{\mu\nu} - \nabla_\nu\delta\overset{\mathrm{L}}{\Gamma}{}^\lambda{}_{\mu\lambda}\right) - T_a{}^{bc}\delta K_{bc}{}^a. \end{cases} \tag{2.359}$$

From the above results, we obtain

$$\delta I_G = \frac{1}{16\pi}\int\left[\begin{array}{c} \left(R_i{}^\mu - \frac{1}{2}V_i{}^\mu R - V_i{}^\mu\Lambda\right)\delta V^i_\mu + 2g^{\rho\lambda}T_{\mu\lambda}{}^\sigma\delta\Gamma^\mu{}_{\rho\sigma} \\ +g^{\mu\nu}\left(\nabla_\lambda\delta\overset{\mathrm{L}}{\Gamma}{}^\lambda{}_{\mu\nu} - \nabla_\nu\delta\overset{\mathrm{L}}{\Gamma}{}^\lambda{}_{\mu\lambda}\right) \end{array}\right]\sqrt{-g}d^4x. \tag{2.360}$$

The last term in the action can be converted into a surface term, so it may be ignored. Using the four-current v^μ introduced earlier, the action for the matter fields read [63]

$$\delta I_{\text{field}} = \int\left[\overline{\psi}\delta\gamma^\mu\nabla_\mu\psi + \overline{\psi}\gamma^\mu\delta\widehat{\Gamma}_\mu\psi\right]\sqrt{-g}d^4x$$

$$= \int\left\{\begin{array}{c} \left[\frac{1}{2}g^{\mu\nu}\overline{\psi}\gamma_i(\nabla_\nu\psi) + T^\mu{}_{\rho\sigma}T_i{}^{\rho\sigma} - \delta_i^\mu T_{\lambda\rho\sigma}T^{\lambda\rho\sigma}\right]\delta V^i_\mu \\ +\frac{1}{8}\left(g^{\rho\nu}v^\mu - g^{\rho\mu}v^\nu\right)\left(g_{\mu\sigma}\delta\overset{\mathrm{L}}{\Gamma}{}^\sigma{}_{\nu\rho} - g_{\nu\sigma}\delta\overset{\mathrm{L}}{\Gamma}{}^\sigma{}_{\mu\rho}\right) \end{array}\right\}\sqrt{-g}d^4x. \tag{2.361}$$

Removing the derivatives of variations of the metric appearing in $\delta\Gamma^\sigma{}_{\nu\rho}$ via partial integration, and equating to zero the coefficients of $\delta g^{\mu\nu}$ and $\delta T^\sigma_{\nu\rho}$ in the variation of the action integral, we obtain

$$0 = \frac{1}{16\pi}\left(R_{\mu\nu} - \frac{1}{2}g_{\mu\nu}R - g_{\mu\nu}\Lambda\right) + \left(\frac{1}{2}\overline{\psi}\gamma_\nu\nabla_\mu\psi - \frac{1}{4}\nabla_\mu v_\nu\right)$$
$$+\nabla_\sigma T_{\mu\nu}{}^{\sigma} + T_{\mu\rho\sigma}T_\nu{}^{\rho\sigma} - g_{\mu\nu}T_{\lambda\rho\sigma}T^{\lambda\rho\sigma} \qquad (2.362)$$

and

$$T_{\rho\sigma\lambda} = 8\pi\tau_{\rho\sigma\lambda}. \qquad (2.363)$$

Eqs. (2.362) have the form of Einstein equations

$$G_{\mu\nu} - g_{\mu\nu}\Lambda = 8\pi\Sigma_{\mu\nu}, \qquad (2.364)$$

where the Einstein tensor and non-symmetric energy-momentum tensors are

$$G_{\mu\nu} = R_{\mu\nu} - \frac{1}{2}g_{\mu\nu}R, \qquad (2.365)$$

$$\Sigma_{\mu\nu} = \Theta_{\mu\nu} + \Upsilon_{\mu\nu}, \qquad (2.366)$$

respectively. Here we identify $\Theta_{\mu\nu}$ as the canonical energy-momentum

$$\Theta^{\mu}{}_{\nu} = \frac{\partial\mathcal{L}_{\text{field}}}{\partial(\nabla_\mu\chi)}\nabla_\nu\chi - \delta^{\mu}_{\nu}\mathcal{L}_{\text{field}}, \qquad (2.367)$$

while $\Upsilon_{\mu\nu}$ is the stress-tensor form of the non-Riemannian manifold. For the case of spinor fields being considered here the explicit form of the energy-momentum components [363] are (after symmetrization of corresponding canonical source terms in the Einstein equation),

$$\Theta_{\mu\nu} = -\left[\overline{\psi}\gamma_\mu\nabla_\nu\psi - (\nabla_\nu\overline{\psi})\gamma_\mu\psi + \overline{\psi}\gamma_\nu\nabla_\mu\psi - (\nabla_\mu\overline{\psi})\gamma_\nu\psi\right] \qquad (2.368)$$

and by using the second field equation (2.363), we determine

$$\Upsilon_{\mu\nu} = \nabla_\sigma T_{\mu\nu}{}^{\sigma} + T_{\mu\rho\sigma}\tau_\nu{}^{\rho\sigma} - g_{\mu\nu}T_{\lambda\rho\sigma}\tau^{\lambda\rho\sigma}, \qquad (2.369)$$

where $\tau_{\mu\nu}{}^{\sigma}$ is the so-called spin - energy potential [210, 211]

$$\tau_{\mu\nu}{}^{\sigma} := \frac{\partial\mathcal{L}_{\text{field}}}{\partial(\nabla_\sigma\chi)}\gamma_{\mu\nu}\chi. \qquad (2.370)$$

Explicitly, the spin energy potential reads $\tau^{\mu\nu\sigma} = \overline{\psi}\gamma^{[\mu}\gamma^{\nu}\gamma^{\sigma]}\psi$. The equation of motion obtained from the variation of the action with respect to $\overline{\psi}$ reads [210, 211]

$$\gamma^{\mu}\nabla_\mu\psi + \frac{3}{8}T_{\mu\nu\sigma}\gamma^{[\mu}\gamma^{\nu}\gamma^{\sigma]}\psi = 0. \qquad (2.371)$$

It is interesting to observe that this generalized curved space-time Dirac equation can be recast into the nonlinear equation of the Heisenberg-Pauli type

$$\gamma^{\mu}\nabla_{\mu}\psi+\frac{3}{8}\varepsilon\left(\overline{\psi}\gamma^{\mu}\gamma_{5}\psi\right)\gamma_{\mu}\gamma_{5}\psi=0. \tag{2.372}$$

Although the gravitational field equation is similar in form to the Einstein field equation, it differs from the original Einstein equations because the curvature tensor, containing space-time torsion, is non-Riemannian. Assuming that the Euler-Lagrange equations for the matter fields are satisfied, we obtain the following conservation laws for the angular - momentum and energy - momentum

$$\begin{aligned} V^{\mu}_{i}V^{\nu}_{j}\Sigma_{[\mu\nu]} &= \nabla_{\nu}\tau_{ij}^{\ \ \nu}, \\ V_{\mu}^{\ k}\nabla_{\nu}\Sigma^{\nu}{}_{\kappa} &= \Sigma^{\nu}{}_{\kappa}T^{k}{}_{\mu\nu}+\tau^{\nu}{}_{ij}R^{ij}{}_{\mu\nu}. \end{aligned} \tag{2.373}$$

To conclude, we have shown that a theory of gravitation can be obtained from a gauge theory of local Poincaré symmetry. Gauge fields were obtained by requiring invariance of the Lagrangian density under local Poincaré transformations. The resulting Einstein-Cartan theory describes a space-time endowed with non-vanishing curvature and torsion. This approach could be a prototype to obtain any theory of gravity.

Chapter 3

Space-Time Deformations

3.1. Deformations and Conformal Transformations

To complete the considerations in Chapter 2, it is worth noticing that also space-time deformations con be related to the generation of gravitational field. In this Chapter, we develop the space-time deformation formalism considering the main quantities which can be related to gravity. The issue to consider a general way to deform the space-time metrics is not new. It has been posed in different ways and is related to several physical problems ranging from the spontaneous symmetry breaking of unification theories up to gravitational waves, considered as space-time perturbations. In cosmology, for example, one faces the problem to describe an observationally lumpy universe at small scales which becomes isotropic and homogeneous at very large scales according to the Cosmological Principle. In this context, it is crucial to find a way to connect background and locally perturbed metrics [173]. For example, McVittie [294] considered a metric which behaves as a Schwarzschild one at short ranges and as a Friedman-Lemaitre-Robertson-Walker metric at very large scales. Gautreau [198] calculated the metric generated by a Schwarzschild mass embedded in a Friedman cosmological fluid trying to address the same problem. On the other hand, the post-newtonian parameterization, as a standard, can be considered as a deformation of a background, asymptotically flat Minkowski metric. In general, the deformation problem has been explicitly posed by Coll and collaborators [142, 143, 279] who conjectured the possibility to obtain any metric from the deformation of a space-time with constant curvature. The problem was solved only for 3-dimensional spaces but a straightforward extension should be to achieve the same result for space-times of any dimension.

In principle, new exact solutions of the Einstein field equations can be obtained by studying perturbations. In particular, dealing with perturbations as Lorentz matrices of scalar fields $\Phi^A{}_C$ reveals particularly useful. Firstly they transform as scalars with respect the coordinate transformations. Secondly, they are dimensionless and, in each point, the matrix $\Phi^A{}_C$ behaves as the element of a group. As we shall see below, such an approach can be related to the conformal transformations giving an "extended" interpretation and a straightforward physical meaning of them (see [9, 387] and references therein for a comprehensive review). Furthermore scalar fields related to space-time deformations have a straightforward physical interpretation which could contribute to explain several fundamental issues as the Higgs mechanism in unification theories, the inflation in cosmology and other pictures where scalar fields play a fundamental role in dynamics. In this Chapter, we are going to discuss the properties of the deforming matrices $\Phi^A{}_C$ and we will derive, from the Einstein equations, the field equations for them, showing how them can parameterize the deformed metrics, according to the boundary and initial conditions and to the energy-momentum tensor. The layout is the following, we define the space-time perturbations in the framework of the metric formalism giving the notion of first and second deformation matrices. In particular, we discuss how deformation matrices can be split in their trace, traceless and skew parts. We derive the contributions of deformation to the geodesic equation and, starting from the curvature Riemann tensor, the general equation of deformations. We discuss the notion of linear perturbations under the standard of deformations. In particular, we recast the equation of gravitational waves and the transverse traceless gauge under the standard of deformations. After we discuss the action of deformations on the Killing vectors. The result consists in achieving a notion of approximate symmetry.

3.2. Generalities in Space-Time Deformations

In order to start our considerations, let us take into account a metric $\boldsymbol{g}$ on a space-time manifold $\mathcal{M}$. Such a metric is assumed to be an exact solution of the Einstein field equations. We can decompose it by a co-tetrad field $\omega^A(x)$

$$\boldsymbol{g} = \eta_{AB}\omega^A\omega^B. \tag{3.1}$$

Let us define now a new tetrad field $\widetilde{\omega} = \Phi^A{}_C(x)\,\omega^C$, with $\Phi^A{}_C(x)$ a matrix of scalar fields. Finally we introduce a space-time $\widetilde{\mathcal{M}}$ with the metric $\widetilde{g}$ defined in the following way

$$\widetilde{\boldsymbol{g}} = \eta_{AB}\Phi^A{}_C\Phi^B{}_D\,\omega^C\omega^D = \gamma_{CD}(x)\omega^C\omega^D, \tag{3.2}$$

where also $\gamma_{CD}(x)$ is a matrix of fields which are scalars with respect to the coordinate transformations.

If $\Phi^A{}_C(x)$ is a Lorentz matrix in any point of $\mathcal{M}$, then

$$\widetilde{g} \equiv g \tag{3.3}$$

otherwise we say that $\widetilde{g}$ is a deformation of g and $\widetilde{\mathcal{M}}$ is a deformed $\mathcal{M}$. If all the functions of $\Phi^A{}_C(x)$ are continuous, then there is a *one - to - one* correspondence between the points of $\mathcal{M}$ and the points of $\widetilde{\mathcal{M}}$.

In particular, if ξ is a Killing vector for g on $\mathcal{M}$, the corresponding vector $\widetilde{\xi}$ on $\widetilde{\mathcal{M}}$ could not necessarily be a Killing vector.

A particular subset of these deformation matrices is given by

$$\Phi^A_C(x) = \Omega(x)\,\delta^A{}_C. \tag{3.4}$$

which define conformal transformations of the metric,

$$\widetilde{g} = \Omega^2(x)g. \tag{3.5}$$

In this sense, the deformations defined by Eq. (3.2) can be regarded as a generalization of the conformal transformations which will be discussed in details in the next Chapter.

We call the matrices $\Phi^A{}_C(x)$ *first deformation matrices*, while we can refer to

$$\gamma_{CD}(x) = \eta_{AB}\Phi^A{}_C(x)\Phi^B{}_D(x). \tag{3.6}$$

as the *second deformation matrices*, which, as seen above, are also matrices of scalar fields. They generalize the Minkowski matrix η_{AB} with constant elements in the definition of the metric. A further restriction on the matrices $\Phi^A{}_C$ comes from the theorem proved by Riemann by which an n-dimensional metric has $n(n-1)/2$ degrees of freedom (see [143] for details). With this definitions in mind, let us consider the main properties of deforming matrices.

3.3. Properties of Deforming Matrices

Let us take into account a four dimensional space-time with Lorentzian signature. A family of matrices $\Phi^A{}_C(x)$ such that

$$\Phi^A{}_C(x) \in GL(4)\,\forall x, \tag{3.7}$$

are defined on such a space-time.

These functions are not necessarily continuous and can connect space-times with different topologies. A singular scalar field introduces a deformed manifold $\widetilde{\mathcal{M}}$ with a space-time singularity.

As it is well known, the Lorentz matrices $\Lambda^A{}_C$ leave the Minkowski metric invariant and then

$$\boldsymbol{g} = \eta_{EF}\Lambda^E{}_A\Lambda^F{}_B\Phi^A{}_C\Phi^B{}_D\,\omega^C\omega^D = \eta_{AB}\Phi^A{}_C\Phi^B{}_D\,\omega^C\omega^D. \tag{3.8}$$

It follows that $\Phi^A{}_C$ give rise to right cosets of the Lorentz group, i.e. they are the elements of the quotient group $GL(4,\boldsymbol{R})/SO(3,1)$. On the other hand, a right-multiplication of $\Phi^A{}_C$ by a Lorentz matrix induces a different deformation matrix.

The inverse deformed metric is

$$\tilde{g}^{ab} = \eta^{CD}\Phi^{-1A}{}_C\Phi^{-1B}{}_D e^a_A e^b_B \tag{3.9}$$

where $\Phi^{-1A}{}_C\Phi^C{}_B = \Phi^A{}_C\Phi^{-1C}{}_B = \delta^A_B$.

Let us decompose now the matrix $\Phi_{AB} = \eta_{AC}\Phi^C{}_B$ in its symmetric and antisymmetric parts

$$\Phi_{AB} = \Phi_{(AB)} + \Phi_{[AB]} = \Omega\eta_{AB} + \Theta_{AB} + \varphi_{AB} \tag{3.10}$$

where $\Omega = \Phi^A{}_A$, Θ_{AB} is the traceless symmetric part and φ_{AB} is the skew symmetric part of the first deformation matrix respectively. Then standard conformal transformations are nothing else but deformations with $\Theta_{AB} = \varphi_{AB} = 0$ [422].

Finding the inverse matrix $\Phi^{-1A}{}_C$ in terms of Ω, Θ_{AB} and φ_{AB} is not immediate, but as above, it can be split in the three terms

$$\Phi^{-1A}{}_C = \alpha\delta^A{}_C + \Psi^A{}_C + \Sigma^A{}_C \tag{3.11}$$

where α, $\Psi^A{}_C$ and $\Sigma^A{}_C$ are respectively the trace, the traceless symmetric part and the antisymmetric part of the inverse deformation matrix. The second deformation matrix, from the above decomposition, takes the form

$$\gamma_{AB} = \eta_{CD}(\Omega\,\delta^C_A + \Theta^C{}_A + \varphi^C{}_A)(\Omega\,\delta^D_B + \Theta^D{}_B + \varphi^D{}_B) \tag{3.12}$$

and then

$$\gamma_{AB} = \Omega^2\,\eta_{AB} + 2\Omega\,\Theta_{AB} + \eta_{CD}\,\Theta^C{}_A\,\Theta^D{}_B + \eta_{CD}\,(\Theta^C{}_A\,\varphi^D{}_B + \varphi^C{}_A\,\Theta^D{}_B) + \eta_{CD}\,\varphi^C{}_A\,\varphi^D{}_B. \tag{3.13}$$

In general, the deformed metric can be split as

$$\tilde{g}_{ab} = \Omega^2 g_{ab} + \gamma_{ab} \tag{3.14}$$

where

$$\gamma_{ab} = \left(2\Omega\,\Theta_{AB} + \eta_{CD}\,\Theta^C{}_A\,\Theta^D{}_B + \eta_{CD}\,(\Theta^C{}_A\,\varphi^D{}_B + \varphi^C{}_A\,\Theta^D{}_B) + \eta_{CD}\,\varphi^C{}_A\,\varphi^D{}_B\right)\omega^A_a\,\omega^B_b \tag{3.15}$$

In particular, if $\Theta_{AB} = 0$, the deformed metric simplifies to

$$\tilde{g}_{ab} = \Omega^2 g_{ab} + \eta_{CD}\,\varphi^C{}_A\,\varphi^D{}_B\omega^A{}_a\omega^B{}_b \tag{3.16}$$

and, if $\Omega = 1$, the deformation of a metric consists in adding to the background metric a tensor γ_{ab}. We have to remember that all these quantities are not independent as, by the theorem mentioned in [143], they have to form at most six independent functions in a four dimensional space-time.

Similarly the controvariant deformed metric can be always decomposed in the following way

$$\tilde{g}^{ab} = \alpha^2 g^{ab} + \lambda^{ab} \tag{3.17}$$

Let us find the relation between γ_{ab} and λ^{ab}. By using $\widetilde{g_{ab}g^{bc}} = \delta^c_a$, we obtain

$$\alpha^2\Omega^2\delta^c_a + \alpha^2\gamma^c_a + \Omega^2\lambda^c_a + \gamma_{ab}\lambda^{bc} = \delta^c_a \tag{3.18}$$

if the deformations are conformal transformations, we have $\alpha = \Omega^{-1}$, so assuming such a condition, one obtain the following matrix equation

$$\alpha^2\gamma^c_a + \Omega^2\lambda^c_a + \gamma_{ab}\lambda^{bc} = 0\,, \tag{3.19}$$

and

$$(\delta^b_a + \Omega^{-2}\gamma^b_a)\lambda^c_b = -\Omega^{-4}\gamma^c_a \tag{3.20}$$

and finally

$$\lambda^c_b = -\Omega^{-4}(\delta + \Omega^{-2}\gamma)^{-1}{}^a_b\gamma^c_a \tag{3.21}$$

where $(\delta + \Omega^{-2}\gamma)^{-1}$ is the inverse tensor of $(\delta^b_a + \Omega^{-2}\gamma^b_a)$.

To each matrix $\Phi^A{}_B$, we can associate a (1,1)-tensor $\phi^a{}_b$ defined by

$$\phi^a{}_b = \Phi^A{}_B \omega^B_b e^a_A \tag{3.22}$$

such that

$$\tilde{g}_{ab} = g_{cd}\phi^c{}_a\phi^d{}_b \tag{3.23}$$

which can be decomposed as in Eq. (3.16). Vice-versa from a (1,1)-tensor $\phi^a{}_b$, we can define a matrix of scalar fields as

$$\phi^A{}_B = \phi^a{}_b \omega^A_a e^b_B. \tag{3.24}$$

The Levi Civita connection corresponding to the metric (3.14) is related to the original connection by the relation

$$\tilde{\Gamma}^c_{ab} = \Gamma^c_{ab} + C^c_{ab} \tag{3.25}$$

(see [422]), where

$$C^c_{ab} = 2\tilde{g}^{cd} g_{d(a}\nabla_{b)}\Omega - g_{ab}\tilde{g}^{cd}\nabla_d\Omega + \frac{1}{2}\tilde{g}^{cd}\left(\nabla_a\gamma_{db} + \nabla_b\gamma_{ad} - \nabla_d\gamma_{ab}\right). \tag{3.26}$$

Therefore, in a deformed space-time, the connection deformation acts like a force that deviates the test particles from the geodesic motion in the unperturbed space-time. As a matter of fact the geodesic equation for the deformed space-time

$$\frac{d^2x^c}{d\lambda^2} + \tilde{\Gamma}^c{}_{ab}\frac{dx^a}{d\lambda}\frac{dx^b}{d\lambda} = 0 \tag{3.27}$$

becomes

$$\frac{d^2x^c}{d\lambda^2} + \Gamma^c{}_{ab}\frac{dx^a}{d\lambda}\frac{dx^b}{d\lambda} = -C^c{}_{ab}\frac{dx^a}{d\lambda}\frac{dx^b}{d\lambda}. \tag{3.28}$$

The deformed Riemann curvature tensor is then

$$\tilde{R}_{abc}{}^d = R_{abc}{}^d + \nabla_b C^d{}_{ac} - \nabla_a C^d{}_{bc} + C^e{}_{ac}C^d{}_{be} - C^e{}_{bc}C^d{}_{ae}, \tag{3.29}$$

while the deformed Ricci tensor obtained by contraction is

$$\tilde{R}_{ab} = R_{ab} + \nabla_d C^d{}_{ab} - \nabla_a C^d{}_{db} + C^e{}_{ab}C^d{}_{de} - C^e{}_{db}C^d{}_{ae} \tag{3.30}$$

and the curvature scalar

$$\tilde{R} = \tilde{g}^{ab}\tilde{R}_{ab} = \tilde{g}^{ab}R_{ab} + \tilde{g}^{ab}\left[\nabla_d C^d{}_{ab} - \nabla_a C^d{}_{db} + C^e{}_{ab}C^d{}_{de} - C^e{}_{db}C^d{}_{ae}\right] \tag{3.31}$$

From the above curvature quantities, we obtain finally the equations for the deformations. In the vacuum case, we simply have

$$\tilde{R}_{ab} = R_{ab} + \nabla_d C^d{}_{ab} - \nabla_a C^d{}_{db} + C^e{}_{ab}C^d{}_{de} - C^e{}_{db}C^d{}_{ae} = 0 \tag{3.32}$$

where R_{ab} must be regarded as a known function. In presence of matter, we consider the equation

$$R_{ab} + \nabla_d C^d{}_{ab} - \nabla_a C^d{}_{db} + C^e{}_{ab} C^d{}_{de} - C^e{}_{db} C^d{}_{ae} = \widetilde{T}_{ab} - \frac{1}{2}\widetilde{g}_{ab}\widetilde{T} \tag{3.33}$$

we are assuming, for the sake of simplicity $8\pi G = c = 1$. This last equation can be improved by considering the Einstein field equations

$$R_{ab} = T_{ab} - \frac{1}{2} g_{ab} T \tag{3.34}$$

and then

$$\nabla_d C^d{}_{ab} - \nabla_a C^d{}_{db} + C^e{}_{ab} C^d{}_{de} - C^e{}_{db} C^d{}_{ae} = \widetilde{T}_{ab} - \frac{1}{2}\widetilde{g}_{ab}\widetilde{T} - \left(T_{ab} - \frac{1}{2} g_{ab} T\right) \tag{3.35}$$

is the most general equation for deformations.

3.4. Metric Deformations as Perturbations

Metric deformations can be used to describe perturbations. To this aim we can simply consider the deformations

$$\Phi^A{}_B = \delta^A{}_B + \varphi^A{}_B \tag{3.36}$$

with

$$|\varphi^A{}_B| \ll 1, \tag{3.37}$$

together with their derivatives

$$|\partial \varphi^A{}_B| \ll 1\,. \tag{3.38}$$

With this approximation, immediately we find the inverse relation

$$(\Phi^{-1})^A{}_B \simeq \delta^A{}_B - \varphi^A{}_B. \tag{3.39}$$

As a remarkable example, we have that gravitational waves are generally described, in linear approximation, as perturbations of the Minkowski metric

$$g_{ab} = \eta_{ab} + \gamma_{ab}. \tag{3.40}$$

In our case, we can extend in a covariant way such an approximation. If φ_{AB} is an antisymmetric matrix, we have

$$\widetilde{g}_{ab} = g_{ab} + \gamma_{ab} \tag{3.41}$$

where the first order terms in $\varphi^A{}_B$ vanish and γ_{ab} is of second order

$$\gamma_{ab} = \eta_{AB} \varphi^A{}_C \varphi^B{}_D \omega^C{}_a \omega^D{}_b. \tag{3.42}$$

Consequently

$$\widetilde{g}^{ab} = g^{ab} + \gamma^{ab} \tag{3.43}$$

where

$$\gamma^{ab} = \eta^{AB}(\varphi^{-1})^C{}_A(\varphi^{-1})^D{}_B e_C{}^a e_D{}^b . \tag{3.44}$$

Let us consider the background metric g_{ab}, solution of the Einstein equations in the vacuum

$$R_{ab} = 0. \tag{3.45}$$

We obtain the equation of perturbations considering only the linear terms in Eq. (3.32) and neglecting the contributions of quadratic terms. We find

$$\widetilde{R}_{ab} = \nabla_d C^d{}_{ab} - \nabla_a C^d{}_{db} = 0\,, \tag{3.46}$$

and, by the explicit form of $C^d{}_{ab}$, this equation becomes

$$\left(\nabla_d \nabla_a \gamma^d_b + \nabla_d \nabla_b \gamma^d_a - \nabla_d \nabla^d \gamma_{ab}\right) - \left(\nabla_a \nabla_d \gamma^d_b + \nabla_a \nabla_b \gamma^d_d - \nabla_a \nabla^d \gamma^d_b\right) = 0\,. \tag{3.47}$$

Imposing the transverse traceless gauge on γ_{ab} , i.e. the standard gauge conditions

$$\nabla^a \gamma_{ab} = 0 \tag{3.48}$$

and

$$\gamma = \gamma^a_a = 0 \tag{3.49}$$

Eq. (3.47) reduces to

$$\nabla_b \nabla^b \gamma_{ac} - 2R^{b\ d}_{ac}\gamma_{bd} = 0\,, \tag{3.50}$$

see also [422]. In our context, this equation is a linearized equation for deformations and it is straightforward to consider perturbations and, in particular, gravitational waves, as small deformations of the metric. This result can be immediately translated into the above scalar field matrix equations. Note that such an equation can be applied to the conformal part of the deformation, when the general decomposition is considered.

As an example, let us take into account the deformation matrix equations applied to the Minkowski metric, when the deformation matrix assumes the form (3.36). In this case, the equations (3.47), become ordinary wave equations for γ_{ab}. Considering the deformation matrices, these equations become, for a tetrad field of constant vectors,

$$\partial^d \partial_d \varphi^C{}_A \varphi_{CB} + 2\partial_d \varphi^C{}_A \partial^d \varphi_{CB} + \varphi^C{}_A \partial^d \partial_d \varphi_{CB} = 0\,. \tag{3.51}$$

The above gauge conditions are now

$$\varphi_{AB}\varphi^{BA} = 0 \tag{3.52}$$

and

$$e^d_D \left[\partial_d \varphi_{CA} \varphi^C_B + \varphi_{CA} \partial_d \varphi^C_B\right] = 0\,. \tag{3.53}$$

This result shows that the gravitational waves can be fully recovered starting from the scalar fields which describe the deformations of the metric. In other words, such scalar fields can assume the meaning of gravitational waves modes.

3.5. Approximate Killing Vectors

Another important issue which can be addressed starting from space-time deformations is related to the symmetries. In particular, they assume a fundamental role in describing when a symmetry is preserved or broken under the action of a given field. In General Relativity, the Killing vectors are always related to the presence of given space-time symmetries [422].

Let us take an exact solution of the Einstein equations, which satisfies the Killing equation

$$(L_\xi g)_{ab} = 0 \tag{3.54}$$

where ξ, being the generator of an infinitesimal coordinate transformation, is a Killing vector. If we take a deformation of the metric with the scalar matrix

$$\Phi^A{}_B = \delta^A{}_B + \varphi^A{}_B \tag{3.55}$$

with

$$|\varphi^A{}_B| \ll 1\,, \tag{3.56}$$

and

$$(L_\xi \tilde{g})_{ab} \neq 0\,, \tag{3.57}$$

being

$$(L_\xi e^A)_a = 0\,, \tag{3.58}$$

we have

$$(L_\xi \varphi)^A{}_B = \xi^a \partial_a \varphi^A{}_B \neq 0\,. \tag{3.59}$$

If there is some region $\mathcal{D}$ of the deformed space-time $\mathcal{M}_{deformed}$ where

$$|(L_\xi \varphi)^A{}_B| \ll 1 \tag{3.60}$$

we say that ξ is an *approximate Killing vector* on $\mathcal{D}$. In other words, these approximate Killing vectors allow to "control" the space-time symmetries under the action of a given deformation. We can calculate the modified connection $\Gamma^{\tilde{c}}_{ab}$ in many alternative ways. Let us introduce the tetrad e_A and cotetrad ω^B satisfying the orthogonality relation

$$i_{e_A}\omega^B = \delta^B_A \tag{3.61}$$

and the non-integrability condition (anholonomy)

$$d\omega^A = \frac{1}{2}\Omega^A_{BC}\omega^B \wedge \omega^C. \tag{3.62}$$

The corresponding connection is

$$\Gamma^A_{BC} = \frac{1}{2}\left(\Omega^A_{BC} - \eta^{AA'}\eta_{BB'}\Omega^{B'}_{A'C} - \eta^{AA'}\eta_{CC'}\Omega^{C'}_{A'B}\right) \tag{3.63}$$

If we deform the metric as in (3.2), we have two alternative ways to write this expression: either writing the "deformation" of the metric in the space of tetrads or "deforming" the tetrad field as in the following expression

$$\hat{g} = \eta_{AB}\Phi^{A}{}_{C}\Phi^{B}{}_{D}\,\omega^{C}\omega^{D} = \gamma_{AB}\,\omega^{A}\omega^{B} = \eta_{AB}\,\hat{\omega}^{A}\hat{\omega}^{B}. \tag{3.64}$$

In the first case, the contribution of the Christoffel symbols, constructed by the metric γ_{AB}, appears

$$\hat{\Gamma}^{A}_{BC} = \frac{1}{2}\left(\Omega^{A}_{BC} - \gamma^{AA'}\gamma_{BB'}\Omega^{B'}_{A'C} - \gamma^{AA'}\gamma_{CC'}\Omega^{C'}_{A'B}\right)$$
$$+\frac{1}{2}\gamma^{AA'}\left(i_{e_C}d\gamma_{BA'} - i_{e_B}d\gamma_{CA'} - i_{e_{A'}}d\gamma_{BC}\right) \tag{3.65}$$

In the second case, using (3.62), we can define the new anholonomy objects $\hat{C}^{A}_{BC}$.

$$d\hat{\omega}^{A} = \frac{1}{2}\hat{\Omega}^{A}_{BC}\hat{\omega}^{B}\wedge\hat{\omega}^{C}. \tag{3.66}$$

After some calculations, we have

$$\hat{\Omega}^{A}_{BC} = \Phi^{A}{}_{E}\Phi^{-1}{}^{D}{}_{B}\Phi^{-1}{}^{F}{}_{C}\Omega^{E}_{DF} + 2\Phi^{A}{}_{F}e^{a}_{G}\left(\Phi^{-1}{}^{G}{}_{[B}\partial_{a}\Phi^{-1}{}^{F}{}_{C]}\right) \tag{3.67}$$

As we are assuming a constant metric in tetradic space, the deformed connection is

$$\hat{\Gamma}^{A}_{BC} = \frac{1}{2}\left(\hat{\Omega}^{A}_{BC} - \eta^{AA'}\eta_{BB'}\hat{\Omega}^{B'}_{A'C} - \eta^{AA'}\eta_{CC'}\hat{\Omega}^{C'}_{A'B}\right). \tag{3.68}$$

Substituting (3.67) in (3.68), the final expression of $\hat{\Gamma}^{A}_{BC}$, as a function of Ω^{A}_{BC}, $\Phi^{A}{}_{B}$, $\Phi^{-1}{}^{D}{}_{C}$ and e^{a}_{G} is

$$\hat{\Gamma}_{ABC} = \Delta^{DEF}_{ABC}\left[\frac{1}{2}\eta_{DG}\Phi^{G}{}_{G'}\Phi^{-1}{}^{E'}{}_{E}\Phi^{-1}{}^{F'}{}_{F}\Omega^{G'}_{E'F'}\right.$$
$$\left.+\eta_{DK}\Phi^{K}{}_{H}e^{a}_{G}\Phi^{-1}{}^{G}{}_{[E}\partial_{|a|}\Phi^{-1}{}^{H}{}_{F]}\right] \tag{3.69}$$

where

$$\Delta^{DEF}_{ABC} = \delta^{D}_{A}\delta^{E}_{C}\delta^{F}_{B} - \delta^{D}_{B}\delta^{E}_{C}\delta^{F}_{A} + \delta^{D}_{C}\delta^{E}_{A}\delta^{F}_{B}. \tag{3.70}$$

Now we have developed all the tools we need to take into account Extended Theories of Gravity. First of all, results in Chapter 2 and 3 work for any theory of gravity stemming out from gauge invariance. being deformations related to conformal transformations, any theory of gravity, conformally related to the Einstein one, can be easily generated and framed in the context of theories where diffeomorphism invariance and covariant structure hold, with these results in mind, we can pass to analyze extensively the generalization of the Einstein approach.

Chapter 4

Extended Theories of Gravity

As widely discussed above, due to the problems of Standard Cosmological Model, and, first of all, to the lack of a definitive Quantum Gravity Theory, alternative theories have been considered in order to attempt, at least, a semi-classical scheme where General Relativity and its positive results could be recovered. One of the most fruitful approaches has been that of *Extended Theories of Gravity* which have become a sort of paradigm in the study of gravitational interaction. They are based on corrections and enlargements of the Einstein theory. The paradigm consists, essentially, in adding higher-order curvature invariants and minimally or non-minimally coupled scalar fields into dynamics which come out from the effective action of Quantum Gravity [88, 308]. This approach is coherent to the fact that these generalized theories emerge, like Einstein's gravity, from the Poincaré gauge invariance and can be framed in the bundle structure discussed above.

Other motivations to modify General Relativity, as discussed, come from the issue of a full recovering of the Mach principle which leads to assume a varying gravitational coupling. The principle states that the local inertial frame is determined by some average of the motion of distant astronomical objects [59]. This fact implies that the gravitational coupling can be scale-dependent and related to some scalar field. As a consequence, the concept of "inertia" and the Equivalence Principle have to be revised as discussed above.

Besides, as mentioned, every unification scheme as Superstrings, Supergravity or Grand Unified Theories, takes into account effective actions where non-minimal couplings to the geometry or higher-order terms in the curvature invariants are present. Such contributions are due to one-loop or higher-loop corrections in the high-curvature regimes near the full (not yet available) Quantum Gravity Regime [308]. Specifically, this scheme was adopted in order to deal with the quantization on curved space-times as discussed in Chapter 1 and the result was that the interactions among quantum scalar fields and background geometry or the gravitational self-interactions yield corrective terms in the Hilbert-Einstein Lagrangian [56]. Moreover, it has been realized that such corrective terms are inescapable in order to obtain the effective action of Quantum Gravity at scales closed to the Planck one [420]. All these approaches are not the "*full quantum gravity*" but are needed as working schemes toward it.

In summary, higher-order terms in curvature invariants (such as R^2, $R^{\mu\nu}R_{\mu\nu}$, $R^{\mu\nu\alpha\beta}R_{\mu\nu\alpha\beta}$, $R\Box R$, or $R\Box^k R$) or non-minimally coupled terms between scalar fields and geometry (such as $\phi^2 R$) have to be added to the effective Lagrangian of gravitational field when quantum corrections are considered. For instance, one can notice that such terms occur in the effective Lagrangian of strings or in Kaluza-Klein theories, when the mechanism of dimensional reduction is used [197].

On the other hand, from a conceptual point of view, there are no *a priori* reason to restrict the gravitational Lagrangian to a linear function of the Ricci scalar R, minimally coupled with matter [290]. Furthermore, the idea that there are no "exact" laws of physics could be taken into serious account: in such a case, the effective Lagrangians of physical interactions are "stochastic" functions. This feature means that the local gauge invariances (*i.e.* conservation laws) are well approximated only in the low energy limit and the fundamental physical constants can vary [36].

Besides fundamental physics motivations, all these theories have acquired a huge interest in cosmology due to the fact that they "naturally" exhibit inflationary behaviors able to overcome the shortcomings of Cosmological Standard Model (based on GR). The related cosmological models seem realistic and capable of matching with the Cosmic Microwave Background Radiation observations [168, 258, 389]. Furthermore, it is possible to show that, via conformal transformations, the higher-order and non-minimally coupled terms always correspond to the Einstein gravity plus one or more than one minimally coupled scalar fields [11, 92, 201, 287, 403, 423].

More precisely, higher-order terms appear always as contributions of order two in the field equations. For example, a term like R^2 gives fourth order equations [350], $R\,\Box R$ gives sixth order equations [11, 18, 48, 68, 201], $R\Box^2 R$ gives eighth order equations [42] and so on. By a conformal transformation, any 2nd-order derivative term corresponds to a scalar field[1]: for example, fourth-order gravity gives Einstein plus one scalar field, sixth-order gravity gives Einstein plus two scalar fields and so on [201, 373].

Furthermore, it is possible to show that the $f(R)$-gravity is equivalent not only to a scalar-tensor one but also to the Einstein theory plus an ideal fluid [90]. This feature results very interesting if we want to obtain multiple inflationary events since an early stage could select "very" large-scale structures (clusters of galaxies today), while a late stage could select "small" large-scale structures (galaxies today) [18, 48, 68]. The philosophy is that each inflationary era is related to the dynamics of a scalar field. Finally, these extended schemes could naturally solve the problem of "graceful exit" bypassing the shortcomings of former inflationary models [17, 258].

In addition to the revision of Standard Cosmology at early epochs (leading to the inflation), a new approach is necessary also at late epochs. Extended Theories of Gravity could play a fundamental role also in this context. In fact, the increasing bulk of data that have been accumulated in the last few years have paved the way to the emergence of a new cosmological model usually referred to as the *Concordance Model* or the ΛCDM model outlined in Chapter 1.

Actually, there is still a different way to face the problem of cosmic acceleration. It is possible that the observed acceleration is not the manifestation of another ingredient in the cosmic pie, but rather the first signal of a breakdown of our understanding of the laws of gravitation (in the infra-red limit).

From this point of view, it is thus tempting to modify the Friedmann equations to see

[1]The dynamics of such scalar fields is usually given by the corresponding Klein-Gordon Equation, which is second order.

whether it is possible to fit the astrophysical data with models comprising only the standard matter. Interesting examples of this kind are the Cardassian expansion [194] and the DGP gravity [169]. Moving in this same framework, it is possible to find alternative schemes where a quintessential behavior is obtained by taking into account effective models coming from fundamental physics giving rise to generalized or higher-order gravity actions [8, 94, 126, 307, 308].

For instance, a cosmological constant term may be recovered as a consequence of a non - vanishing torsion field thus leading to a model which is consistent with both SNeIa Hubble diagram and Sunyaev - Zel'dovich data coming from clusters of galaxies [96]. SNeIa data could also be efficiently fitted including higher-order curvature invariants in the gravity Lagrangian [97, 264–266]. It is worth noticing that these alternative models provide naturally a cosmological component with negative pressure whose origin is related to the geometry of the Universe thus overcoming the problems linked to the physical significance of the scalar field.

It is evident, the high number of cosmological models which are viable candidates to explain the observed accelerated expansion. This abundance of models is, from one hand, the signal of the fact that we have a limited number of cosmological tests to discriminate among rival theories, and, from the other hand, that a urgent degeneracy problem has to be faced. To this aim, it is useful to remark that both the SNeIa Hubble diagram and the angular size - redshift relation of compact radio sources [135, 136] are distance based methods to probe cosmological models so then systematic errors and biases could be iterated. From this point of view, it is interesting to search for tests based on time-dependent observables.

For example, one can take into account the *lookback time* to distant objects since this quantity can discriminate among different cosmological models. The lookback time is observationally estimated as the difference between the present day age of the Universe and the age of a given object at redshift z. Such an estimate is possible if the object is a galaxy observed in more than one photometric band since its color is determined by its age as a consequence of stellar evolution. It is thus possible to get an estimate of the galaxy age by measuring its magnitude in different bands and then using stellar evolutionary codes to choose the model that reproduces the observed colors at best.

Coming to the weak-field-limit approximation, which essentially means considering Solar System scales, Extended Theories of Gravity are expected to reproduce General Relativity which, in any case, is firmly tested only in this limit [436]. This fact is matter of debate since several relativistic theories *do not* reproduce exactly the Einstein results in the Newtonian approximation but, in some sense, generalize them. As it was firstly noticed by Stelle [394], a R^2-theory gives rise to Yukawa-like corrections in the Newtonian potential. Such a feature could have interesting physical consequences. For example, some authors claim to explain the flat rotation curves of galaxies by using such terms [354]. Others [291] have shown that a conformal theory of gravity is nothing else but a fourth-order theory containing such terms in the Newtonian limit. Besides, indications of an apparent, anomalous, long-range acceleration revealed from the data analysis of Pioneer 10/11, Galileo, and Ulysses spacecrafts could be framed in a general theoretical scheme by taking corrections to the Newtonian potential into account [22, 50].

In general, any relativistic theory of gravitation yields corrections to the Newton potential (see for example [337]) which, in the post-Newtonian (PPN) formalism, could be a test for the same theory [436]. Furthermore the newborn *gravitational lensing astronomy* [360] is giving rise to additional tests of gravity over small, large, and very large scales which can provide direct measurements for the variation of the Newton coupling [255], the potential of galaxies, clusters of galaxies and several other features of self-gravitating systems.

Such data will be, very likely, capable of confirming or ruling out the physical consistency of General Relativity or of any Extended Theories of Gravity. In summary, the general features of Extended Theories of Gravity are that the Einstein field equations result to be modified in two senses: *i*) geometry can be non-minimally coupled to some scalar field, and/or *ii*) higher than second order derivative terms in the metric come out. In the former case, we generically deal with scalar-tensor theories of gravity; in the latter, we deal with higher-order theories. However combinations of non-minimally coupled and higher-order terms can emerge as contributions in effective Lagrangians. In this case, we deal with higher-order-scalar-tensor theories of gravity.

Considering a mathematical point of view, the problem of reducing more general theories to Einstein standard form has been extensively treated; one can see that, through a "Legendre" transformation on the metric, higher-order theories, under suitable regularity conditions on the Lagrangian, take the form of the Einstein one in which a scalar field (or more than one) is the source of the gravitational field (see for example [187,289,290,382]); on the other side, as discussed above, it has been studied the mathematical equivalence between models with variable gravitational coupling with the Einstein standard gravity through suitable conformal transformations (see [147,149,162,163,404]).

In any case, the debate on the physical meaning of conformal transformations is far to be solved [see [183,387] and references therein for a comprehensive review]. Several authors claim for a true physical difference between Jordan frame (higher-order theories and/or variable gravitational coupling) since there are experimental and observational evidences which point out that the Jordan frame could be suitable to better match solutions with data. Others state that the true physical frame is the Einstein one according to the energy theorems [289]. However, the discussion is open and no definitive statement has been formulated up to now.

The problem should be faced from a more general point of view and the Palatini approach to gravity could be useful to this goal. The Palatini approach in gravitational theories was firstly introduced and analyzed by Einstein himself [177]. It was, however, called the Palatini approach as a consequence of an historical misunderstanding [69,188].

The fundamental idea of the Palatini formalism is to consider the (usually torsion-less) connection Γ, entering the definition of the Ricci tensor, to be independent of the metric g defined on the space-time $\mathcal{M}$. The Palatini formulation for the standard Hilbert-Einstein theory results to be equivalent to the purely metric theory: this follows from the fact that the field equations for the connection Γ, firstly considered to be independent of the metric, give the Levi-Civita connection of the metric g. As a consequence, there is no reason to impose the Palatini variational principle in the standard Hilbert-Einstein theory instead of the metric variational principle.

However, the situation completely changes if we consider the Extended Theories of Gravity, depending on functions of curvature invariants, as $f(R)$, or non-minimally coupled to some scalar field. In these cases, the Palatini and the metric variational principle provide different field equations and the theories thus derived differ [189,289]. The relevance of

Palatini approach, in this framework, has been recently proven in relation to cosmological applications [79, 94, 126, 268, 269, 307, 308, 421].

It has also been studied the crucial problem of the Newtonian potential in alternative theories of Gravity and its relations with the conformal factor [295]. From a physical point of view, considering the metric g and the connection Γ as independent fields means to decouple the metric structure of space-time and its geodesic structure (being, in general, the connection Γ not the Levi-Civita connection of g). The chronological structure of space-time is governed by g while the trajectories of particles, moving in the space-time, are governed by Γ.

This decoupling enriches the geometric structure of space-time and generalizes the purely metric formalism. This metric-affine structure of space-time is naturally translated, by means of the same (Palatini) field equations, into a bi-metric structure of space-time. Beside the *physical* metric g, another metric h is involved. This new metric is related, in the case of $f(R)$-gravity, to the connection. As a matter of fact, the connection Γ results to be the Levi-Civita connection of h and thus provides the geodesic structure of space-time.

If we consider the case of non-minimally coupled interaction in the gravitational Lagrangian (scalar-tensor theories), the new metric h is related to the non-minimal coupling. The new metric h can be thus related to a different geometric and physical aspect of the gravitational theory. Thanks to the Palatini formalism, the non-minimal coupling and the scalar field, entering the evolution of the gravitational fields, are separated from the metric structure of space-time. The situation mixes when we consider the case of higher-order-scalar-tensor theories. Due to these features, the Palatini approach could greatly contribute to clarify the physical meaning of conformal transformation [9].

4.1. The Effective Action and the Field Equations

With the above considerations in mind, let us start with a general class of higher-order-scalar-tensor theories in four dimensions [2] given by the action

$$\mathcal{A} = \int d^4x\sqrt{-g}\left[F(R,\Box R,\Box^2 R,\ldots,\Box^k R,\phi) - \frac{\varepsilon}{2}g^{\mu\nu}\phi_{;\mu}\phi_{;\nu} + \mathcal{L}_m\right], \tag{4.1}$$

where F is an unspecified function of curvature invariants and of a scalar field ϕ. The term $\mathcal{L}_m$, as above, is the minimally coupled ordinary matter contribution. We shall use physical units $8\pi G = c = \hbar = 1$; ε is a constant which specifies the theory. Actually its values can be $\varepsilon = \pm 1, 0$ fixing the nature and the dynamics of the scalar field which can be a standard scalar field, a phantom field or a field without dynamics (see [181, 347] for details).

In the metric approach, the field equations are obtained by varying (4.1) with respect to

[2]For the aims of this Chapter, we do not need more complicated invariants like $R_{\mu\nu}R^{\mu\nu}$, $R_{\mu\nu\alpha\beta}R^{\mu\nu\alpha\beta}$, $C_{\mu\nu\alpha\beta}C^{\mu\nu\alpha\beta}$ which are also possible. these terms will be considered in Chapter 5 and 6.

$g_{\mu\nu}$. We get

$$
\begin{aligned}
G^{\mu\nu} = & \frac{1}{\mathcal{G}}\left[T^{\mu\nu} + \frac{1}{2}g^{\mu\nu}(F - \mathcal{G}R) + (g^{\mu\lambda}g^{\nu\sigma} - g^{\mu\nu}g^{\lambda\sigma})\mathcal{G}_{;\lambda\sigma}\right. \\
& + \frac{1}{2}\sum_{i=1}^{k}\sum_{j=1}^{i}(g^{\mu\nu}g^{\lambda\sigma} + g^{\mu\lambda}g^{\nu\sigma})(\Box^{j-i})_{;\sigma}\left(\Box^{i-j}\frac{\partial F}{\partial \Box^{i}R}\right)_{;\lambda} \\
& \left. - g^{\mu\nu}g^{\lambda\sigma}\left((\Box^{j-1}R)_{;\sigma}\Box^{i-j}\frac{\partial F}{\partial \Box^{i}R}\right)_{;\lambda}\right],
\end{aligned}
\tag{4.2}
$$

where $G^{\mu\nu}$ is the above Einstein tensor and

$$
\mathcal{G} \equiv \sum_{j=0}^{n}\Box^{j}\left(\frac{\partial F}{\partial \Box^{j}R}\right). \tag{4.3}
$$

The differential Eqs. (4.2) are of order $(2k+4)$. The stress-energy tensor is due to the kinetic part of the scalar field and to the ordinary matter:

$$
T_{\mu\nu} = T^{m}_{\mu\nu} + \frac{\varepsilon}{2}[\phi_{;\mu}\phi_{;\nu} - \frac{1}{2}\phi^{\alpha}_{;}\phi_{;\alpha}]. \tag{4.4}
$$

The (eventual) contribution of a potential $V(\phi)$ is contained in the definition of F. We shall indicate by a capital F a Lagrangian density containing also the contribution of a potential $V(\phi)$ and by $F(\phi)$, $f(R)$, or $f(R,\Box R)$ a function of such fields without potential.

By varying with respect to the scalar field ϕ, we obtain the Klein-Gordon equation

$$
\varepsilon\Box\phi = -\frac{\partial F}{\partial\phi}. \tag{4.5}
$$

Several approaches can be used to deal with such equations. For example, as we said, by a conformal transformation, it is possible to reduce an Extended Theories of Gravity to a (multi) scalar-tensor theory of gravity [11, 92, 147, 149, 201, 337, 423].

The simplest extension of General Relativity is achieved assuming

$$
F = f(R), \qquad \varepsilon = 0, \tag{4.6}
$$

in the action (4.1); $f(R)$ is an arbitrary (analytic) function of the Ricci curvature scalar R. We are considering here the simplest case of fourth-order gravity but we could construct such kind of theories also using other invariants in $R_{\mu\nu}$ or $R^{\alpha}_{\beta\mu\nu}$. The standard Hilbert-Einstein action is, of course, recovered for $f(R) = R$. Varying with respect to $g_{\alpha\beta}$, we get the field equations

$$
f'(R)R_{\alpha\beta} - \frac{1}{2}f(R)g_{\alpha\beta} = f'(R)^{;\mu\nu}\left(g_{\alpha\mu}g_{\beta\nu} - g_{\alpha\beta}g_{\mu\nu}\right), \tag{4.7}
$$

which are fourth-order equations due to the term $f'(R)^{;\mu\nu}$; the prime indicates the derivative with respect to R. Eq. (4.7) is also the equation for $T_{\mu\nu} = 0$ when the matter term is absent.

By a suitable manipulation, the above equations can be rewritten as:

$$G_{\alpha\beta} = \frac{1}{f'(R)}\left\{\frac{1}{2}g_{\alpha\beta}\left[f(R)-Rf'(R)\right]+f'(R)_{;\alpha\beta}-g_{\alpha\beta}\Box f'(R)\right\}, \tag{4.8}$$

where the gravitational contribution due to higher-order terms can be simply reinterpreted as a stress-energy tensor contribution. This means that additional and higher-order terms in the gravitational action act, in principle, as a stress-energy tensor, related to the form of $f(R)$. Considering also the standard perfect-fluid matter contribution, we have

$$\begin{aligned} G_{\alpha\beta} &= \frac{1}{f'(R)}\left\{\frac{1}{2}g_{\alpha\beta}\left[f(R)-Rf'(R)\right]+f'(R)_{;\alpha\beta}-g_{\alpha\beta}\Box f'(R)\right\}+\frac{T^{m}_{\alpha\beta}}{f'(R)} \\ &= T^{curv}_{\alpha\beta}+\frac{T^{m}_{\alpha\beta}}{f'(R)}, \end{aligned} \tag{4.9}$$

where $T^{curv}_{\alpha\beta}$ is an effective stress-energy tensor constructed by the extra curvature terms. In the case of General Relativity, $T^{curv}_{\alpha\beta}$ identically vanishes while the standard, minimal coupling is recovered for the matter contribution. The peculiar behavior of $f(R)=R$ is due to the particular form of the Lagrangian itself which, even though it is a second order Lagrangian, can be non-covariantly rewritten as the sum of a first order Lagrangian plus a pure divergence term. The Hilbert-Einstein Lagrangian can be in fact recast as follows:

$$L_{HE} = \mathcal{L}_{HE}\sqrt{-g} = \left[p^{\alpha\beta}(\Gamma^{\rho}_{\alpha\sigma}\Gamma^{\sigma}_{\rho\beta}-\Gamma^{\rho}_{\rho\sigma}\Gamma^{\sigma}_{\alpha\beta})+\nabla_{\sigma}(p^{\alpha\beta}u^{\sigma}{}_{\alpha\beta})\right] \tag{4.10}$$

where:

$$p^{\alpha\beta} = \sqrt{-g}g^{\alpha\beta} = \frac{\partial\mathcal{L}}{\partial R_{\alpha\beta}} \tag{4.11}$$

Γ is the Levi-Civita connection of g and $u^{\sigma}_{\alpha\beta}$ is a quantity constructed out with the variation of Γ [429]. Since $u^{\sigma}_{\alpha\beta}$ is not a tensor, the above expression is not covariant; however a standard procedure has been studied to recast covariance in the first order theories [190]. This clearly shows that the field equations should consequently be second order and the Hilbert-Einstein Lagrangian is thus degenerate.

From the action (4.1), it is possible to obtain another interesting case by choosing

$$F = F(\phi)R - V(\phi), \qquad \varepsilon = -1. \tag{4.12}$$

In this case, we get

$$\mathcal{A} = \int d^4x\sqrt{-g}\left[F(\phi)R+\frac{1}{2}g^{\mu\nu}\phi_{;\mu}\phi_{;\nu}-V(\phi)\right] \tag{4.13}$$

$V(\phi)$ and $F(\phi)$ are generic functions describing respectively the potential and the coupling of a scalar field ϕ. The Brans-Dicke theory of gravity is a particular case of the action (4.13) for $V(\phi)$=0 [98]. The variation with respect to $g_{\mu\nu}$ gives the second-order field equations

$$F(\phi)G_{\mu\nu} = F(\phi)\left[R_{\mu\nu}-\frac{1}{2}Rg_{\mu\nu}\right] = -\frac{1}{2}T^{\phi}_{\mu\nu}-g_{\mu\nu}\Box_g F(\phi)+F(\phi)_{;\mu\nu}, \tag{4.14}$$

here $\Box_g$ is the d'Alembert operator with respect to the metric g The energy-momentum tensor relative to the scalar field is

$$T^{\phi}_{\mu\nu} = \phi_{;\mu}\phi_{;\nu} - \frac{1}{2}g_{\mu\nu}\phi_{;\alpha}\phi^{\alpha}_{;} + g_{\mu\nu}V(\phi) \tag{4.15}$$

The variation with respect to ϕ provides the Klein - Gordon equation, i.e. the field equation for the scalar field:

$$\Box_g\phi - RF_{\phi}(\phi) + V_{\phi}(\phi) = 0 \tag{4.16}$$

where $F_{\phi} = dF(\phi)/d\phi$, $V_{\phi} = dV(\phi)/d\phi$. This last equation is equivalent to the Bianchi contracted identity [99]. Standard fluid matter can be treated as above.

4.2. Conformal Transformations

Let us now introduce conformal transformations to show that any higher-order or scalar-tensor theory, in absence of ordinary matter, e.g. a perfect fluid, is conformally equivalent to an Einstein theory plus minimally coupled scalar fields. If standard matter is present, conformal transformations allow to transfer non-minimal coupling to the matter component [289]. The conformal transformation on the metric $g_{\mu\nu}$ is

$$\tilde{g}_{\mu\nu} = e^{2\omega}g_{\mu\nu} \tag{4.17}$$

in which $e^{2\omega}$ is the conformal factor. This is nothing else but a particular case of space-time deformations discussed in Chapter 3. Under this transformation, the Lagrangian in (4.13) becomes

$$\sqrt{-g}\left(FR + \frac{1}{2}g^{\mu\nu}\phi_{;\mu}\phi_{;\nu} - V\right)$$
$$= \sqrt{-\tilde{g}}e^{-2\omega}\left(F\tilde{R} - 6F\Box_{\tilde{g}}\omega + -6F\omega_{;\alpha}\omega^{\alpha}_{;} + \frac{1}{2}\tilde{g}^{\mu\nu}\phi_{;\mu}\phi_{;\nu} - e^{-2\omega}V\right) \tag{4.18}$$

in which $\tilde{R}$ and $\Box_{\tilde{g}}$ are the Ricci scalar and the d'Alembert operator relative to the metric $\tilde{g}$. Requiring the theory in the metric $\tilde{g}_{\mu\nu}$ to appear as a standard Einstein theory, the conformal factor has to be related to F, that is

$$e^{2\omega} = -2F. \tag{4.19}$$

where F must be negative in order to restore physical coupling. Using this relation and introducing a new scalar field $\tilde{\phi}$ and a new potential $\tilde{V}$, defined respectively by

$$\tilde{\phi}_{;\alpha} = \sqrt{\frac{3F_{\phi}{}^2 - F}{2F^2}}\phi_{;\alpha}, \quad \tilde{V}(\tilde{\phi}(\phi)) = \frac{V(\phi)}{4F^2(\phi)}, \tag{4.20}$$

we see that the Lagrangian (4.18) becomes

$$\sqrt{-g}\left(FR+\frac{1}{2}g^{\mu\nu}\phi_{;\mu}\phi_{;\nu}-V\right)=\sqrt{-\tilde{g}}\left(-\frac{1}{2}\tilde{R}+\frac{1}{2}\tilde{\phi}_{;\alpha}\tilde{\phi}^{\alpha}_{;}-\tilde{V}\right)$$

which is the usual Hilbert-Einstein Lagrangian plus the standard Lagrangian relative to the scalar field $\tilde{\phi}$. Therefore, every non-minimally coupled scalar-tensor theory, in absence of ordinary matter, e.g. perfect fluid, is conformally equivalent to an Einstein theory, being the conformal transformation and the potential suitably defined by (4.19) and (4.20). The converse is also true: for a given $F(\phi)$, such that $3F_{\phi}{}^{2}-F>0$, we can transform a standard Einstein theory into a non-minimally coupled scalar-tensor theory. This means that, in principle, if we are able to solve the field equations in the framework of the Einstein theory in presence of a scalar field with a given potential, we should be able to get the solutions for the scalar-tensor theories, assigned by the coupling $F(\phi)$, via the conformal transformation (4.19) with the constraints given by (4.20). Following the standard terminology, the "Einstein frame" is the framework of the Einstein theory with the minimal coupling and the "Jordan frame" is the framework of the non-minimally coupled theory [289].

In the context of alternative theories of gravity, as previously discussed, the gravitational contribution to the stress-energy tensor of the theory can be reinterpreted by means of a conformal transformation as the stress-energy tensor of a suitable scalar field and then as "matter" like terms. Performing the conformal transformation (4.17) in the field equations (4.8), we get:

$$\begin{aligned}\tilde{G}_{\alpha\beta} &= \frac{1}{f'(R)}\left\{\frac{1}{2}g_{\alpha\beta}\left[f(R)-Rf'(R)\right]+f'(R)_{;\alpha\beta}-g_{\alpha\beta}\Box f'(R)\right\}\\ &\quad +2\left(\omega_{;\alpha;\beta}+g_{\alpha\beta}\Box\omega-\omega_{;\alpha}\omega_{;\beta}+\frac{1}{2}g_{\alpha\beta}\omega_{;\gamma}\omega^{;\gamma}\right). \qquad (4.21)\end{aligned}$$

We can then choose the conformal factor to be

$$\omega=\frac{1}{2}\ln|f'(R)|, \qquad (4.22)$$

which has now to be substituted into (4.9). Rescaling ω in such a way that

$$k\phi=\omega, \qquad (4.23)$$

and $k=\sqrt{1/6}$, we obtain the Lagrangian equivalence

$$\sqrt{-g}f(R)=\sqrt{-\tilde{g}}\left(-\frac{1}{2}\tilde{R}+\frac{1}{2}\tilde{\phi}_{;\alpha}\tilde{\phi}^{\alpha}_{;}-\tilde{V}\right) \qquad (4.24)$$

and the Einstein equations in standard form

$$\tilde{G}_{\alpha\beta}=\phi_{;\alpha}\phi_{;\beta}-\frac{1}{2}\tilde{g}_{\alpha\beta}\phi_{;\gamma}\phi^{;\gamma}+\tilde{g}_{\alpha\beta}V(\phi), \qquad (4.25)$$

with the potential

$$V(\phi) = \frac{e^{-4k\phi}}{2}\left[\mathcal{P}(\phi) - \mathcal{N}\left(e^{2k\phi}\right)e^{2k\phi}\right] = \frac{1}{2}\frac{f(R) - Rf'(R)}{f'(R)^2}. \tag{4.26}$$

Here $\mathcal{N}$ is the inverse function of $\mathcal{P}'(\phi)$ and $\mathcal{P}(\phi) = \int \exp(2k\phi)d\mathcal{N}$. However, the problem is completely solved if $\mathcal{P}'(\phi)$ can be analytically inverted. In summary, a fourth-order theory is conformally equivalent to the standard second-order Einstein theory plus a scalar field (see also [187, 290]).

This procedure can be extended to more general theories. If the theory is assumed to be higher than fourth order, we may have Lagrangian densities of the form [11, 69, 201],

$$\mathcal{L} = \mathcal{L}(R, \Box R, \ldots, \Box^k R). \tag{4.27}$$

Every $\Box$ operator introduces two further terms of derivation into the field equations. For example a theory like

$$\mathcal{L} = R\Box R, \tag{4.28}$$

is a sixth-order theory and the above approach can be pursued by considering a conformal factor of the form

$$\omega = \frac{1}{2}\ln\left|\frac{\partial\mathcal{L}}{\partial R} + \Box\frac{\partial\mathcal{L}}{\partial\Box R}\right|. \tag{4.29}$$

In general, increasing two orders of derivation in the field equations (*i.e.* for every term $\Box R$), corresponds to adding a scalar field in the conformally transformed frame [11, 201]. A sixth-order theory can be reduced to an Einstein theory with two minimally coupled scalar fields; a $2n$-order theory can be, in principle, reduced to an Einstein theory plus $(n-1)$-scalar fields. On the other hand, these considerations can be directly generalized to higher-order-scalar-tensor theories in any number of dimensions as shown in [287].

As concluding remarks, we can say that conformal transformations work at three levels: *i*) on the Lagrangian of the given theory; *ii*) on the field equations; *iii*) on the solutions. The table below summarizes the situation for fourth-order gravity, non-minimally coupled scalar-tensor theories and standard Hilbert-Einstein theory. Clearly, direct and inverse transformations correlate all the steps of the table but no absolute criterion, at this point of the discussion, is able to select which is the "physical" framework since, at least from a mathematical point of view, all the frames are equivalent [289]. This point is up to now unsolved several if wide discussions are present in literature [9].

$\mathcal{L}_{FOG}$	$\longleftrightarrow$	$\mathcal{L}_{NMC}$	$\longleftrightarrow$	$\mathcal{L}_{HE}$
$\updownarrow$		$\updownarrow$		$\updownarrow$
FOG Eqs.	$\longleftrightarrow$	NMC Eqs.	$\longleftrightarrow$	Einstein Eqs.
$\updownarrow$		$\updownarrow$		$\updownarrow$
FOG Solutions	$\longleftrightarrow$	NMC Solutions	$\longleftrightarrow$	Einstein Solutions

4.3. The Intrinsic Conformal Structure

As we said, the Palatini approach, considering g and Γ as independent fields, is "intrinsically" bi-metric and capable of disentangling the geodesic structure from the chronological structure of a given manifold. Starting from these considerations, conformal transformations assume a fundamental role in defining the affine connection which is merely "Levi-Civita" only for the Hilbert-Einstein theory.

In this section, we work out examples showing how conformal transformations assume a fundamental physical role in relation to the Palatini approach in Extended Theories of Gravity [9].

Let us start from the case of fourth-order gravity where Palatini variational principle is straightforward in showing the differences with Hilbert-Einstein variational principle, involving only metric. Besides, cosmological applications of $f(R)$ gravity have shown the relevance of Palatini formalism, giving physically interesting results with singularity - free solutions [421]. This last nice feature is not present in the standard metric approach.

An important remark is in order at this point. The Ricci scalar entering in $f(R)$ is $R \equiv R(g,\Gamma) = g^{\alpha\beta}R_{\alpha\beta}(\Gamma)$ that is a *generalized Ricci scalar* and $R_{\mu\nu}(\Gamma)$ is the Ricci tensor of a torsion-less connection Γ, which, *a priori*, has no relations with the metric g of space-time. The gravitational part of the Lagrangian is controlled by a given real analytical function of one real variable $f(R)$, while $\sqrt{-g}$ denotes a related scalar density of weight 1. Field equations, deriving from the Palatini variational principle, are:

$$f'(R)R_{(\mu\nu)}(\Gamma) - \frac{1}{2}f(R)g_{\mu\nu} = T^{m}_{\mu\nu} \tag{4.30}$$

$$\nabla^{\Gamma}_{\alpha}(\sqrt{-g}f'(R)g^{\mu\nu}) = 0 \tag{4.31}$$

where ∇^{Γ} is the covariant derivative with respect to Γ. As above, we assume $8\pi G = 1$. We shall use the standard notation denoting by $R_{(\mu\nu)}$ the symmetric part of $R_{\mu\nu}$, *i.e.* $R_{(\mu\nu)} \equiv \frac{1}{2}(R_{\mu\nu} + R_{\nu\mu})$.

In order to get (4.31), one has to additionally assume that $\mathcal{L}_m$ is functionally independent of Γ; however it may contain metric covariant derivatives $\overset{g}{\nabla}$ of fields. This means that the matter stress-energy tensor $T^{m}_{\mu\nu} = T^{m}_{\mu\nu}(g,\Psi)$ depends on the metric g and some matter fields denoted here by Ψ, together with their derivatives (covariant derivatives with respect to the Levi-Civita connection of g). From (4.31) one sees that $\sqrt{-g}f'(R)g^{\mu\nu}$ is a symmetric twice contravariant tensor density of weight 1. As previously discussed in [9, 189], this naturally leads to define a new metric $h_{\mu\nu}$, such that the following relation holds:

$$\sqrt{-g}f'(R)g^{\mu\nu} = \sqrt{-h}h^{\mu\nu}\,. \tag{4.32}$$

This *ansatz* is suitably made in order to impose Γ to be the Levi-Civita connection of h and the only restriction is that $\sqrt{-g}f'(R)g^{\mu\nu}$ should be non-degenerate. In the case of Hilbert-Einstein Lagrangian, it is $f'(R) = 1$ and the statement is trivial.

The above Eq. (4.32) imposes that the two metrics h and g are conformally equivalent. The corresponding conformal factor can be easily found to be $f'(R)$ (in dim $\mathcal{M} = 4$) and the conformal transformation results to be ruled by:

$$h_{\mu\nu} = f'(R)g_{\mu\nu} \tag{4.33}$$

Therefore, as it is well known, Eq. (4.31) implies that $\Gamma = \Gamma_{LC}(h)$ and $R_{(\mu\nu)}(\Gamma) =$

$R_{\mu\nu}(h) \equiv R_{\mu\nu}$. Field equations can be supplemented by the scalar-valued equation obtained by taking the trace of (4.30), (we define $\tau = \mathrm{tr}\hat{T}$)

$$f'(R)R - 2f(R) = g^{\alpha\beta}T^{m}_{\alpha\beta} \equiv \tau^{m} \tag{4.34}$$

which controls solutions of (4.31). We shall refer to this scalar-valued equation as the *structural equation* of the space-time. In the vacuum case (and space-times filled with radiation, such that $\tau^m = 0$) this scalar-valued equation admits constant solutions, which are different from zero only if one add a cosmological constant. This means that the universality of Einstein field equations holds [189], corresponding to a theory with cosmological constant [352].

In the case of interaction with matter fields, the structural equation (4.33), if explicitly solvable, provides an expression of $R = F(\tau)$, where F is a generic function, and consequently both $f(R)$ and $f'(R)$ can be expressed in terms of τ. The matter content of space-time thus rules the bi-metric structure of space-time and, consequently, both the geodesic and metric structures which are intrinsically different. This behavior generalizes the vacuum case and corresponds to the case of a time-varying cosmological constant. In other words, due to these features, conformal transformations, which allow to pass from a metric structure to another one, acquire an intrinsic physical meaning since "select" metric and geodesic structures which, for a given Extended Theories of Gravity, in principle, *do not* coincide.

Let us now try to extend the above formalism to the case of non-minimally coupled scalar-tensor theories. The effort is to understand if and how the bi-metric structure of space-time behaves in this cases and which could be its geometric and physical interpretation.

We start by considering scalar-tensor theories in the Palatini formalism, calling A_1 the action functional. After, we take into account the case of decoupled non-minimal interaction between a scalar-tensor theory and a $f(R)$ theory, calling A_2 this action functional. We finally consider the case of non-minimal-coupled interaction between the scalar field ϕ and the gravitational fields (g,Γ), calling A_3 the corresponding action functional. Particularly significant is, in this case, the limit of low curvature R. This resembles the physical relevant case of present values of curvatures of the Universe and it is important for cosmological applications.

The action (4.13) for scalar-tensor gravity can be generalized, in order to better develop the Palatini approach, as:

$$A_1 = \int \sqrt{-g}\ [F(\phi)R + \frac{\varepsilon}{2}\overset{g}{\nabla}_{\mu}\phi\overset{g\,\mu}{\nabla}\phi - V(\phi) + \mathcal{L}_m(\Psi,\overset{g}{\nabla}\Psi)]d^4x\,. \tag{4.35}$$

As above, the values of $\varepsilon = \pm 1$ selects between standard scalar field theories and quintessence (phantom) field theories. The relative "signature" can be selected by conformal transformations. Field equations for the gravitational part of the action are, respectively for the metric g and the connection Γ:

$$F(\phi)[R_{(\mu\nu)} - \frac{1}{2}Rg_{\mu\nu}] = T^{\phi}_{\mu\nu} + T^{m}_{\mu\nu}\nabla^{\Gamma}_{\alpha}(\sqrt{-g}F(\phi)g^{\mu\nu}) = 0 \tag{4.36}$$

$R_{(\mu\nu)}$ is the same defined in (4.30). For matter fields we have the following field equations:

$$\varepsilon\Box\phi = -V_\phi(\phi) + F_\phi(\phi)R\frac{\delta\mathcal{L}_m}{\delta\Psi} = 0\,. \tag{4.37}$$

In this case, the structural equation of space-time implies that:

$$R = -\frac{\tau^\phi + \tau^{\mathrm{m}}}{F(\phi)} \tag{4.38}$$

which expresses the value of the Ricci scalar curvature in terms of the traces of the stress-energy tensors of standard matter and scalar field (we have to require $F(\phi) \neq 0$). The bi-metric structure of space-time is thus defined by the ansatz:

$$\sqrt{-g}F(\phi)g^{\mu\nu} = \sqrt{-h}h^{\mu\nu} \tag{4.39}$$

such that g and h result to be conformally related being

$$h_{\mu\nu} = F(\phi)g_{\mu\nu}\,. \tag{4.40}$$

The conformal factor is exactly the interaction factor. From (4.38), it follows that in the vacuum case $\tau^\phi = 0$ and $\tau^m = 0$: this theory is equivalent to the standard Einstein one without matter. On the other hand, for $F(\phi) = F_0$ we recover the Einstein theory plus a minimally coupled scalar field: this means that the Palatini approach intrinsically gives rise to the conformal structure (4.40) of the theory which is trivial in the Einstein, minimally coupled case. Beside fundamental physics motivations, these theories have acquired a huge interest in Cosmology due to the fact that they "naturally" exhibit inflationary behaviors able to overcome the shortcomings of Cosmological Standard Model (based on General Relativity). The related cosmological models seem realistic and capable of matching with the Cosmic Microwave Background Radiation observations [56, 196].

As a further step, let us generalize the previous results considering the case of a non-minimal coupling in the framework of $f(R)$ theories. The action functional can be written as:

$$A_2 = \int \sqrt{-g}\,\left[F(\phi)f(R) + \frac{\varepsilon}{2}\overset{g}{\nabla}_\mu\phi\overset{g\,\mu}{\nabla}\phi - V(\phi) + \mathcal{L}_m(\Psi,\overset{g}{\nabla}\Psi)\right]d^4x \tag{4.41}$$

where $f(R)$ is, as usual, any analytical function of the Ricci scalar R. Field equations (in the Palatini formalism) for the gravitational part of the action are:

$$F(\phi)[f'(R)R_{(\mu\nu)} - \frac{1}{2}f(R)g_{\mu\nu}] = T^\phi_{\mu\nu} + T^m_{\mu\nu}\nabla^\Gamma_\alpha(\sqrt{-g}F(\phi)f'(R)g^{\mu\nu}) = 0\,. \tag{4.42}$$

For scalar and matter fields we have, otherwise, the following field equations:

$$\varepsilon\Box\phi = -V_\phi(\phi) + \sqrt{-g}F_\phi(\phi)f(R)\frac{\delta\mathcal{L}_m}{\delta\Psi} = 0 \tag{4.43}$$

where the non-minimal interaction term enters into the modified Klein-Gordon equations.

In this case the structural equation of space-time implies that:

$$f'(R)R - 2f(R) = \frac{\tau^{\phi} + \tau^{m}}{F(\phi)}. \tag{4.44}$$

We remark again that this equation, if solved, expresses the value of the Ricci scalar curvature in terms of traces of the stress-energy tensors of standard matter and scalar field (we have to require again that $F(\phi) \neq 0$). The bi-metric structure of space-time is thus defined by the ansatz:

$$\sqrt{-g}F(\phi)f'(R)g^{\mu\nu} = \sqrt{-h}h^{\mu\nu} \tag{4.45}$$

such that g and h result to be conformally related by:

$$h_{\mu\nu} = F(\phi)f'(R)g_{\mu\nu}. \tag{4.46}$$

Once the structural equation is solved, the conformal factor depends on the values of the matter fields (ϕ, Ψ) or, more precisely, on the traces of the stress-energy tensors and the value of ϕ. From equation (4.44), it follows that in the vacuum case, *i.e.* both $\tau^{\phi} = 0$ and $\tau^{m} = 0$, the universality of Einstein field equations still holds as in the case of minimally interacting $f(R)$ theories [189]. The validity of this property is related to the decoupling of the scalar field and the gravitational field.

Let us finally consider the case where the gravitational Lagrangian is a general function of ϕ and R. The action functional can thus be written as:

$$A_3 = \int \sqrt{-g}\; [K(\phi,R) + \frac{\varepsilon}{2} \overset{g}{\nabla}_{\mu} \phi \overset{g\,\mu}{\nabla} \phi - V(\phi) + \mathcal{L}_m(\Psi, \overset{g}{\nabla}\Psi)]d^4x \tag{4.47}$$

Field equations for the gravitational part of the action are:

$$\left[\frac{\partial K(\phi,R)}{\partial R}\right] R_{(\mu\nu)} - \frac{1}{2}K(\phi,R)g_{\mu\nu} = T^{\phi}_{\mu\nu} + T^{m}_{\mu\nu}\nabla^{\Gamma}_{\alpha}\left(\sqrt{-g}\left[\frac{\partial K(\phi,R)}{\partial R}\right] g^{\mu\nu}\right) = 0. \tag{4.48}$$

For matter fields, we have:

$$\varepsilon\Box\phi = -V_{\phi}(\phi) + \left[\frac{\partial K(\phi,R)}{\partial\phi}\right] \frac{\delta L_{\text{mat}}}{\delta\Psi} = 0. \tag{4.49}$$

The structural equation of space-time can be expressed as:

$$\frac{\partial K(\phi,R)}{\partial R} R - 2K(\phi,R) = \tau^{\phi} + \tau^{m} \tag{4.50}$$

This equation, if solved, expresses again the form of the Ricci scalar curvature in terms of traces of the stress-energy tensors of matter and scalar field (we have to impose regularity conditions and, for example, $K(\phi,R) \neq 0$). The bi-metric structure of space-time is thus defined by the ansatz:

$$\sqrt{-g}\frac{\partial K(\phi,R)}{\partial R}g^{\mu\nu} = \sqrt{-h}h^{\mu\nu} \tag{4.51}$$

such that g and h result to be conformally related by

$$h_{\mu\nu} = \frac{\partial K(\phi,R)}{\partial R}g_{\mu\nu} \tag{4.52}$$

Again, once the structural equation is solved, the conformal factor depends just on the values of the matter fields and (the trace of) their stress energy tensors. In other words, the evolution, the definition of the conformal factor and the bi-metric structure is ruled by the values of traces of the stress-energy tensors and by the value of the scalar field ϕ. In this case, the universality of Einstein field equations does not hold anymore in general. This is evident from (4.50) where the strong coupling between R and ϕ avoids the possibility, also in the vacuum case, to achieve simple constant solutions.

We consider, furthermore, the case of small values of R, corresponding to small curvature space-times. This limit represents, as a good approximation, the present epoch of the observed Universe under suitably regularity conditions. A Taylor expansion of the analytical function $K(\phi,R)$ can be performed:

$$K(\phi,R) = K_0(\phi) + K_1(\phi)R + o(R^2) \tag{4.53}$$

where only the first leading term in R is considered and we have defined:

$$K_0(\phi) = K(\phi,R)_{R=0} K_1(\phi) = \left(\frac{\partial K(\phi,R)}{\partial R}\right)_{R=0}. \tag{4.54}$$

Substituting this expression in (4.50) and (4.52) we get (neglecting higher order approximations in R) the structural equation and the bi-metric structure in this particular case. From the structural equation, we get:

$$R = \frac{1}{K_1(\phi)}[-(\tau^{\phi} + \tau^{m}) - 2K_0(\phi)] \tag{4.55}$$

such that the value of the Ricci scalar is always determined, in this first order approximation, in terms of $\tau^{\phi}, \tau^{\mathrm{m}}, \phi$. The bi-metric structure is, otherwise, simply defined by means of the first term of the Taylor expansion, which is

$$h_{\mu\nu} = K_1(\phi)g_{\mu\nu}. \tag{4.56}$$

It reproduces, as expected, the scalar-tensor case (4.40). In other words, scalar-tensor theories can be recovered in a first order approximation of a general theory where gravity and non-minimal couplings are any (compare (4.55) with (4.44)). This fact agrees with the above considerations where Lagrangians of physical interactions can be considered as stochastic functions with local gauge invariance properties [36].

Finally we have to say that there are also bi-metric theories which cannot be conformally related (see for example the summary of alternative theories given in [436]) and torsion field should be taken into account, if one wants to consider the most general point of view

[100, 211]. We will not take into account these general theories in this review.

4.4. Cosmological Solutions in the Einstein and Jordan Frames

The above discussion tells us that, for a given Extended Theories of Gravity, Palatini approach intrinsically define a bi-metric structure where geodesic and chronological structures of space-time do not coincide a priori. This fact is extremely relevant in the interpretation of conformal transformations since the interpretation of physical results in the metrics $h_{\mu\nu}$ and $g_{\mu\nu}$ (or alternatively $\bar{g}_{\mu\nu}$ and $g_{\mu\nu}$)is something different since, in the Palatini formalism h and g are entangled. This means that g provides the chronological structure while h is related to the geodesic structure as the affine connection is assumed to be $\Gamma = \Gamma_{LC}(h)$. This feature assume a crucial role at the level of the solutions which can be worked out in the two dynamics, first of all in cosmology. In fact, a bad interpretation of the geodesic structure of a given space-time can lead to misunderstand the results and the interpretation of observations. In this section, we want to show how the "same" theory, conformally transformed, can give rise to completely different cosmological solutions. For example, in the Einstein frame we can have solutions with cosmological constant which is the same at every epoch while in the Jordan frame a self-interacting potential and a non-minimal coupling come out. This fact leads to a completely different interpretation of data. The shortcoming is unambiguously solved only if the structure of affine connections is completely controlled as in the Palatini approach.

In order to support these statements, let us take into account scalar-tensor theories in the the Friedmann-Robertson-Walker cosmology. A part the interest of such theories discussed in the Introduction, they are remarkable since, as we have seen, represent the low-curvature limit of general non-minimally coupled higher-order theories whose interpretation is straightforward in the Palatini approach.

Let us assume now that the space-time manifold is described by a Friedmann-Robertson-Walker metric. The Lagrangian density (4.13) takes the form

$$L_t = 6F(\phi)a\dot{a}^2 + 6F_\phi(\phi)a^2\dot{a}\dot{\phi} - 6F(\phi)aK + \frac{1}{2}a^3\dot{\phi}^2 - a^3V(\phi). \tag{4.57}$$

With the subscript t, we mean that the time–coordinate considered is the cosmic time t: this remark is important for the forthcoming discussion. Here a is the scale factor of the universe and K is spatial curvature constant. The Euler-Lagrange equations relative to (4.57) are then

$$\begin{cases} \dfrac{2\ddot{a}}{a} + \dfrac{\dot{a}^2}{a^2} + \dfrac{2F_\phi\dot{a}\dot{\phi}}{Fa} + \dfrac{F_\phi\ddot{\phi}}{F} + \dfrac{K}{a^2} + \dfrac{F_{\phi\phi}\dot{\phi}^2}{F} - \dfrac{\dot{\phi}^2}{4F} + \dfrac{V}{2F} = 0 \\ \ddot{\phi} + \dfrac{3\dot{a}\dot{\phi}}{a} + \dfrac{6F_\phi\dot{a}^2}{a^2} + \dfrac{6F_\phi\ddot{a}}{a} + \dfrac{6F_\phi K}{a^2} + V_\phi = 0 \end{cases} \tag{4.58}$$

which correspond to the (generalized) second order Einstein equation and to the Klein–Gordon equation in the Friedmann-Robertson-Walker case. The energy function relative to

(4.57) is

$$E_t = \frac{\partial L_t}{\partial \dot{a}}\dot{a} + \frac{\partial L_t}{\partial \dot{\phi}}\dot{\phi} - L_t = 6Fa\dot{a}^2 + 6F_\phi a^2\dot{a}\dot{\phi} + 6FaK + \frac{1}{2}a^3\dot{\phi}^2 + a^3V = 0 \tag{4.59}$$

which is the first order generalized Einstein equation. Performing the conformal transformation defined by (4.17), (4.19), (4.20) on the Friedmann-Robertson-Walker metric, one should obtain the corresponding expression for the Lagrangian and the corresponding equations of the Einstein-cosmology from the nonstandard coupled Lagrangian (4.57) and from the generalized Einstein and Klein–Gordon equations, respectively. Unfortunately we see that the presence of the conformal factor (4.19) implies that the transformed line element which is obtained is no longer expressed in the "cosmic time form". Actually the scale factor of the Einstein theory can be defined as the scale factor of the non-minimally coupled theory multiplied by the conformal factor, but the time coordinate of the Einstein theory has to be redefined if we require to have the cosmic time as well. Absorbing the conformal factor in the redefinition of time, we obtain the transformation on the time coordinate. Therefore, the transformation from the Jordan frame to the Einstein frame in the cosmological case is given by

$$\begin{cases} \bar{a} = \sqrt{-2F(\phi)}\,a \\ \dfrac{d\bar{\phi}}{dt} = \sqrt{\dfrac{3F_\phi{}^2 - F}{2F^2}}\dfrac{d\phi}{dt} \\ d\bar{t} = \sqrt{-2F(\phi)}\,dt. \end{cases} \tag{4.60}$$

From the Palatini point of view, these transformations are "natural" due to the intrinsic different geodesic structure of the two frames. Furthermore, the system of Eqs. (4.58),(4.59) and the relations (4.60) to pass from the Jordan frame to the Einstein frame are immediately recovered from the Palatini field equations (4.30)-(4.31) and (4.37), linked together by the structural equation (4.44). Moreover in the Palatini formalism, the redefinition of cosmic time in the two frames (*i.e.* considering h or g as the physical metric) naturally follows from (4.39) and reproduces (4.60).In other words, Palatini field equations give, at once, dynamics of fields and, being endowed with a bi-metric structure, the relation between the Jordan frame and the Einstein frame.

Using the first and the third of (4.60), the scale factor $\bar{a}$ in the Einstein frame depends only on $\bar{t}$. The factor $F(\phi)$, which modifies the geodesic structure, is absorbed into the definition of the cosmic time in the Einstein frame. The second of (4.60) corresponds to the first of relations (4.20) under the given assumption of homogeneity and isotropy. Under transformation (4.60) we have that

$$\begin{aligned} \frac{1}{\sqrt{-2F}}L_t &= \frac{1}{\sqrt{-2F}}\left(6Fa\dot{a}^2 + 6F_\phi a^2\dot{a}\dot{\phi} - 6FaK + \frac{1}{2}a^3\dot{\phi}^2 - a^3V\right) \\ &= -3\bar{a}\dot{\bar{a}}^2 + 3K\bar{a} + \frac{1}{2}\bar{a}^3\dot{\bar{\phi}}^2 - \bar{a}^3\bar{V}(\bar{\phi}) = \bar{L}_{\bar{t}} \end{aligned} \tag{4.61}$$

in which the dot over barred quantities means the derivative with respect to $\bar{t}$; L_t is given by

(4.57) and $\overline{L}_{\bar t}$ coincides with the "point–like" Lagrangian obtained from the Hilbert-Einstein action plus a scalar field under the assumption of homogeneity and isotropy. In this way, the invariance of the homogeneus and isotropic action under (4.60) is restored, being L_t and $\overline{L}_{\bar t}$ equivalent by the (4.61). The same correspondence as (4.61) exists between the energy function E_t and $\overline{E}_{\bar t}$, that is, there is correspondence between the two first order Einstein equations in the two frames. It is interesting to note that the relation (4.61) reflects the Palatini bi-metric structure: the Lagrangians are equivalent only if the time is conformally transformed and Levi-Civita connection is restored in the new metric.

We focus now our attention on the way in which the Euler-Lagrange equations transform under (4.60). The Euler-Lagrange equations relative to (4.61) are the usual second order Einstein equation and Klein–Gordon equation

$$\begin{cases} \dfrac{2\ddot{\bar{a}}}{\bar{a}}+\dfrac{\dot{\bar{a}}^2}{\bar{a}}+\dfrac{K}{\bar{a}^2}+\dfrac{1}{2}\dot{\bar{\phi}}^2-\bar{V}=0 \\ \ddot{\bar{\phi}}+\dfrac{3\dot{\bar{a}}\dot{\bar{\phi}}}{\bar{a}}+\bar{V}_{\bar{\phi}}=0. \end{cases} \tag{4.62}$$

Under (4.60) it is straightforward to verify that they become

$$\begin{cases} \dfrac{2\ddot{a}}{a}+\dfrac{\dot{a}^2}{a^2}+\dfrac{2F_\phi\dot{a}\dot{\phi}}{Fa}+\dfrac{F_\phi\ddot{\phi}}{F}+\dfrac{K}{a^2}+\dfrac{F_{\phi\phi}\dot{\phi}^2}{F}-\dfrac{\dot{\phi}^2}{4F}+\dfrac{V}{2F}=0 \\ \ddot{\phi}+\dfrac{3\dot{a}\dot{\phi}}{a}+\left(\dfrac{6F_\phi F_{\phi\phi}-F_\phi}{3F_\phi{}^2-F}\right)\dfrac{\dot{\phi}^2}{2}+\dfrac{2F_\phi V}{3F_\phi{}^2-F}-\dfrac{FV_\phi}{3F_\phi{}^2-F}=0 \end{cases} \tag{4.63}$$

which do not coincide with the Euler–Lagrange equations given by (4.58). Using the first of (4.58), the second of (4.58) can be written as

$$\begin{aligned} &\frac{F-3F_\phi{}^2}{F}\ddot{\phi}+\frac{3(F-3F_\phi{}^2)}{F}\frac{\dot{a}\dot{\phi}}{a}+\left(\frac{F_\phi-6F_{\phi\phi}F_\phi}{F}\right)\frac{\dot{\phi}^2}{2} \\ &+\frac{F_\phi\dot{\phi}^2}{4F}-\frac{2F_\phi V}{F}+V_\phi+\frac{3F_\phi\dot{a}^2}{a^2}+\frac{3F_\phi K}{a^2}+\frac{3F_\phi{}^2\dot{a}\dot{\phi}}{a}=0, \end{aligned} \tag{4.64}$$

which becomes, taking into account (4.59)

$$\begin{aligned} &\frac{F-3F_\phi{}^2}{F}\ddot{\phi}+\frac{3(F-3F_\phi{}^2)}{F}\frac{\dot{a}\dot{\phi}}{a}+\frac{\dot{\phi}^2}{2F}\frac{d}{d\phi}(F-3F_\phi{}^2) \\ &+\frac{F_\phi\dot{\phi}^2}{4F}-\frac{2F_\phi V}{F}+V_\phi+\frac{F_\phi}{2a^3F}E_t=0. \end{aligned} \tag{4.65}$$

Comparing (4.65) with the second of (4.63), we see that they coincide if $F-3F_\phi{}^2\neq 0$ and $E_t=0$. The quantity $F-3F_\phi{}^2$ is proportional to the Hessian determinant of L_t with respect to $(\dot{a},\ \dot{\phi})$; this Hessian has to be different from zero in order to avoid pathologies in the dynamics [99], while $E_t=0$ corresponds to the first order Einstein equation. Clearly, such pathologies are naturally avoided in the Palatini approach where the cosmological equations of motion are derived from the field equations (4.30)-(4.31) and (4.37). It is possible to see more clearly at the problem of the cosmological conformal equivalence, formulated in the context of the "point–like" Lagrangian, if we use, as time–coordinate, the conformal time η, connected to the cosmic time t by the usual relation

$$a^2(\eta)d\eta^2 = dt^2. \tag{4.66}$$

We can see that the use of η makes much easier the treatment of all the problems we have discussed till now. The crucial point is the following: given the form of the Friedmann-Robertson-Walker line element expressed in conformal time η one does not face the problem of redefining time after performing a conformal transformation, since in this case, the expansion parameter appears in front of all the terms of the line element. From this point of view, the conformal transformation which connects Einstein and Jordan frame is given by

$$\begin{cases} \bar{a} = \sqrt{-2F(\phi)}\,a \\ \\ \dfrac{d\bar{\phi}}{d\eta} = \sqrt{\dfrac{3F_\phi{}^2 - F}{2F^2}}\,\dfrac{d\phi}{d\eta} \end{cases} \tag{4.67}$$

where a, ϕ, $\bar{a}$, $\bar{\phi}$ are assumed as functions of η.

The Hilbert-Einstein "point–like" Lagrangian is given by

$$\bar{L}_\eta = -3\bar{a}'^2 + 3K\bar{a}^2 + \frac{1}{2}\bar{a}^2\bar{\phi}'^2 - \bar{a}^4\bar{V}(\bar{\phi}) \tag{4.68}$$

in which the prime means the derivative with respect to η, and the subscript η means that the time–coordinate considered is the conformal time. Under transformation (4.67), it becomes

$$\begin{aligned} \bar{L}_\eta &= -3\bar{a}'^2 + 3K\bar{a}^2 + \frac{1}{2}\bar{a}^2\bar{\phi}'^2 - \bar{a}^4\bar{V}(\bar{\phi}) \\ &= 6F(\phi)a' + 6F_\phi(\phi)aa'F'(\phi) - 6F(\phi)Ka^2 + \frac{1}{2}a^2\phi'^2 - a^4V(\phi) = L_\eta \end{aligned} \tag{4.69}$$

which corresponds to the "point–like" Lagrangian obtained from the Lagrangian density in (4.13) under the hypotheses of homogeneity and isotropy, using the conformal time as time coordinate. This means that the Euler–Lagrange equations relative to (4.68), which coincides with the second order Einstein equation and the Klein–Gordon equation in conformal time, correspond to the Euler–Lagrange equations relative to (4.69), under the transformation (4.67). Moreover, the energy function $\bar{E}_\eta$ relative to (4.68) corresponds to the energy function E_η relative to (4.69), so that there is correspondence between the first order Einstein equations. Furthermore, in order to have full coherence between the two formulations, it is easy to verify that, both in the Jordan frame and in the Einstein frame, the Euler–Lagrange equations, written using the conformal time, correspond to the Euler–Lagrange equations written using the cosmic time except for terms in the energy function; for it, one gets the relation

$$E_\eta = aE_t \tag{4.70}$$

which holds in both the frames; thus the first order Einstein equation is preserved under the transformation from η to t and there is full equivalence between the two formulations. We want to point out that for the two Lagrangians L_η and L_t the same relation as (4.70) holds. On the other hand, such results naturally hold if one takes into account the relation (4.40) derived from the second Palatini equation(4.31)

When ordinary matter is present the standard Einstein (cosmological) "point–like" Lagrangian is

$$\overline{L}_{tot} = \overline{L}_{\bar{t}} + \overline{L}_{mat}, \tag{4.71}$$

in which $\overline{L}_{\bar{t}}$ is given by (4.61) and $\overline{L}_{mat}$ is the Lagrangian relative to perfect fluid matter. Using the contracted Bianchi identity, it can be seen that $\overline{L}_{mat}$ can be written as [89]

$$\overline{L}_{mat} = -D\bar{a}^{3(1-\gamma)}, \tag{4.72}$$

where D is connected to the total amount of matter. In writing (4.71) and (4.72) we have chosen the cosmic time as time–coordinate. Under the transformation (4.60) we have, besides relation (4.61), that (4.72) corresponds to

$$\overline{L}_{mat} = (\sqrt{-2F})^{3(1-\gamma)} L_{mat}, \tag{4.73}$$

where, analogously to (4.72)

$$L_{mat} = Da^{3(1-\gamma)}. \tag{4.74}$$

Then we have that, using (4.60), (4.71), it becomes

$$\frac{1}{\sqrt{-2F}} L_{tot} = \frac{1}{\sqrt{-2F}} [L_t + (\sqrt{-2F})^{(4-3\gamma)} L_{mat}] \tag{4.75}$$

in which we have defined the total "point–like" Lagrangian after the conformal transformation as

$$L_{tot} = L_t + (\sqrt{-2F})^{(4-3\gamma)} L_{mat}, \tag{4.76}$$

(cfr. (4.61)); the transformation of $\overline{L}_{tot}$ under (4.60) has to be written following the expression (4.75) and consequently the "point–like" Lagrangian L_{tot} has to be defined as in (4.76).

Summarizing, the perfect fluid-matter, which minimally interact in the Jordan frame, results non-minimally interacting in the conformally transformed Einstein frame unless $\gamma = \frac{4}{3}$ (radiation), since the standard matter Lagrangian term is coupled with the scalar field in a way which depends on the coupling F. Such a coupling between the matter and the scalar field is an effect of the transformation, therefore depending on the coupling. Also this interaction which emerges passing from the Jordan frame to the Einstein frame, is immediately recovered considering the Palatini structural equation (4.44) and follows directly from (4.45) which express the relation between the different metrics (and consequently between the two frames).

4.4.1. Some Relevant Examples

The exact identification of the frame is crucial when the solutions are matched with data. We are going to give some examples where the nature of solutions drastically changes considering the Einstein frame or the Jordan frame without taking into account the problem of transformations of physical quantities between them. The ambiguity is removed in the Palatini approach since, due to the intrinsic bi-metric structure, the two frames are given together by the same dynamics.

i) Let us consider a model in the Einstein frame with a scalar field, a constant potential and zero curvature. The Lagrangian is given by

$$\bar{L}_{\bar{t}} = -3\bar{a}\dot{\bar{a}}^2 + \frac{1}{2}\bar{a}^3\dot{\bar{\phi}}^2 - \bar{a}^3\Lambda; \tag{4.77}$$

the Euler–Lagrange equations and the energy condition are

$$\begin{cases} \dfrac{2\ddot{\bar{a}}}{\bar{a}} + \dfrac{\dot{\bar{a}}^2}{\bar{a}^2} + \dfrac{1}{2}\dot{\bar{\phi}}^2 - \Lambda = 0 \\[2ex] \ddot{\bar{\phi}} + \dfrac{3\dot{\bar{a}}\dot{\bar{\phi}}}{\bar{a}} = 0. \end{cases} \tag{4.78}$$

$$\frac{\dot{\bar{a}}^2}{\bar{a}^2} - \frac{1}{3}\left(\frac{1}{2}\dot{\bar{\phi}}^2 + \Lambda\right) = 0. \tag{4.79}$$

The system can be easily solved giving the solution

$$\begin{cases} \bar{a} = \left[c_1 e^{\sqrt{3\Lambda}\bar{t}} - \dfrac{\dot{\bar{\phi}}_0^2}{8\Lambda c_1^2} e^{-\sqrt{3\Lambda}\bar{t}}\right]^{\frac{1}{3}} \\[3ex] \bar{\phi} = \bar{\phi}_0 + \sqrt{\dfrac{2}{3}} \ln \dfrac{1 - \dfrac{\dot{\bar{\phi}}_0}{2c_1\sqrt{2\Lambda}} e^{-\sqrt{3\Lambda}\bar{t}}}{1 + \dfrac{\dot{\bar{\phi}}_0}{2c_1\sqrt{2\Lambda}} e^{-\sqrt{3\Lambda}\bar{t}}} \end{cases} \tag{4.80}$$

Three integration constants appear in the solution, since Eq. (4.79) corresponds to a constraint on the value of the first integral $\bar{E}_{\bar{t}}$. We have that, in the limit of $\bar{t} \to +\infty$, the behavior of $\bar{a}$ is exponential with characteristic time given by $\sqrt{\frac{\Lambda}{3}}$, as we would expect, and $\bar{\phi}$ goes to a constant. Looking at the second of (4.20), we have that such a model in the Einstein frame corresponds, in the Jordan frame, to the class of models with (arbitrarily given) coupling F and potential V connected by the relation

$$\frac{V}{4F^2} = \Lambda, \tag{4.81}$$

the solution of which can be obtained from (4.80) via the transformation (4.60). We can thus fix the potential V and obtain, from (4.81), the corresponding coupling. This can be used as a method to find the solutions of non-minimally coupled models with given potentials, the coupling being determined by (4.81). In other words, a single model in the Einstein frame corresponds to a family of models in the Jordan frame, but giving "a priori" the bi-metric structure of the theory by the Palatini approach, the model is only one. As an example, let us take into account the case

$$V = \lambda\phi^4, \quad \lambda > 0 \tag{4.82}$$

which correspond to a "chaotic inflationary" potential. The corresponding coupling is quadratic in ϕ

$$F = k_0\phi^2 \tag{4.83}$$

in which

$$k_0 = -\frac{1}{2}\sqrt{\frac{\lambda}{\Lambda}}. \tag{4.84}$$

Substituting (4.80) into (4.60), we get

$$\begin{cases} a = \dfrac{\bar{a}}{\phi\sqrt{-2k_0}} \\[2ex] d\phi = \phi\sqrt{\dfrac{2k_0}{12k_0 - 1}}\, d\bar{\phi} \\[2ex] dt = \dfrac{d\bar{t}}{\phi\sqrt{-2k_0}}. \end{cases} \tag{4.85}$$

As we see from these relations, it has to be $k_0 < 0$. Integrating the second of (4.65), we have the conformal relation between the scalar fields, *i.e.* ϕ in terms of $\bar{\phi}$

$$\phi = \alpha_0 e^{\sqrt{\frac{2k_0}{12k_0-1}}\,\bar{\phi}}. \tag{4.86}$$

Substituting (4.86) in the first of (4.85) and taking into account the second of (4.80), we have the solutions a and ϕ as functions of $\bar{t}$

$$\begin{cases} \phi = \phi_0 \left[\dfrac{1 - \dfrac{\bar{\phi}_0}{2c_1\sqrt{2\Lambda}} e^{-\sqrt{3\Lambda}\bar{t}}}{1 + \dfrac{\bar{\phi}_0}{2c_1\sqrt{2\Lambda}} e^{-\sqrt{3\Lambda}\bar{t}}} \right]^{\sqrt{\frac{4k_0}{3(12k_0-1)}}} \\[4ex] a = \dfrac{1}{\phi_0\sqrt{-2k_0}} \left[c_1 e^{\sqrt{3\Lambda}\bar{t}} - \dfrac{\bar{\phi}_0^2}{8\Lambda c_1^2} e^{-\sqrt{3\Lambda}\bar{t}} \right]^{\frac{1}{3}} \left[\dfrac{1 + \dfrac{\bar{\phi}_0}{2c_1\sqrt{2\Lambda}} e^{-\sqrt{3\Lambda}\bar{t}}}{1 - \dfrac{\bar{\phi}_0}{2c_1\sqrt{2\Lambda}} e^{-\sqrt{3\Lambda}\bar{t}}} \right]^{\sqrt{\frac{4k_0}{3(12k_0-1)}}} \end{cases} \tag{4.87}$$

in which $\phi_0 = \alpha_0 e^{\sqrt{\frac{2k_0}{12k_0-1}}\bar{\phi}_0}$. Substituting (4.86) in the third of (4.85), taking into account (4.80), we get

$$dt = \frac{d\bar{t}}{\bar{\phi}_0\sqrt{-2k_0}}\left[\frac{1+\dfrac{\bar{\phi}_0}{2c_1\sqrt{2\Lambda}}e^{-\sqrt{3\Lambda}\bar{t}}}{1-\dfrac{\bar{\phi}_0}{2c_1\sqrt{2\Lambda}}e^{-\sqrt{3\Lambda}\bar{t}}}\right]^{\sqrt{\frac{4k_0}{3(12k_0-1)}}}. \tag{4.88}$$

We obtain $\bar{t}$ as a function of t integrating (4.88) and then considering the inverse function; Eq. (4.88) could be easily integrated if the exponent $\sqrt{\frac{4k_0}{3(12k_0-1)}}$ would be equal to ± 1, but this corresponds to a value of $k_0 = \frac{3}{32}$ which is positive and thus it turns out to be not physically acceptable. In general, (4.88) is not of easy solution. We can analyze its asymptotic behavior, obtaining

$$\frac{dt}{d\bar{t}} \overset{\bar{t}\to+\infty}{\longrightarrow} \frac{1}{\bar{\phi}_0\sqrt{-2k_0}} \tag{4.89}$$

that is, asymptotically,

$$t - t_0 \simeq \frac{\bar{t}}{\bar{\phi}_0\sqrt{-2k_0}}. \tag{4.90}$$

Substituting (4.90) in the asymptotic expression of (4.87), we obtain the asymptotic behavior of the solutions (since from (4.89) one has $t \overset{\bar{t}\to+\infty}{\longrightarrow} +\infty$)

$$\begin{cases} a \simeq \dfrac{c_1^{1/3}}{\bar{\phi}_0\sqrt{-2k_0}}e^{\bar{\phi}_0\sqrt{\frac{-2\Lambda k_0}{3}}(t-t_0)} \\ \phi \simeq \phi_0. \end{cases} \tag{4.91}$$

Thus we have that, asymptotically, $a(t)$ is exponential, and $\phi(t)$ is constant; the coupling F is asymptotically constant too, so that, fixing the arbitrary constant of integration to obtain the finite transformation of $\bar{a}$, $\bar{\phi}$ (that is, fixing the units, see [382]), once k_0 is fixed, it is possible to recover asymptotically the Einstein gravity from the Jordan frame.

As a remark we would like to notice that the asymptotic expression (4.91) of $a(t)$ and $\phi(t)$ are solutions of the Einstein equations and Klein–Gordon equation with zero curvature and F and V given by (4.82), (4.83). They have not been obtained as solutions of the asymptotic limits of these equations. It means then that they are, in any case, particular solutions of the given non-minimally coupled model.

ii) Another interesting case is the Ginzburg–Landau potential

$$V = \lambda(\phi^2 - \mu^2)^2, \quad \lambda > 0. \tag{4.92}$$

The corresponding coupling is given by

$$F = k_0(\phi^2 - \mu^2), \tag{4.93}$$

in which k_0 is given by (4.84) when $\phi^2 > \mu^2$ while is given by (4.84) with opposite sign when $\phi^2 < \mu^2$, in order to have $F < 0$. With this coupling, the corresponding conformal transformation turns out to be singular for $\phi^2 = \mu^2$, thus with this method it is not possible to solve this model for ϕ equal to the Ginzburg–Landau mass μ. The explicit function $\phi = \phi(\bar{\phi})$ is obtained inverting the integral

$$\bar{\phi} - \bar{\phi}_0 = \int \frac{[3\sqrt{\frac{\lambda}{\Lambda}}\phi^2 + \frac{1}{2}(\phi^2 - \mu^2)]^{\frac{1}{2}}}{[\frac{\lambda}{4\Lambda}]^{\frac{1}{4}}(\phi^2 - \mu^2)} d\phi; \tag{4.94}$$

and it is possible to carry analogous considerations as in the previous case, concluding that asymptotically the behavior of $a(t)$ is exponential and that of $\phi(t)$ is constant.

iii) Another interesting case is

$$V = \lambda\phi^2, \;\; \lambda > 0; \quad F = k_0\phi^2, \;\; k_0 < 0 \tag{4.95}$$

in the Jordan frame. The coupling is the same as in (4.83)) and the conformal transformation is given by (4.85). To obtain the corresponding potential in the Einstein frame we have to substitute (4.86) in the relation

$$\bar{V}(\bar{\phi}) = \frac{\lambda}{4k_0^2\phi^2(\bar{\phi})}, \tag{4.96}$$

that is

$$\bar{V}(\bar{\phi}) = \frac{\lambda}{4k_0^2\phi_0^2} e^{-2\sqrt{\frac{2k_0}{12k_0-1}}\bar{\phi}}, \tag{4.97}$$

which gives, in the Einstein frame, power–law solutions [36, 283]. A general remark concerns the relation between the Hubble parameter in the Einstein and in the Jordan frame. It is

$$\bar{H} = \frac{\dot{\bar{a}}}{\bar{a}} = \frac{1}{(-2F)}\left(-\frac{\dot{F}}{\sqrt{-2F}} + \sqrt{-2F}\frac{\dot{a}}{a}\right) = \frac{\dot{F}}{2F\sqrt{-2F}} + \frac{H}{\sqrt{-2F}}, \tag{4.98}$$

in which we have used the relations (4.60). Relation (4.98) is useful to study the asymptotic behavior of the Hubble parameter: if we require an asymptotic de Sitter–behavior in both the Einstein and Jordan frame (for example, in order to reproduce quintessential accelerated behavior), we have to require $\bar{H} \overset{t\to+\infty}{\longrightarrow} \bar{C}$ and $H \overset{t\to+\infty}{\longrightarrow} C$ where $\bar{C}$ and C are constants, from (4.98), we obtain a differential equation for the coupling F as a function of t ($t >> 0$), given by

$$\dot{F} + 2CF - 2\bar{C}F\sqrt{-2F} = 0. \tag{4.99}$$

Its solution is

$$F = -\frac{C^2}{2C^2}\left[\frac{1}{1-F_0 e^{Ct}}+1\right]^2, \tag{4.100}$$

in which F_0 is the integration constant; this is the time–behavior that F has to assume on the solution $\phi(t)$, in order to have a de Sitter asymptotical accelerated behavior in both frames. It easy to verify that both the couplings in the examples i) and ii) satisfy (4.100) asymptotically.

4.5. Summary

It is clear that several mathematical and physical issues allow to take into account alternative and extended theories of gravity with respect to General Relativity.

Such schemes, also if do not fully address the problem of Quantum Gravity, constitute effective approaches which, from one side, gives a semiclassical limit to the problem and, from another side, could potentially solve astrophysical and cosmological riddles as Dark Energy and Dark Matter. As we have shown, the field equations that are derived from such theories can be higher-order than the second in derivatives or non minimally coupled to some scalar field. This fact means that Extended Theories bring further gravitational degrees of freedom (given by scalar fields or curvature invariants) which could account for the bulk of dark components. This "geometric" approach means that one should change the r.h.s. of field equations in order to address dynamical issues at small and large scales and not the l.h.s., where new "sources" should be present. The second approach is genuinely "material" but cannot leave aside the issue to "detect" new ingredients for Dark Energy and Dark Matter at a fundamental level. The forthcoming answers coming from LHC will be crucial in this sense.

Finally, it is worth while to stress the role of conformal transformations and Palatini formalism. The former approach con be seen as a tool to "reduce" the generalized schemes to the Einstein formalism. The latter allows a physical interpretation of the results being endowed with a bimetric structure [79]. Both of them can be framed in the conformal-affine structure of gauge theories as we have shown in Chapters 2 and 3.

Chapter 5

Probing the Post-Minkowskian Limit

Beside addressing theoretical issues related to Quantum Gravity adn observational issues related to Dark Energy and Dark Matter, Extended Theories of Gravity should reproduce consistently weak field limits along the lines traced by General Relativity. In fact, such a theory is considered self-consistent since successfully gives the Minkowski space-time of Special Relativity, when gravitational field goes to zero, and the Newton gravity as soon as gravitational field is weak and the velocities are small compared to the speed of light. These situations are partially reproduced starting from Extended Theories of Gravity since dynamics is now endowed with further degrees of freedom. this means that, beside the standard massless modes of General Relativity, further massive and ghost modes emerge in the post-Minkowskian limit and generalized Poisson equations are reproduced in the post-Newtonian limit.

Probing these limits is crucial to retain or reject Extended Theories of Gravity. For example the detection of massive scalar modes or the deviation from the Newtonian gravity at certain astrophysical scales could constitute foundamental probes to confirm the approach we have presented, leaving apart from dark side issues. In the Chapter 5 and 6, we shall face these problems and discuss possible forthcoming experimental measurements in this sense.

5.1. Gravitational Waves in Extended Gravity

Recently, the data analysis of interferometric gravitational waves detectors has been started (for the current status of gravitational waves interferometers see [2,4,23,375,438]) and the scientific community aims at a first direct detection of gravitational waves in next years. The design and the construction of a number of sensitive detectors for gravitational waves is underway today. There are some laser interferometers like the VIRGO detector, built in Cascina, near Pisa, Italy, by a joint Italian-French collaboration, the GEO 600 detector built in Hannover, Germany, by a joint Anglo-German collaboration, the two LIGO detectors built in the United States (one in Hanford, Washington and the other in Livingston, Louisiana) by a joint Caltech-MIT collaboration, and the TAMA 300 detector, in Tokyo, Japan.

Many detectors are currently in operation too, and several interferometers are in a phase of planning and proposal stages (for the current status of gravitational waves experiments see [224–226]). The results of these detectors will have a fundamental impact on astrophysics and gravitational physics and will be important for a better knowledge of the Universe and either to confirm or rule out the physical consistency of General Relativity or any other theory of gravitation [436]. At a fundamental level, detecting new gravitational modes could be a sort of *experimentum crucis* in order to discriminate among theories since this fact would be the "signature" that General Relativity should be enlarged or modified [44, 58, 101].

5.2. The Post Minkowskian Limit of Extended Gravity: The Case of $f(R)$-Gravity

Before discussing the further gravitational modes related to Extended Theories of gravity, let us formally develop, the post-Minkowskian limit for higher order gravity models [388]. As we shall see, massive modes naturally emerge beside the standard massless mode of General Relativity.

The post-Minkowskian limit of whatever gravity theory arises when the regime of small field is considered without any prescription on the propagation velocity of the field. This case has to be clearly distinguished with respect to the Newtonian limit which, differently, requires both the small velocity and the field approximations.

In literature such a distinction is after not clearly remarked and several cases of pathological analysis can be accounted. The post-Minkowskian limit of General Relativity naturally provides massless waves as the propagation modes of gravity. We can now develop an analogous study considering instead of the Hilbert-Einstein Lagrangian, a general function of the Ricci scalar. Actually, in order to perform the post-Minkowskian limit of field equations, one has to implement the field equations with a small perturbation on the Minkowski background $\eta_{\mu\nu}$. It is reasonable to assume that the $f(R)$ - Lagrangian is an analytic function in term of the Ricci scalar (i.e. Taylor expandable around the Ricci scalar value $R = R_0 = 0$). In such a case, field Eqs (4.7), at the first order of approximation in term of the perturbation, become :

$$f'_0\left[R^{(1)}_{\mu\nu} - \frac{R^{(1)}}{2}\eta_{\mu\nu}\right] - 2f''_0\left[R^{(1)}_{,\mu\nu} - \eta_{\mu\nu}\Box_\eta R^{(1)}\right] = \mathcal{X}\, T^{(0)}_{\mu\nu}\,. \tag{5.1}$$

where we are assuming $g_{\mu\nu} \simeq \eta_{\mu\nu} + h_{\mu\nu}$. At zero-order one gets again $f(0) = 0$ while $T_{\mu\nu}$ is fixed at zero-order in (5.1). The explicit expression of the Ricci tensor and scalar at the first order in metric perturbations are

$$\begin{aligned} R^{(1)}_{\mu\nu} &= h^{\sigma}_{(\mu,\sigma)\sigma} - \frac{1}{2}\Box h_{\mu\nu} - \frac{1}{2}h_{,\mu\nu} \\ R^{(1)} &= h^{,\sigma\tau}_{\sigma\tau} - \Box h \end{aligned} \tag{5.2}$$

where $(,)$ is the symmetrization and $h =^{\sigma}_{\sigma}$. Introducing the constant $\lambda^2 = -\frac{f'_0}{3f''_0}$ we obtain the field equations

$$\begin{aligned} h^{\sigma}_{(\mu,\nu)\sigma} - \frac{1}{2}\Box_\eta h_{\mu\nu} - \frac{1}{2}h_{,\mu\nu} - \frac{1}{2}(h_{\sigma\tau}{}^{,\sigma\tau} - \Box_\eta h)\eta_{\mu\nu} \\ + \frac{1}{3\lambda^2}(\partial^2_{\mu\nu} - \eta_{\mu\nu}\Box_\eta)(h_{\sigma\tau}{}^{,\sigma\tau} - \Box_\eta h) = \frac{\mathcal{X}}{f'(R)}T^{(0)}_{\mu\nu} \end{aligned} \tag{5.3}$$

and by choosing the harmonic gauge: $\tilde{h}_{\mu\nu} = h_{\mu\nu} - \frac{h}{2}\eta_{\mu\nu}$ with the condition $\tilde{h}^{\mu\nu}{}_{,\mu} = 0$, one obtains that the field equations and the trace equation respectively read

$$\begin{cases} \Box_\eta \tilde{h}_{\mu\nu} + \frac{1}{3\lambda^2}(\eta_{\mu\nu}\Box_\eta - \partial^2_{\mu\nu})\Box_\eta \tilde{h} = -\frac{2\mathcal{X}}{f'_0}T^{(0)}_{\mu\nu} \\ \Box_\eta \tilde{h} + \frac{1}{\lambda^2}\Box^2_\eta \tilde{h} = -\frac{2\mathcal{X}}{f'_0}T^{(0)} \end{cases} \tag{5.4}$$

In order to deduce the analytic solutions of (5.4), we can adopt momentum representation. This approach simplifies the equations and, above all, allows to directly observe what the physical properties of our problem are. In such picture we have:

$$\begin{cases} k^2\tilde{h}_{\mu\nu}(k) + \frac{1}{3\lambda^2}(k_\mu k_\nu - k^2\eta_{\mu\nu})k^2\tilde{h}(k) = \frac{2\mathcal{X}}{f'_0}T^{(0)}_{\mu\nu}(k) \\ k^2\tilde{h}(k)(1 - \frac{k^2}{\lambda^2}) = \frac{2\mathcal{X}}{f'_0}T^{(0)}(k) \end{cases} \tag{5.5}$$

where

$$\begin{cases} \tilde{h}_{\mu\nu}(k) = \int \frac{d^4x}{(2\pi)^2}\tilde{h}_{\mu\nu}(x)\, e^{-ikx} \\ T^{(0)}_{\mu\nu}(k) = \int \frac{d^4x}{(2\pi)^2}T^{(0)}_{\mu\nu}(x)\, e^{-ikx} \end{cases} \tag{5.6}$$

are the Fourier transforms of the perturbation $\tilde{h}_{\mu\nu}(x)$ and of the matter tensor $T^{(0)}_{\mu\nu}$. We have defined, as usual, $kx = \omega t - \mathbf{k}\cdot\mathbf{x}$ and $k^2 = \omega^2 - \mathbf{k}^2$. On the other hand $\tilde{h}(k)$ and $T^{(0)}(k)$ are the traces of $\tilde{h}_{\mu\nu}(k)$ and $T^{(0)}_{\mu\nu}(k)$. In the momentum space one can easily recognize the solutions of (5.5); the expression for $\tilde{h}_{\mu\nu}(k)$ turns out to be

$$\tilde{h}_{\mu\nu}(k) = \frac{2\mathcal{X}}{f'_0}\frac{T^{(0)}_{\mu\nu}(k)}{k^2} + \frac{2\mathcal{X}}{3f'_0}\frac{k_\mu k_\nu - k^2\eta_{\mu\nu}}{k^2(k^2 - \lambda^2)}T^{(0)}(k), \tag{5.7}$$

which fulfils the condition $\tilde{h}^{\mu\nu}{}_{,\mu} = 0$ ($\tilde{h}^{\mu\nu}(k)\, k_\mu = 0$). The true perturbation variable $h_{\mu\nu}(k)$ can ne obtained inverting the relation with the tilded variables, in particular inserting the matter functions $S^{(0)}_{\mu\nu}(k) = T^{(0)}_{\mu\nu}(k) - \frac{1}{2}\eta_{\mu\nu}T^{(0)}(k)$ and $S^{(0)}(k) = \eta^{\mu\nu}S^{(0)}_{\mu\nu}(k)$, one obtains:

$$h_{\mu\nu}(k) = \frac{2\mathcal{X}}{f'_0}\frac{S^{(0)}_{\mu\nu}(k)}{k^2} - \frac{\mathcal{X}}{3f'_0}\frac{k^2\eta_{\mu\nu} + 2k_\mu k_\nu}{k^2(k^2 - \lambda^2)}S^{(0)}(k), \tag{5.8}$$

which represents a wavelike solution, in the momentum space, with a massless and a massive contribution since the pole in the denominator of the second term, whose mass is directly related with the pole itself. The explicit wavelike solution can be obtained returning the the configuration space inverting the Fourier transform of $h_{\mu\nu}$.

Let us stress again that field equations, for a generic $f(R)$ - model, can be rewritten by isolating the Einstein tensor in the l.h.s. [94, 102, 103]. In such a case, higher than second order differential contributions, in term of the metric tensor, are considered in the r.h.s. as a source component of the space-time dynamics as well as the energy momentum tensor of ordinary matter, that is

$$G_{\mu\nu} = T^{(m)}_{\mu\nu} + T^{(curv)}_{\mu\nu}\,, \tag{5.9}$$

where we have restored the notation in Chapter 4 and

$$\begin{cases} T^{(m)}_{\mu\nu} & = \frac{\mathcal{X}\,T_{\mu\nu}}{f'(R)}\,, \\ T^{(curv)}_{\mu\nu} & = \frac{1}{2}g_{\mu\nu}\frac{f(R)-Rf'(R)}{f'(R)} + \frac{f'(R)_{;\mu\nu}-g_{\mu\nu}\Box f'(R)}{f'(R)}\,. \end{cases} \tag{5.10}$$

Actually if we consider the perturbed metric and develop the Einstein tensor up to the first order in perturbations, we have

$$G_{\mu\nu} \sim G^{(1)}_{\mu\nu} = h^{\sigma}_{(\mu,\nu)\sigma} - \frac{1}{2}\Box_\eta h_{\mu\nu} - \frac{1}{2}h_{,\mu\nu} - \frac{1}{2}(h_{\sigma\tau}{}^{,\sigma\tau} - \Box_\eta h)\eta_{\mu\nu} \tag{5.11}$$

while the curvature tensor gives the contributions

$$T^{(curv)}_{\mu\nu} \sim \frac{1}{3\lambda^2}(\eta_{\mu\nu}\Box_\eta - \partial^2_{\mu\nu})(h_{\sigma\tau}{}^{,\sigma\tau} - \Box_\eta h) \tag{5.12}$$

This expression allow to recognize that, in the space of momenta, such a quantity has a pole-like term which implies the introduction of a massive degree of freedom. In fact, inserting these two expressions into the the field Eqs. (5.9) and considering the Ricci tensor and scalar to first order, we obtain

$$\Box_\eta h_{\mu\nu}(x) = -\frac{2\mathcal{X}}{f_0}\left[S^{(0)}_{\mu\nu}(x) + \Sigma^{\lambda}_{\mu\nu}(x)\right] \tag{5.13}$$

where $\Sigma^{\lambda}_{\mu\nu}(x)$ is related to the curvature tensor and is defined as

$$\Sigma^{\lambda}_{\mu\nu}(x) = -\frac{1}{6}\int\frac{d^4k}{(2\pi)^2}\frac{k^2\eta_{\mu\nu} + 2k_\mu k_\nu}{k^2-\lambda^2}S^{(0)}(k)\,e^{ikx}\,. \tag{5.14}$$

The general solution for the metric perturbation $h_{\mu\nu}(x)$ can be rewritten as

$$\begin{aligned} h_{\mu\nu}(x) &= \frac{2\mathcal{X}}{f'_0}\int\frac{d^4k}{(2\pi)^2}\frac{S^{(0)}_{\mu\nu}(k)}{k^2}e^{ikx} \\ &\quad -\frac{\mathcal{X}}{3f'_0}\int\frac{d^4k}{(2\pi)^2}\frac{k^2\eta_{\mu\nu} + 2k_\mu k_\nu}{k^2(k^2-\lambda^2)}S^{(0)}(k)\,e^{ikx}\,, \end{aligned} \tag{5.15}$$

which has, in the second term, a pole whose properties can be easily evaluated in vacuum. In fact, in such a case (i.e. $T_{\mu\nu} = 0$), the Eqs. (5.4) are

$$\begin{cases} k^2\,[\tilde{h}_{\mu\nu}(k) + \frac{1}{3\lambda^2}(k_\mu k_\nu - k^2\eta_{\mu\nu})\tilde{h}(k)] = 0 \\ \\ k^2\tilde{h}(k)(1 - \frac{k^2}{\lambda^2}) = 0 \end{cases} \tag{5.16}$$

showing that the allowed solutions are of two types :

$$\begin{cases} \omega = \pm|\mathbf{k}| \\ \\ h_{\mu\nu}(x) = \int \frac{d^4k}{(2\pi)^2} h_{\mu\nu}(k)\, e^{ikx} \quad \text{with} \;\; h(k) = 0 \end{cases} \tag{5.17}$$

and

$$\begin{cases} \omega \;=\; \pm\sqrt{\mathbf{k}^2 + \lambda^2} \\ \\ h_{\mu\nu}(x) \;=\; -\int \frac{d^4k}{(2\pi)^2}\left[\frac{\lambda^2\eta_{\mu\nu} + 2k_\mu k_\nu}{6\lambda^2}\right] h(k)\, e^{ikx} \quad \text{with} \;\; h(k) \neq 0 \end{cases} \tag{5.18}$$

It is evident, that the first solution represents a massless graviton according to the standard prescriptions of General Relativity while the second one gives a massive degree of freedom with $m^2 \doteq \lambda^2$. In this sense we can further rewrite Eqs. (5.4) introducing $\phi \doteq \Box\tilde{h}$ so that the system can be rearranged in the following way

$$\begin{cases} \Box_\eta \tilde{h}_{\mu\nu} \;=\; -\frac{2\chi}{f'_0} T^{(0)}_{\mu\nu} + \left[\frac{\partial^2_{\mu\nu} - \eta_{\mu\nu}\Box_\eta}{3m^2}\right]\phi \\ \\ (\Box_\eta + m^2)\phi \;=\; -\frac{2\chi}{f'_0} m^2 T^{(0)} \end{cases} \tag{5.19}$$

which suggests that the $f(R)$ contributions act in the post-Minkowskian limit as a massive scalar field whose mass depends on f'_0, f''_0, calculated on the background metric.

It is important to stress that the massive contributions in the gravitational waves spectrum is strictly related to the trace equation. In fact, in the case of $f(R)$-gravity the trace equation establishes a constraint for the Ricci scalar under the form of a dynamical equation. This relation tell us that the Ricci scalar is not univocally fixed as in General Relativity. In oder words, the fact that the $f(R)$-Lagrangian is non-linear in R gives rise to a further massive scalar mode directly related to the analytic form of $f(R)$ and its derivatives. As we will see in the next Section, this feature is standard in any higher-order-theory of gravity.

5.3. The General Theory: Ghost, Massless and Massive Modes

The above considerations can be generalized assuming a generic fourth-order gravitational model constructed starting from curvature invariants. Specifically, we will consider the action

$$S = \int d^4x\sqrt{-g}f(R,P,Q) \tag{5.20}$$

where

$$P \equiv R_{ab}R^{ab} \qquad Q \equiv R_{abcd}R^{abcd} \tag{5.21}$$

Varying with respect to the metric, one gets the field equations:

$$\begin{aligned} FG_{\mu\nu} &= \frac{1}{2}g_{\mu\nu}(f - R\,F) - (g_{\mu\nu}\Box - \nabla_\mu\nabla_\nu)F - 2\left(f_P R^a_\mu R_{a\nu} + f_Q\, R_{abc\mu}R^{abc}{}_\nu\right) \\ &-g_{\mu\nu}\nabla_a\nabla_b(f_P R^{ab}) - \Box(f_P R_{\mu\nu}) + 2\nabla_a\nabla_b\left(f_P\, R^a{}_{(\mu}\delta^b{}_{\nu)} + 2f_Q\, R^a{}_{(\mu\nu)}{}^b\right) \end{aligned} \tag{5.22}$$

where we have set

$$F \equiv \frac{\partial f}{\partial R}, \quad f_P \equiv \frac{\partial f}{\partial P}, \quad f_Q \equiv \frac{\partial f}{\partial Q} \tag{5.23}$$

and $\Box = g^{ab}\nabla_a\nabla_b$ is the d'Alembert operator while the notation $T_{(ij)} = \frac{1}{2}(T_{ij} + T_{ji})$ denotes symmetrization with respect to the indices (i, j).

Taking the trace of Eq. (5.22), we find:

$$\Box\left(F + \frac{f_P}{3}R\right) = \frac{1}{3}\left(2f - RF - 2\nabla_a\nabla_b((f_P + 2f_Q)R^{ab}) - 2(f_P P + f_Q Q)\right) \tag{5.24}$$

Expanding the third term on the r.h.s. of (5.24) and using the purely geometrical identity $G^{ab}{}_{;b} = 0$ we get:

$$\begin{aligned} \Box\left(F + \frac{2}{3}(f_P + f_Q)R\right) &= \frac{1}{3}[2f - RF - 2R^{ab}\nabla_a\nabla_b(f_P + 2f_Q) - R\Box(f_P + 2f_Q) \\ &\quad -2(f_P P + f_Q Q)] \end{aligned} \tag{5.25}$$

If we define

$$\Phi \equiv F + \frac{2}{3}(f_P + f_Q)R \tag{5.26}$$

$$\frac{dV}{d\Phi} \equiv \text{RHS of (5.25)} \tag{5.27}$$

then we get a Klein-Gordon equation for the scalar field Φ:

$$\Box\Phi = \frac{dV}{d\Phi} \tag{5.28}$$

In order to find the various modes of the gravitational waves, we need to linearize gravity around the Minkowski background, that is as above

$$g_{\mu\nu} = \eta_{\mu\nu} + h_{\mu\nu} \tag{5.29}$$

$$\Phi = \Phi_0 + \delta\Phi \tag{5.30}$$

Then from Eq. (5.26) we get

$$\delta\Phi = \delta F + \frac{2}{3}(\delta f_P + \delta f_Q)R_0 + \frac{2}{3}(f_{P0} + f_{Q0})\delta R \tag{5.31}$$

where $R_0 \equiv R(\eta_{\mu\nu}) = 0$ and similarly $f_{P0} = \frac{\partial f}{\partial P}|_{\eta_{\mu\nu}}$ (note that 0 indicates evaluation with respect the Minkowski metric) which is either constant or zero. By δR we denote the first order perturbation on the Ricci scalar which, along with the perturbed parts of the Riemann and Ricci tensors, are given by:

$$\delta R_{\mu\nu\rho\sigma} = \frac{1}{2}\left(\partial_\rho\partial_\nu h_{\mu\sigma} + \partial_\sigma\partial_\mu h_{\nu\rho} - \partial_\sigma\partial_\nu h_{\mu\rho} - \partial_\rho\partial_\mu h_{\nu\sigma}\right) \tag{5.32}$$

$$\delta R_{\mu\nu} = \frac{1}{2}\left(\partial_\sigma\partial_\nu h^\sigma{}_\mu + \partial_\sigma\partial_\mu h^\sigma_\nu - \partial_\mu\partial_\nu h - \Box h_{\mu\nu}\right) \tag{5.33}$$

$$\delta R = \partial_\mu\partial_\nu h^{\mu\nu} - \Box h \tag{5.34}$$

where as above $h = \eta^{\mu\nu}h_{\mu\nu}$. The first term of Eq. (5.31) is

$$\delta F = \frac{\partial F}{\partial R}|_0\ \delta R + \frac{\partial F}{\partial P}|_0\ \delta P + \frac{\partial F}{\partial Q}|_0\ \delta Q \tag{5.35}$$

However, since δP and δQ are second order, we get $\delta F \simeq F_{,R0}\ \delta R$ and

$$\delta\Phi = \left(F_{,R0} + \frac{2}{3}(f_{P0} + f_{Q0})\right)\delta R \tag{5.36}$$

Finally, from Eq. (5.25) we get the Klein-Gordon equation for the scalar perturbation $\delta\Phi$

$$\begin{aligned}\Box\delta\Phi &= \frac{1}{3}\frac{F_0}{F_{,R0} + \frac{2}{3}(f_{P0} + f_{Q0})}\delta\Phi - \frac{2}{3}\delta R^{ab}\partial_a\partial_b(f_{P0} + 2f_{Q0}) - \frac{1}{3}\delta R\Box(f_{P0} + 2f_{Q0}) \\ &= m_s^2\delta\Phi \end{aligned}\tag{5.37}$$

The last two terms in the first line are actually are zero since the terms f_{P0}, f_{Q0} are constants and we have defined the scalar mass as $m_s^2 \equiv \frac{1}{3}\frac{F_0}{F_{,R0} + \frac{2}{3}(f_{P0} + f_{Q0})}$.

Perturbing the field Eqs. (5.22) we get:

$$\begin{aligned} F_0\left(\delta R_{\mu\nu} - \frac{1}{2}\eta_{\mu\nu}\delta R\right) &= -(\eta_{\mu\nu}\Box - \partial_\mu\partial_\nu)(\delta\Phi - \frac{2}{3}(f_{P0} + f_{Q0})\delta R) \\ -\eta_{\mu\nu}\partial_a\partial_b(f_{P0}\delta R^{ab}) &- \Box(f_{P0}\delta R_{\mu\nu}) + 2\partial_a\partial_b(f_{P0}\ \delta R^a{}_{(\mu}\delta^b{}_{\nu)} + 2f_{Q0}\ \delta R^a{}_{(\mu\nu)}{}^b) \end{aligned}\tag{5.38}$$

As above, it is convenient to work in Fourier space so that $\partial_\gamma h_{\mu\nu} \rightarrow ik_\gamma h_{\mu\nu}$ and $\Box h_{\mu\nu} \rightarrow -k^2 h_{\mu\nu}$. Then the above equations, become

$$\begin{aligned} F_0\left(\delta R_{\mu\nu} - \frac{1}{2}\eta_{\mu\nu}\delta R\right) &= (\eta_{\mu\nu}k^2 - k_\mu k_\nu)(\delta\Phi - \frac{2}{3}(f_{P0} + f_{Q0})\delta R) \\ &+\eta_{\mu\nu}k_a k_b(f_{P0}\delta R^{ab}) + k^2(f_{P0}\delta R_{\mu\nu}) \\ &-2k_a k_b(f_{P0}\ \delta R^a{}_{(\mu}\delta^b{}_{\nu)}) - 4k_a k_b(f_{Q0}\ \delta R^a{}_{(\mu\nu)}{}^b) \end{aligned}\tag{5.39}$$

We can rewrite the metric perturbation as

$$h_{\mu\nu} = \bar{h}_{\mu\nu} - \frac{\bar{h}}{2}\eta_{\mu\nu} + \eta_{\mu\nu}h_f \tag{5.40}$$

and use our gauge freedom to demand that the usual conditions $\partial_\mu \bar{h}^{\mu\nu} = 0$ and $\bar{h} = 0$ hold. The first of these conditions implies that $k_\mu \bar{h}^{\mu\nu} = 0$ while the second gives

$$h_{\mu\nu} = \bar{h}_{\mu\nu} + \eta_{\mu\nu}h_f \tag{5.41}$$

$$h = 4h_f \tag{5.42}$$

With these assumptions in mind we have:

$$\delta R_{\mu\nu} = \frac{1}{2}\left(2k_\mu k_\nu h_f + k^2\eta_{\mu\nu}h_f + k^2\bar{h}_{\mu\nu}\right) \tag{5.43}$$

$$\delta R = 3k^2 h_f \tag{5.44}$$

$$k_\alpha k_\beta\, \delta R^{\alpha}{}_{(\mu\nu)}{}^{\beta} = -\frac{1}{2}\left((k^4\eta_{\mu\nu} - k^2 k_\mu k_\nu)h_f + k^4\bar{h}_{\mu\nu}\right) \tag{5.45}$$

$$k_a k_b\, \delta R^{a}{}_{(\mu}\delta^{b}{}_{\nu)} = \frac{3}{2}k^2 k_\mu k_\nu h_f \tag{5.46}$$

Using Eqs. (5.40)-(5.46) into (5.39) and after some algebra, we get:

$$\frac{1}{2}\left(k^2 - k^4\frac{f_{P0} + 4f_{Q0}}{F_0}\right)\bar{h}_{\mu\nu} = (\eta_{\mu\nu}k^2 - k_\mu k_\nu)\frac{\delta\Phi}{F_0} + (\eta_{\mu\nu}k^2 - k_\mu k_\nu)h_f \tag{5.47}$$

Defining $h_f \equiv -\frac{\delta\Phi}{F_0}$ we find the equation for the perturbations:

$$\left(k^2 + \frac{k^4}{m^2_{spin2}}\right)\bar{h}_{\mu\nu} = 0 \tag{5.48}$$

where we have defined $m^2_{spin2} \equiv -\frac{F_0}{f_{P0} + 4f_{Q0}}$, while from Eq. (5.37) we get:

$$\Box h_f = m_s^2 h_f \tag{5.49}$$

From equation (5.48) it is easy to see that we have a modified dispersion relation which corresponds to a massless spin-2 field ($k^2 = 0$) and a massive spin-2 ghost mode $k^2 = \frac{F_0}{\frac{1}{2}f_{P0} + 2f_{Q0}} \equiv -m^2_{spin2}$ with mass m^2_{spin2}. To see this, note that the propagator for $\bar{h}_{\mu\nu}$ can be rewritten as

$$G(k) \propto \frac{1}{k^2} - \frac{1}{k^2 + m^2_{spin2}} \tag{5.50}$$

Clearly the second term has the opposite sign, which indicates the presence of a ghost, and this agrees with the results found in the literature for this class of theories [137, 315, 392].

Also, as a sanity check, we can see that for the Gauss-Bonnet term $\mathcal{L}_{GB} = Q - 4P + R^2$ we have $f_{P0} = -4$ and $f_{Q0} = 1$. Then, Eq. (5.48) simplifies to $k^2\bar{h}_{\mu\nu} = 0$ and, in this case, we have no ghosts as expected.

The solution to Eqs. (5.48) and (5.49) can be written in terms of plane waves

$$\bar{h}_{\mu\nu} = A_{\mu\nu}(\overrightarrow{p})\cdot exp(ik^{\alpha}x_{\alpha})+cc \tag{5.51}$$

$$h_f = a(\overrightarrow{p})\cdot exp(iq^{\alpha}x_{\alpha})+cc \tag{5.52}$$

where

$$\begin{aligned} k^{\alpha} &\equiv (\omega_{m_{spin2}}, \overrightarrow{p}) \qquad \omega_{m_{spin2}} = \sqrt{m^2_{spin2}+p^2} \\ q^{\alpha} &\equiv (\omega_{m_s}, \overrightarrow{p}) \qquad \omega_{m_s} = \sqrt{m_s^2+p^2}. \end{aligned} \tag{5.53}$$

and where m_{spin2} is zero (non-zero) in the case of massless (massive) spin-2 mode and the polarization tensors $A_{\mu\nu}(\overrightarrow{p})$ can be found in Ref. [418] (see equations (21)-(23)). In Eqs. (5.48) and (5.51) the equation and the solution for the standard waves of General Relativity [300] have been obtained, while Eqs. (5.49) and (5.52) are respectively the equation and the solution for the massive mode (see also [104]).

The fact that the dispersion law for the modes of the massive field h_f is not linear has to be emphasized. The velocity of every "ordinary" (i.e. which arises from General Relativity) mode $h_{\mu\nu}$ is the light speed c, but the dispersion law (the second of Eq. (5.53)) for the modes of h_f is that of a massive field which can be discussed like a wave-packet [104]. Also, the group-velocity of a wave-packet of h_f, centered in $\overrightarrow{p}$, is

$$\overrightarrow{v_G} = \frac{\overrightarrow{p}}{\omega}, \tag{5.54}$$

which is exactly the velocity of a massive particle with mass m and momentum $\overrightarrow{p}$.

From the second of Eqs. (5.53) and Eq. (5.54) it is straightforward to obtain:

$$v_G = \frac{\sqrt{\omega^2 - m^2}}{\omega}. \tag{5.55}$$

Then, the constant speed of the wave-packet, is

$$m = \sqrt{(1-v_G^2)}\omega. \tag{5.56}$$

Now, before starting the analysis, it has to be discussed if there are phenomenological limitations to the mass of the gravitational waves [146]. A strong limitation arises from the fact that the gravitational waves needs a frequency which falls in the frequency-range for both of earth based and space based gravitational antennas, that is the interval $10^{-4}Hz \leq f \leq 10KHz$ [2,4,23,226,375,401,438]. For a massive gravitational waves, from [105] it is:

$$2\pi f = \omega = \sqrt{m^2+p^2}, \tag{5.57}$$

where p is the momentum. Thus, it needs

$$0eV \leq m \leq 10^{-11}eV. \tag{5.58}$$

A stronger limitation is given by requirements of cosmology and Solar System tests. In this case, it is

$$0eV \leq m \leq 10^{-33}eV. \tag{5.59}$$

However, these light scalars have to be discussed as coherent massive states.

5.4. Polarization States of Gravitational Waves

Considering the above equations, we can note that there are two conditions for Eq. (5.37) that depend on the value of k^2. In fact we can have a $k^2 = 0$ mode that corresponds to a massless spin-2 field with two independent polarizations plus a scalar mode, while if we have $k^2 \neq 0$ we have a further massive spin-2 ghost mode and there are five independent polarization tensors plus a scalar mode. First, let us consider the case where the spin-2 field is massless.

Taking $\overrightarrow{p}$ in the z direction, a gauge in which only A_{11}, A_{22}, and $A_{12} = A_{21}$ are different to zero can be chosen. The condition $h = 0$ gives $A_{11} = -A_{22}$. In this frame, we may take the bases of polarizations defined as [1]

$$e_{\mu\nu}^{(+)} = \frac{1}{\sqrt{2}}\begin{pmatrix} 1 & 0 & 0 \\ 0 & -1 & 0 \\ 0 & 0 & 0 \end{pmatrix}, \qquad e_{\mu\nu}^{(\times)} = \frac{1}{\sqrt{2}}\begin{pmatrix} 0 & 1 & 0 \\ 1 & 0 & 0 \\ 0 & 0 & 0 \end{pmatrix}$$

$$e_{\mu\nu}^{(s)} = \frac{1}{\sqrt{2}}\begin{pmatrix} 0 & 0 & 0 \\ 0 & 0 & 0 \\ 0 & 0 & 1 \end{pmatrix} \tag{5.60}$$

Now, putting these matrices in Eq. (5.40), it results

$$h_{\mu\nu}(t,z) = A^{+}(t-z)e_{\mu\nu}^{(+)} + A^{\times}(t-z)e_{\mu\nu}^{(\times)} + h_s(t - v_{G}z)e_{\mu\nu}^{s} \tag{5.61}$$

The terms $A^{+}(t-z)e_{\mu\nu}^{(+)} + A^{\times}(t-z)e_{\mu\nu}^{(\times)}$ describe the two standard polarizations of gravitational waves which arise from General Relativity, while the term $h_s(t - v_{G}z)\eta_{\mu\nu}$ is the massive field arising from the generic high order $f(R)$ theory.

When the spin-2 field is massive, we have that the bases of the six polarizations are defined by

$$e_{\mu\nu}^{(+)} = \frac{1}{\sqrt{2}}\begin{pmatrix} 1 & 0 & 0 \\ 0 & -1 & 0 \\ 0 & 0 & 0 \end{pmatrix}, \qquad e_{\mu\nu}^{(\times)} = \frac{1}{\sqrt{2}}\begin{pmatrix} 0 & 1 & 0 \\ 1 & 0 & 0 \\ 0 & 0 & 0 \end{pmatrix}$$

$$e_{\mu\nu}^{(B)} = \frac{1}{\sqrt{2}}\begin{pmatrix} 0 & 0 & 1 \\ 0 & 0 & 0 \\ 1 & 0 & 0 \end{pmatrix}, \qquad e_{\mu\nu}^{(C)} = \frac{1}{\sqrt{2}}\begin{pmatrix} 0 & 0 & 0 \\ 0 & 0 & 1 \\ 0 & 1 & 0 \end{pmatrix}$$

[1] The polarizations are defined in our 3-space, not in a space-time with extra dimensions. Each polarization mode is orthogonal to one other and it is normalized $e_{\mu\nu}e^{\mu\nu} = 2\delta$. Note that other modes are not traceless, in contrast to the ordinary plus and cross polarization modes in General Relativity.

$$e^{(D)}_{\mu\nu} = \frac{\sqrt{2}}{3}\begin{pmatrix} \frac{1}{2} & 0 & 0 \\ 0 & \frac{1}{2} & 0 \\ 0 & 0 & -1 \end{pmatrix}, \qquad e^{(s)}_{\mu\nu} = \frac{1}{\sqrt{2}}\begin{pmatrix} 0 & 0 & 0 \\ 0 & 0 & 0 \\ 0 & 0 & 1 \end{pmatrix}$$

and the amplitude can be written in terms of the 6 polarization states as

$$\begin{aligned} h_{\mu\nu}(t,z) &= A^+(t - v_{G_{s2}}z)e^{(+)}_{\mu\nu} + A^\times(t - v_{G_{s2}}z)e^{(\times)}_{\mu\nu} + B^B(t - v_{G_{s2}}z)e^{(B)}_{\mu\nu} \\ &\quad + C^C(t - v_{G_{s2}}z)e^{(C)}_{\mu\nu} + D^D(t - v_{G_{s2}}z)e^{(D)}_{\mu\nu} + h_s(t - v_G z)e^{s}_{\mu\nu}. \end{aligned} \tag{5.62}$$

where $v_{G_{s2}}$ is the group velocity of the massive spin-2 field. It is given by

$$v_{G_{s2}} = \frac{\sqrt{\omega^2 - m_{s2}^2}}{\omega}. \tag{5.63}$$

The first two polarizations are the same as in the massless case, inducing tidal deformations on the x-y plane. In Fig. 5.4., we illustrate how each gravitational waves polarization affects test masses arranged on a circle.

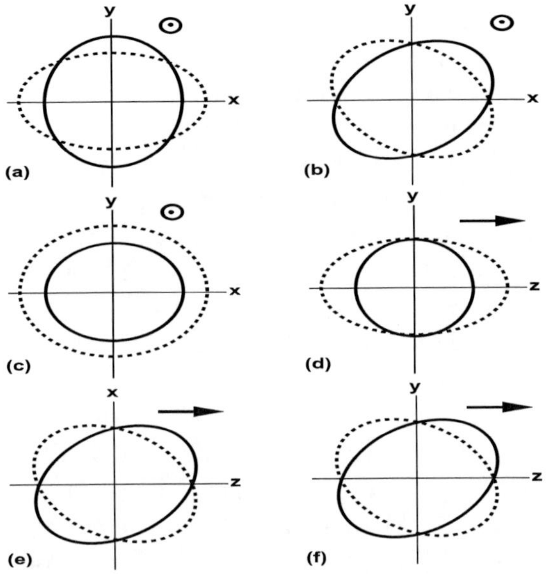

Figure 5.1. The polarization modes of weak gravitational waves permitted. Shown is the displacement that each mode induces on a sphere of test particles at the moments of different phases by π. The wave propagates out of the plane in (a), (b), (c), and it propagates in the plane in (d), (e) and (f). Where in (a) and (b) we have respectively the plus mode and cross mode, in (c) the scalar mode, in (d), (e) and (f) the D, B and C mode.

The presence of the ghost mode may seem as a pathology of the theory from a purely quantum-mechanical approach. There are several reasons to consider such a mode as problematic if we wish to pursuit the particle picture interpretation of the metric perturbations. The ghost mode can be viewed as either a particle state of positive energy and "negative" probability density, or a positive probability density state with a negative energy. In the first case, allowing the presence of such a particle will quickly induce the violation of unitarity and then affecting the consistency of the theory. The negative energy scenario leads to a theory where there is no minimum of energy and the system thus becomes unstable. The vacuum can decay into pairs of ordinary and ghost gravitons leading to a catastrophic instability.

A way out of such problems is to impose a very weak coupling of the ghost with the rest of the particles in the theory, such that the decay rate of the vacuum will become comparable to the inverse of the Hubble scale. The present vacuum state will then appear to be sufficiently stable. This is not a viable option in our theory, since the ghost state comes in the gravitational sector, which is bound to couple to all kinds of matter present and it seems physically and mathematically unlikely for the ghost graviton to couple differently than the ordinary massless graviton does. Another option is to assume that this picture does not hold up to arbitrarily high energies and that at some cutoff scale M_{cutoff} the theory gets modified appropriately as to ensure a ghost-free behavior and a stable ground state. This can happen, for example, if we assume that Lorentz invariance is violated at M_{cutoff}, thereby restricting any potentially harmful decay rates [180].

However, there is no guaranty that theories of modified gravity such as the one investigated here are supposed to hold up to arbitrary energies. Such models are plagued at the quantum level by the same problems as ordinary General Relativity, i.e. they are non-renormalizable. It is therefore not necessary for them to be considered as genuine candidates for a quantum gravity theory and the corresponding ghost particle interpretation becomes rather ambiguous. At the purely classical level, the perturbation $h_{\mu\nu}$ should be viewed as nothing more than a tensor representing the "stretching" of space-time away from flatness. A ghost mode then makes sense as just another way of propagating this perturbation of the space-time geometry, one which carries the opposite sign in the propagator than an ordinary massive graviton would.

Viewed in this way, the presence of the massive ghost graviton will induce on an interferometer the same effects as an ordinary massive graviton transmitting the perturbation, but with the opposite sign in the displacement. Tidal stretching from a polarized wave on the polarization plane will be turned into shrinking and vice-versa. This signal will, at the end, be a superposition of the displacements coming from the ordinary massless spin-2 graviton and the massive ghost. Since these induce two competing effects, this will lead to a less pronounced signal than the one we would expect if the ghost mode was absent, setting in this way less severe constraints on the theory. However, the presence of the new modes will also affect the total energy density carried by the gravitational waves and this may also appear as a candidate signal in stochastic backgrounds, as we will see in the following.

5.5. Potential Detection by Interferometers

Here we consider the response of a single detector to a gravitational waves propagating in a certain direction. A perturbed metric $h_{\mu\nu}$, which represents the gravitational waves propagating in three-dimensional space, is decomposed as Eq. (5.61). Although a detector output actually depends on the gravitational waves amplitude determined by a specific theoretical model, we can discuss the detector response to each gravitational waves polarization without specifying a certain theoretical model. The angular pattern function of a detector to gravitational waves is given by

$$\begin{aligned} F_A(\hat{\boldsymbol{\Omega}}) &= \boldsymbol{D} : \boldsymbol{e}_A(\hat{\boldsymbol{\Omega}}), \\ \boldsymbol{D} &= \frac{1}{2}\left[\hat{\boldsymbol{u}} \otimes \hat{\boldsymbol{u}} - \hat{\boldsymbol{v}} \otimes \hat{\boldsymbol{v}}\right], \end{aligned} \tag{5.64}$$

here $A = +, \times, B, C, D$, and s and these symbols : denotes contraction between tensors, $\boldsymbol{D}$ is a so-called detector tensor, which describes the response of a laser-interferometric detector and maps the gravitational metric perturbation to a gravitational waves signal from the detector. The unit vectors $\hat{\boldsymbol{u}}$ and $\hat{\boldsymbol{v}}$ are orthogonal to each other and are directed to each detector arm, which form an orthonormal coordinate system with the unit vector $\hat{\boldsymbol{w}}$, as shown in Fig. 5.2. $\hat{\boldsymbol{\Omega}}$ is the unit vector directed at the gravitational waves propagation direction. Note that the detector tensor, Eq. (5.64), is valid only when the arm length of the detector is much smaller than the wavelength of gravitational waves that we consider. This is relevant for our purpose to deal with the ground-based laser interferometers.

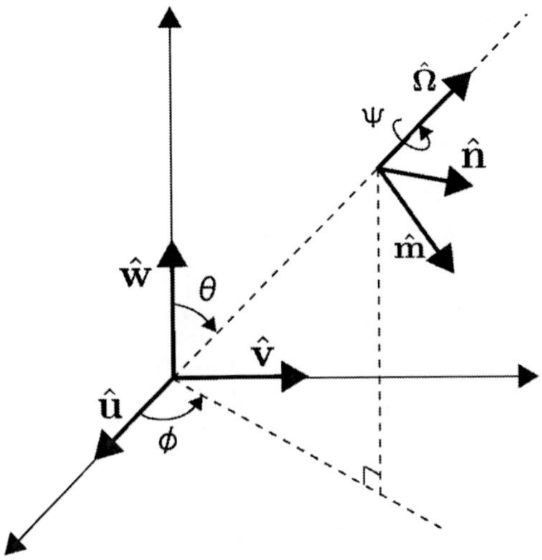

Figure 5.2. Coordinate systems.

Suppose that an orthonormal coordinate system for the detector is

$$\begin{cases} \hat{\boldsymbol{u}} = (1,0,0) \\ \hat{\boldsymbol{v}} = (0,1,0) \\ \hat{\boldsymbol{w}} = (0,0,1) \end{cases},$$

and the gravitational waves coordinate system rotated by angles (θ, ϕ) is

$$\begin{cases} \hat{\boldsymbol{u}}' = (\cos\theta\cos\phi, \cos\theta\sin\phi, -\sin\theta) \\ \hat{\boldsymbol{v}}' = (-\sin\phi, \cos\phi, 0) \\ \hat{\boldsymbol{w}}' = (\sin\theta\cos\phi, \sin\theta\sin\phi, \cos\theta) \end{cases}.$$

The most general choice of the coordinates is obtained by the rotation with respect to the angle ψ around the gravitational waves propagating axis,

$$\begin{cases} \hat{\boldsymbol{m}} = \hat{\boldsymbol{u}}' \cos\psi + \hat{\boldsymbol{v}}' \sin\psi \\ \hat{\boldsymbol{n}} = -\hat{\boldsymbol{v}}' \sin\psi + \hat{\boldsymbol{u}}' \cos\psi \\ \hat{\boldsymbol{\Omega}} = \hat{\boldsymbol{w}}' \end{cases} .$$

The coordinate system $(\hat{\boldsymbol{u}}, \hat{\boldsymbol{v}}, \hat{\boldsymbol{w}})$ is related to the coordinate system $(\hat{\boldsymbol{m}}, \hat{\boldsymbol{n}}, \hat{\boldsymbol{\Omega}})$ by the rotation angles (ϕ, θ, ψ), shown in Fig. 5.2. Using the unit vectors $\hat{\boldsymbol{m}}$, $\hat{\boldsymbol{n}}$, and $\hat{\boldsymbol{\Omega}}$, the polarization tensors can be written as

$$\begin{aligned}
\boldsymbol{e}_+ &= \frac{1}{\sqrt{2}}(\hat{\boldsymbol{m}} \otimes \hat{\boldsymbol{m}} - \hat{\boldsymbol{n}} \otimes \hat{\boldsymbol{n}}) , \\
\boldsymbol{e}_\times &= \frac{1}{\sqrt{2}}(\hat{\boldsymbol{m}} \otimes \hat{\boldsymbol{n}} + \hat{\boldsymbol{n}} \otimes \hat{\boldsymbol{m}}) , \\
\boldsymbol{e}_B &= \frac{1}{\sqrt{2}}\left(\hat{\boldsymbol{m}} \otimes \hat{\boldsymbol{\Omega}} + \hat{\boldsymbol{\Omega}} \otimes \hat{\boldsymbol{m}}\right) , \\
\boldsymbol{e}_C &= \frac{1}{\sqrt{2}}\left(\hat{\boldsymbol{n}} \otimes \hat{\boldsymbol{\Omega}} + \hat{\boldsymbol{\Omega}} \otimes \hat{\boldsymbol{n}}\right) . \\
\boldsymbol{e}_D &= \frac{\sqrt{3}}{2}\left(\frac{\hat{m}}{2} \otimes \frac{\hat{m}}{2} + \frac{\hat{n}}{2} \otimes \frac{\hat{n}}{2} + \hat{\boldsymbol{\Omega}} \otimes \hat{\boldsymbol{\Omega}}\right) , \\
\boldsymbol{e}_s &= \frac{1}{\sqrt{2}}\left(\hat{\boldsymbol{\Omega}} \otimes \hat{\boldsymbol{\Omega}}\right) ,
\end{aligned}$$

Then, from Eqs. (5.64), the angular pattern functions for each polarization result in

$$\begin{aligned}
F_+(\theta, \phi, \psi) &= \frac{1}{\sqrt{2}}(1 + \cos^2\theta) \cos 2\phi \cos 2\psi \\
&\quad - \cos\theta \sin 2\phi \sin 2\psi , \\
F_\times(\theta, \phi, \psi) &= -\frac{1}{\sqrt{2}}(1 + \cos^2\theta) \cos 2\phi \sin 2\psi \\
&\quad - \cos\theta \sin 2\phi \cos 2\psi , \\
F_B(\theta, \phi, \psi) &= \sin\theta\, (\cos\theta \cos 2\phi \cos\psi - \sin 2\phi \sin\psi) , \\
F_C(\theta, \phi, \psi) &= \sin\theta\, (\cos\theta \cos 2\phi \sin\psi + \sin 2\phi \cos\psi) , \\
F_D(\theta, \phi) &= \frac{\sqrt{3}}{32} \cos 2\phi \left(6 \sin^2\theta + (\cos 2\theta + 3) \cos 2\psi\right) , \\
F_s(\theta, \phi) &= \frac{1}{\sqrt{2}} \sin^2\theta \cos 2\phi .
\end{aligned}$$

We plot the angular pattern functions for each polarization in Fig. 5.3. These results are consistent with those obtained in [3].

5.6. The Stochastic Background of Gravitational Waves

The above theoretical results could be probed against the stochastic background of gravitational waves.Depending on its origin, the stochastic background can be broadly divided into

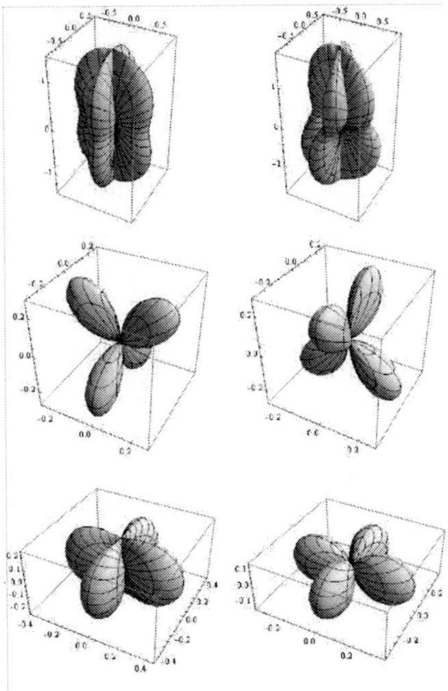

Figure 5.3. Plots along the panel lines from left to right of angular pattern functions of a detector for each polarization. From left plus mode F_+, cross mode $F_\times$, B mode F_B, C mode F_C, D mode F_D,and scalar mode F_s. The angular pattern function of the F_B and F_C mode is the same except for a rotation.

two classes: the astrophysically generated background due to the incoherent superposition of gravitational radiation emitted by large populations of astrophysical sources that cannot be resolved individually [184], and the primordial gravitational wave background generated by processes taking place in the early stages of the Universe. A primordial component of such background is especially interesting, since it would carry unique information about the state of the primordial Universe. The production physical process has been analyzed, for example, in [14, 15, 204] but only for the first two tensorial components of Eq. (5.61) for standard General Relativity. Actually the process can be improved considering all the components. Before starting with the analysis, it has to be emphasized that, considering a stochastic background of gravitational waves, it can be described and characterized by a dimensionless spectrum (see the definition [13, 14, 204, 288])

$$\Omega^A_{gw}(f) = \frac{1}{\rho_c}\frac{d\rho^A_{gw}}{d\ln f}, \tag{5.65}$$

where

$$\rho_c \equiv \frac{3H_0^2}{8\pi G} \tag{5.66}$$

is the (today) critical energy density of the Universe, H_0 the today observed Hubble expansion rate, and $d\rho_{sgw}$ is the energy density of the part of the gravitational radiation contained in the frequency range f to $f+df$.

$$\rho_{gw} = \int_0^\infty df\,\tilde{\rho}_{gw}(f)\,. \tag{5.67}$$

where $\tilde{\rho}_{GW}$ is the GWs energy density per unit frequency. $\Omega_{gw}(f)$ is related to $S_h(f)$ by [15,288]

$$\Omega^A_{gw}(f) = \left(\frac{4\pi^2}{3H_0^2}\right) f^3 S^A_h(f)\,. \tag{5.68}$$

Note that the above definition is different from that in the literature [15,288], by a factor of 2, since it is defined for each polarization. It is convenient to represent the energy density with the form $h_0^2\Omega_{gw}(f)$ by parametrizing the Hubble constant as $H_0 = 100h_0\,\mathrm{km\,s^{-1}\,Mpc^{-1}}$. Then, the gravitational waves stochastic background energy density of all modes can be written as

$$\Omega^A_{gw} \equiv \Omega^+_{gw} + \Omega^\times_{gw} + \Omega^B_{gw} + \Omega^C_{gw} + \Omega^D_{gw} + \Omega^s_{gw} \tag{5.69}$$

we can split Ω^A_{gw} as a part arising from General Rrelativity

$$\Omega^{GR}_{gw} = \Omega^+_{gw} + \Omega^\times_{gw}\,, \qquad \Omega^+_{gw} = \Omega^\times_{gw} \tag{5.70}$$

a part from higher-order-gravity

$$\Omega^{HOG}_{gw} = \Omega^B_{gw} + \Omega^C_{gw} + \Omega^D_{gw}\,, \qquad \Omega^B_{gw} = \Omega^C_{gw} = \Omega^D_{gw} \tag{5.71}$$

and a scalar part Ω^s_{gw}.

We are considering now standard units and study only the modes which arise from higher order theory.

The existence of a relic stochastic background of gravitational waves is a consequence of general assumptions. Essentially it derives from basic principles of Quantum Field Theory and General Relativity. The strong variations of gravitational field in the early Universe amplifies the zero-point quantum fluctuations and produces relic gravitational waves. It is well known that the detection of relic gravitational waves is the only way to learn about the evolution of the very early Universe, up to the bounds of the Planck epoch and the initial singularity [13,14,204,288]. It is very important to stress the unavoidable and fundamental character of such a mechanism. It directly derives from the inflationary scenario [206,426], which fit well the WMAP data in particularly good agreement with almost exponential inflation and spectral index ≈ 1, [45,380].

A remarkable fact about the inflationary scenario is that it contains a natural mechanism which gives rise to perturbations for any field. It is important for our aims that such a mechanism provides also a distinctive spectrum for relic scalar gravitational waves. These perturbations in inflationary cosmology arise from the most basic quantum mechanical effect: the uncertainty principle. In this way, the spectrum of relic gravitational waves that we could detect today is nothing else but the adiabatically-amplified zero-point fluctuations [14,204]. The calculation for a simple inflationary model can be performed for the scalar field component of Eq. (5.61). Let us assume that the early Universe is described an inflationary de Sitter phase emerging in a radiation dominated phase [13, 14, 204]. The conformal metric element is

$$ds^2 = a^2(\eta)[d\eta^2 - d\vec{x}^2 - h_{\mu\nu}(\eta, \vec{x})dx^\mu dx^\nu], \tag{5.72}$$

where, for a purely GW the metric perturbation (5.61) reduces to

$$h_{\mu\nu} = h_A e^{(A)}_{\mu\nu}. \tag{5.73}$$

where $A = +, \times, B, C, D$, and s. Let us assume a phase transition between a de Sitter and a radiation-dominated phase [14, 204], we have: η_1 is the inflation-radiation transition conformal time and η_0 is the value of conformal time today. If we express the scale factor in terms of comoving time $cdt = a(t)d\eta$, we have

$$a(t) \propto \exp(H_{ds}t), \qquad a(t) \propto \sqrt{t} \tag{5.74}$$

for the de Sitter and radiation phases respectively. In order to solve the horizon and flatness problems, the condition $\dfrac{a(\eta_0)}{a(\eta_1)} > 10^{27}$ has to be satisfied. The relic scalar-tensor GWs are the weak perturbations $h_{\mu\nu}(\eta, \vec{x})$ of the metric (5.73) which can be written in the form

$$h_{\mu\nu} = e^{(A)}_{\mu\nu}(\hat{k})X(\eta)\exp(i\vec{k}\cdot\vec{x}), \tag{5.75}$$

in terms of the conformal time η where $\vec{k}$ is a constant wavevector. From Eq. (5.75), the component is

$$\Phi(\eta, \vec{k}, \vec{x}) = X(\eta)\exp(i\vec{k}\cdot\vec{x}). \tag{5.76}$$

Assuming $Y(\eta) = a(\eta)X(\eta)$, from the Klein-Gordon equation in the Friedmann-Robertson-Walker metric, one gets

$$Y'' + \left(|\vec{k}|^2 - \frac{a''}{a}\right)Y = 0 \tag{5.77}$$

where the prime $'$ denotes derivative with respect to the conformal time. The solutions of Eq. (5.77) can be expressed in terms of Hankel functions in both the inflationary and radiation dominated eras, that is:
For $\eta < \eta_1$

$$X(\eta) = \frac{a(\eta_1)}{a(\eta)}[1 + iH_{ds}\omega^{-1}]\exp(-ik(\eta - \eta_1)), \tag{5.78}$$

for $\eta > \eta_1$

$$X(\eta) = \frac{a(\eta_1)}{a(\eta)}[\alpha\exp(-ik(\eta - \eta_1)) + \beta\exp(ik(\eta - \eta_1))], \tag{5.79}$$

where $\omega = ck/a$ is the angular frequency of the wave (which is function of the time being $k = |\vec{k}|$ constant), α and β are time-independent constants which we can obtain demanding that both X and $dX/d\eta$ are continuous at the boundary $\eta = \eta_1$ between the inflationary and the radiation dominated eras. By this constraint, we obtain

$$\alpha = 1 + i\frac{\sqrt{H_{ds}H_0}}{\omega} - \frac{H_{ds}H_0}{2\omega^2}, \qquad \beta = \frac{H_{ds}H_0}{2\omega^2} \tag{5.80}$$

In Eqs. (5.80), $\omega = ck/a(\eta_0)$ is the angular frequency as observed today, $H_0 = c/\eta_0$ is the Hubble expansion rate as observed today. Such calculations are referred in literature as the Bogoliubov coefficient methods [14, 204].

In an inflationary scenario, every classical or macroscopic perturbation is damped out by the inflation, i.e. the minimum allowed level of fluctuations is that required by the uncertainty principle. The solution (5.78) corresponds to a de Sitter vacuum state. If the period of inflation is long enough, the today observable properties of the Universe should be indistinguishable from the properties of a Universe started in the de Sitter vacuum state. In the radiation dominated phase, the eigenmodes which describe particles are the coefficients of α, and these which describe antiparticles are the coefficients of β. Thus, the number of particles created at angular frequency ω in the radiation dominated phase is given by

$$N_\omega = |\beta_\omega|^2 = \left(\frac{H_{ds}H_0}{2\omega^2}\right)^2. \tag{5.81}$$

Now it is possible to write an expression for the energy density of the stochastic scalar-tensor relic gravitons background in the frequency interval $(\omega, \omega + d\omega)$ for each mode as

$$d\rho^A_{gw} = \hbar\omega\left(\frac{\omega^2 d\omega}{2\pi^2 c^3}\right)N_\omega = \frac{\hbar H^2_{ds}H^2_0}{8\pi^2 c^3}\frac{d\omega}{\omega} = \frac{\hbar H^2_{ds}H^2_0}{8\pi^2 c^3}\frac{df}{f}, \tag{5.82}$$

where f, as above, is the frequency in standard comoving time. Eq. (5.82) can be rewritten in terms of the today and de Sitter value of energy density being

$$H_0^2 = \frac{8\pi G\rho_c}{3c^2}, \qquad H^2_{ds} = \frac{8\pi G\rho_{ds}}{3c^2}. \tag{5.83}$$

Introducing the Planck density $\rho_{Planck} = \dfrac{c^7}{\hbar G^2}$ the spectrum is given by

$$\Omega^A_{gw}(f) = \frac{1}{\rho_c}\frac{d\rho_{gw}}{d\ln f} = \frac{f}{\rho_c}\frac{d\rho_{gw}}{df} = \frac{8}{9}\frac{\rho_{ds}}{\rho_{Planck}}. \tag{5.84}$$

At this point, some comments are in order. First of all, such a calculation works for a simplified model that does not include the matter dominated era. If also such an era is also included, the redshift at the equivalence epoch has to be considered. Taking into account also results in [15], we get

$$\Omega^A_{gw}(f) = \frac{8}{9}\frac{\rho_{ds}}{\rho_{Planck}}(1 + z_{eq})^{-1}, \tag{5.85}$$

for the waves which, at the epoch in which the Universe becomes matter dominated, have a frequency higher than H_{eq}, the Hubble parameter at equivalence. This situation corresponds to frequencies $f > (1 + z_{eq})^{1/2} H_0$. The redshift correction in Eq. (5.85) is needed since the today observed Hubble parameter H_0 would result different without a matter dominated contribution. At lower frequencies, the spectrum is given by [14, 204]

$$\Omega_{gw}(f) \propto f^{-2}. \tag{5.86}$$

As a further consideration, let us note that the results (5.84) and (5.85), which are not frequency dependent, do not work correctly in all the range of physical frequencies. For waves with frequencies less than today observed H_0, the notion of energy density has no sense, since the wavelength becomes longer than the Hubble scale of the Universe. In analogous way, at high frequencies, there is a maximal frequency above which the spectrum rapidly drops to zero. In the above calculation, the simple assumption that the phase transition from the inflationary to the radiation dominated epoch is instantaneous has been made. In the physical Universe, this process occurs over some time scale $\Delta\tau$, being

$$f_{max} = \frac{a(t_1)}{a(t_0)}\frac{1}{\Delta\tau}, \tag{5.87}$$

which is the redshifted rate of the transition. In any case, Ω^A_{gw} drops rapidly. The two cutoffs at low and high frequencies for the spectrum guarantee that the total energy density of the relic gravitons is finite. These results can be quantitatively constrained considering the recent WMAP release. Nevertheless, since the spectrum falls off $\propto f^{-2}$ at low frequencies, this means that today, at LIGO-VIRGO and LISA frequencies, one gets for the General Relativity part [288], [71]

$$\Omega^{GR}_{gw}(f)h^2_{100} < 2 \times 10^{-6}. \tag{5.88}$$

for the higher-order-gravity part

$$\Omega^{HOG}_{gw}(f)h^2_{100} < 6.7 \times 10^{-9}. \tag{5.89}$$

and for the scalar part

$$\Omega^{s}_{gw}(f)h^2_{100} < 2.3 \times 10^{-12}. \tag{5.90}$$

It is interesting to calculate the corresponding strain at $\approx 100Hz$, where interferometers like VIRGO and LIGO reach a maximum in sensitivity [224,225]. The well known equation for the characteristic amplitude [288] adapted to one of the components of the gravitational waves can be used[2]:

$$h_A(f) \simeq 8.93 \times 10^{-19}\left(\frac{1Hz}{f}\right)\sqrt{h^2_{100}\Omega_{gw}(f)}, \tag{5.91}$$

and then we obtain for the General Relativity modes

$$h_{GR}(100Hz) < 1.3 \times 10^{-23}. \tag{5.92}$$

[2]The difference between our result and Eq. (19) in Ref. [288] is due to the fact that the latter did their calculation assuming the two polarization modes of General Relativity while we handle each mode separately, hence the $\frac{1}{\sqrt{2}}$ difference.

while for the higher-order modes

$$h_{HOG}(100Hz) < 7.3 \times 10^{-25}. \tag{5.93}$$

and for scalar modes

$$h_s(100Hz) < 2 \times 1.410^{-26}. \tag{5.94}$$

Then, since we expect a sensitivity of the order of 10^{-22} for the above interferometers at $\approx 100Hz$, we need to gain at least three orders of magnitude. Let us analyze the situation also at smaller frequencies. The sensitivity of the VIRGO interferometer is of the order of 10^{-21} at $\approx 10Hz$ and in that case it is for the General Relativity modes

$$h_{GR}(100Hz) < 1.3 \times 10^{-22}. \tag{5.95}$$

while for the higher-order modes

$$h_{HOG}(100Hz) < 7.3 \times 10^{-24}. \tag{5.96}$$

and for scalar modes

$$h_s(100Hz) < 1.4 \times 10^{-25}. \tag{5.97}$$

Still, these effects are below the sensitivity threshold to be observed. The sensitivity of the LISA interferometer will be of the order of 10^{-22} at $\approx 10^{-3}Hz$ (see [226]) and in that case it is

$$h_{GR}(100Hz) < 1.3 \times 10^{-18}. \tag{5.98}$$

while for the higher-order modes

$$h_{HOG}(100Hz) < 7.3 \times 10^{-20}. \tag{5.99}$$

and for scalar modes

$$h_s(100Hz) < 1.4 \times 10^{-21}. \tag{5.100}$$

This means that a stochastic background of relic gravitational waves could be, in principle, detected by the LISA interferometer, including the additional modes.

5.7. The Gravitational Stochastic Background "Tuned" by Extended Gravity

As we show in the above Chapters 3 and 4 the deformations can be described as generalized conformal transformations and this fact gives a straightforward physical interpretation of conformal transformations because conformally related metrics can be seen as the "background" and the "perturbed" metrics. Then space-time metric deformations can be immediately recast in terms of perturbation theory allowing a completely covariant approach to the problem of gravitational waves. In this section, we want to face the problem of how the gravitational waves stochastic background and $f(R)$ gravity can be related showing, vice-versa, that a revealed stochastic gravitational waves signal could be a powerful probe for a given effective theory of gravity. Our goal is to show that the conformal treatment of gravitational waves can be used to parameterize in a natural way $f(R)$ theories [104, 106–108]. gravitational waves are the perturbations $h_{\mu\nu}$ of the metric $g_{\mu\nu}$ which transform as 3-tensors. Following [429], the gravitational wave-equations in the transverse-traceless gauge are

$$\Box h_i^j = 0 \tag{5.101}$$

where $\Box \equiv (-g)^{-1/2}\partial_\mu(-g)^{1/2}g^{\mu\nu}\partial_\nu$ is the usual d'Alembert operator and these equations are derived from the Einstein field equations deduced from the Hilbert-Lagrangian density $\mathcal{L} = R$. Clearly matter perturbations do not appear in (5.101) since scalar and vector perturbations do not couple with tensor perturbations in Einstein equations. The Latin indexes run from 1 to 3 the Greek ones from 0 to 3. Our task is now to derive the analog of Eqs. (5.101) assuming a generic theory of gravity given by the action

$$\mathcal{A} = \frac{1}{2k}\int d^4x\sqrt{-g}f(R) \tag{5.102}$$

where, for the sake of simplicity, we have discarded matter contributions. A conformal analysis will help to this goal. In fact, assuming the conformal transformation

$$\widetilde{g}_{\mu\nu} = e^{2\Phi}g_{\mu\nu} \qquad \text{with} \qquad e^{2\Phi} = f'(R) \tag{5.103}$$

where the prime indicates the derivative with respect to the Ricci scalar R and Φ is the "conformal scalar field", we obtain the conformally equivalent Hilbert-Einstein action

$$\mathcal{A} = \frac{1}{2k}\int \sqrt{-\widetilde{g}}d^4x\left[\widetilde{R} + \mathcal{L}\left(\Phi, \Phi_{;\mu}\right)\right] \tag{5.104}$$

where $\mathcal{L}\left(\Phi, \Phi_{;\mu}\right)$ is the conformal scalar field contribution derived from

$$\widetilde{R}_{\mu\nu} = R_{\mu\nu} + 2\left(\Phi_{;\mu}\Phi_{;\nu} - g_{\mu\nu}\Phi_{;\delta}\Phi^{;\delta} - \Phi_{;\mu\nu} - \frac{1}{2}g_{\mu\nu}\Phi^{;\delta}{}_{;\delta}\right) \tag{5.105}$$

and

$$\widetilde{R} = e^{-2\Phi}\left(R - 6\Box\Phi - 6\Phi_{;\delta}\Phi^{;\delta}\right) \tag{5.106}$$

In any case, as we will see, the $\mathcal{L}\left(\Phi, \Phi_{;\mu}\right)$-term does not affect the gravitational waves tensor equations so it will not be considered any longer[3].

Starting from the action (5.104) and deriving the Einstein-like conformal equations, the gravitational waves equations are

$$\widetilde{\Box}\widetilde{h}_i^j = 0 \tag{5.107}$$

expressed in the conformal metric $\widetilde{g}_{\mu\nu}$. Since no scalar perturbation couples to the tensor part of gravitational waves, we have

$$\widetilde{h}_i^j = \widetilde{g}^{lj}\delta\widetilde{g}_{il} = e^{-2\Phi}g^{lj}e^{2\Phi}\delta g_{il} = h_i^j \tag{5.108}$$

which means that h_i^j is a conformal invariant.

[3] Actually a scalar component in gravitational radiation is often considered [288] but here we are taking into account only the genuine tensor part of stochastic background.

As a consequence, the plane-wave amplitude $h_i^j = h(t)e_i^j \exp(ik_ix^i)$, where e_i^j is the polarization tensor, are the same in both metrics. In any case, the d'Alembert operator transforms as

$$\widetilde{\Box} = e^{-2\Phi}\left(\Box + 2\Phi^{;\lambda}\partial_{;\lambda}\right) \tag{5.109}$$

and this means that the background is changing while the tensor wave amplitude not.

In order to study the cosmological stochastic background, the operator (5.109) can be specified for a Friedmann-Robertson-Walker metric and then Eq. (5.107) becomes

$$\ddot{h} + \left(3H + 2\dot{\Phi}\right)\dot{h} + k^2a^{-2}h = 0 \tag{5.110}$$

being $\Box = \frac{\partial}{\partial t^2} + 3H\frac{\partial}{\partial t}$, $a(t)$ the scale factor and k the wave number.

It is worth stressing that Eq. (5.110) applies to any $f(R)$ theory whose conformal transformation can be defined as $e^{2\Phi} = f'(R)$. The solution, i.e. the gravitational waves amplitude, depends on the specific cosmological background (i.e. $a(t)$) and the specific theory of gravity (i.e. $\Phi(t)$). For example, if we assume power law behaviors for $a(t)$ and $\Phi(t) = \frac{1}{2}\ln f'(R(t))$, that is

$$\Phi(t) = f'(R) = f'_0\,(t/t_0)^m\,, \quad a(t) = a_0\,(t/t_0)^n \tag{5.111}$$

it is easy show that general relativity is recovered for $m = 0$ while

$$n = \frac{m^2 + m - 2}{m + 2} \tag{5.112}$$

is the relation between the parameters for a generic $f(R) = f_0R^s$ where $s = 1 - \frac{m}{2}$ with $n \neq 1$. Eq. (5.110) can be recast in the form

$$\ddot{h} + (3n + m)t^{-1}\dot{h} + k^2a_0\,(t_0/t))^{2n}\,h = 0 \tag{5.113}$$

whose general solution is

$$h(t) = \left(\frac{t}{t_0}\right)^{\beta}\left[C_1J_\alpha(x) + C_2J_{-\alpha}(x)\right] \tag{5.114}$$

J_α's are Bessel functions and

$$\alpha = \frac{1 - 3n - m}{2(n-1)}\,, \quad \beta = \frac{1 - 3n - m}{2}\,, \quad x = \frac{kt^{1-n}}{1-n} \tag{5.115}$$

while t_0, C_1, C_2 are constants related to the specific values of n and m. In Fig. (5.4), some examples are given. The plots are labelled by the set of parameters $\{m, n, s\}$ which assign the time evolution of $\Phi(t)$ and $a(t)$ with respect to a given power-law theory $f(R) = f_0R^s$.

The time units are in terms of the Hubble radius H^{-1}; $n = 1/2$ is a radiation-like evolution; $n = 2/3$ is a dust-like evolution, $n = 2$ labels power-law inflationary phases and $n = -5$ is a pole-like inflation. From Eq. (5.112), a singular case is for $m = -2$ and $s = 2$. It is clear that the conformally invariant plane-wave amplitude evolution of the tensor gravitational waves strictly depends on the background.

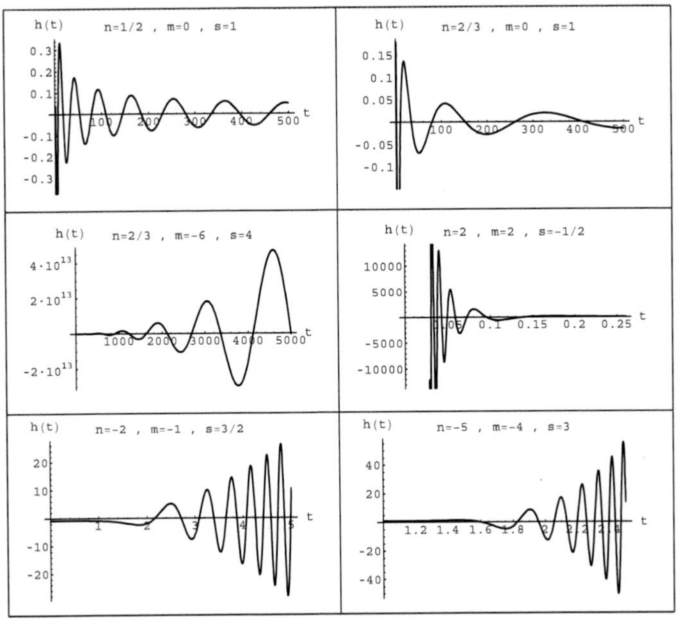

Figure 5.4. Evolution of the gravitational waves amplitude for some power-law behaviors of $a(t)$, $\Phi(t)$ and $f(R)$. The scales of time and amplitude strictly depend on the cosmological background giving a "signature" for the model.

Let us now take into account the issue of the production of gravitational waves contributing to the stochastic background. Several mechanism can be considered as cosmological populations of astrophysical sources [137], vacuum fluctuations, phase transitions [288] and so on. In principle, we could seek for contributions due to every high-energy physical process in the early phases of the Universe evolution.

It is important to distinguish processes coming from transitions like inflation, where the Hubble flow emerges in the radiation dominated phase and process, like the early star formation rates, where the production takes place during the dust dominated era. In the first case, stochastic gravitational waves background is strictly related to the cosmological model. This is the case we are considering here which is, furthermore, also connected to the specific theory of gravity. In particular, one can assume that the main contribution to the stochastic background comes from the amplification of vacuum fluctuations at the transition between an inflationary phase and the radiation dominated era. However, in any inflationary model, we can assume that the gravitational waves generated as zero-point fluctuation during the inflation undergo adiabatically damped oscillations ($\sim 1/a$) until they reach the Hubble radius H^{-1}. This is the particle horizon for the growth of perturbations. On the other hand, any other previous fluctuation is smoothed away by the inflationary expansion. The gravitational waves freeze aut for $a/k \gg H^{-1}$ and reenter the H^{-1} radius after the reheating in the Friedmann era (see also [14, 204]). The reenter in the radiation-dominated or in the dust-dominated era depends on the scale of the gravitational waves. After the reenter, Gws can be detected by their Sachs-Wolfe effect on the temperature anisotropy $\triangle T/T$

at the decoupling [351]. When Φ acts as the inflaton [389] we have $\dot{\Phi} \ll H$ during the inflation. Considering also the conformal time $d\eta = dt/a$, Eq. (5.110) reads

$$h'' + 2\frac{\chi'}{\chi}h' + k^2 h = 0 \tag{5.116}$$

where $\chi = ae^{\Phi}$ and derivation is with respect to η. Inflation means that $a(t) = a_0 \exp(Ht)$ and then $\eta = \int dt/a = (aH)^{-1}$ and $\chi'/\chi = -\eta^{-1}$. The exact solution of (5.116) is

$$h(\eta) = k^{-3/2}\sqrt{2/k}\left[C_1\left(\sin k\eta - \cos k\eta\right) + C_2\left(\sin k\eta + \cos k\eta\right)\right] \tag{5.117}$$

Inside the H^{-1} radius we have $k\eta \gg 1$. Furthermore considering the absence of gravitons in the initial vacuum state, we have only negative-frequency modes and then the adiabatic behavior is

$$h = k^{1/2}\sqrt{2/\pi}\frac{1}{aH}C\exp(-ik\eta)\,. \tag{5.118}$$

At the first horizon crossing $(aH = k)$, the averaged amplitude $A_h = (k/2\pi)^{3/2}\,|h|$ of the perturbation is

$$A_h = \frac{1}{2\pi^2}C \tag{5.119}$$

when the scale a/k grows larger than the Hubble radius H^{-1}, the growing mode of evolution is constant, that is it is frozen. This situation corresponds to the limit $-k\eta \ll 1$ in Eq. (5.117). Since Φ acts as the inflaton field, it is $\Phi \sim 0$ at reenter (after the end of inflation). Then the amplitude A_h of the wave is preserved until the second horizon crossing after which it can be observed, in principle, as an anisotropy perturbation on the Cosmic Microwave Background Radiation. It can be shown that $\triangle T/T \lesssim A_h$ as an upper limit to A_h since other effects can contribute to the background anisotropy . From this consideration, it is clear that the only relevant quantity is the initial amplitude C in Eq. (5.118) which is conserved until the reenter. Such an amplitude directly depends on the fundamental mechanism generating perturbations. Inflation gives rise to processes capable of producing perturbations as zero-point energy fluctuations. Such a mechanism depends on the adopted theory of gravitation and then $(\triangle T/T)$ could constitute a further constraint to select a suitable $f(R)$-theory.

Considering a single graviton in the form of a monochromatic wave, its zero-point amplitude is derived through the commutation relations:

$$[h(t,x), \pi_h(t,y)] = i\delta^3(x-y) \tag{5.120}$$

calculated at a fixed time t, where the amplitude h is the field and π_h is the conjugate momentum operator. Writing the Lagrangian for h

$$\widetilde{\mathcal{L}} = \frac{1}{2}\sqrt{-\widetilde{g}}\widetilde{g}^{\mu\nu}h_{;\mu}h_{;\nu} \tag{5.121}$$

in the conformal Friedmann-Robertson-Walker metric $\widetilde{g}_{\mu\nu}$ (h is conformally invariant), we obtain

$$\pi_h = \frac{\partial \widetilde{\mathcal{L}}}{\partial \dot{h}} = e^{2\Phi} a^3 \dot{h} \tag{5.122}$$

The Eq. (5.120) becomes

$$\left[h(t,x), \dot{h}(y,y)\right] = i \frac{\delta^3(x-y)}{a^3 e^{2\Phi}} \tag{5.123}$$

and the fields h and $\dot{h}$ can be expanded in terms of creation and annihilation operators

$$h(t,x) = \frac{1}{(2\pi)^{3/2}} \int d^3k \left[h(t)e^{-ikx} + h^*(t)e^{+ikx}\right], \tag{5.124}$$

$$\dot{h}(t,x) = \frac{1}{(2\pi)^{3/2}} \int d^3k \left[\dot{h}(t)e^{-ikx} + \dot{h}^*(t)e^{+ikx}\right]. \tag{5.125}$$

The commutation relations is conformal time are then

$$\left[hh'^* - h^*h'\right] = \frac{i(2\pi)^3}{a^3 e^{2\Phi}} \tag{5.126}$$

Inserting (5.118) and (5.119), we obtain $C = \sqrt{2}\pi^2 H e^{-\Phi}$ where H and Φ are calculated at the first horizon-crossing and then

$$A_h = \frac{\sqrt{2}}{2} H e^{-\Phi} \tag{5.127}$$

which means that the amplitude of gravitational waves produced during inflation directly depends on the given $f(R)$ theory being $\Phi = \frac{1}{2}\ln f'(R)$. Explicitly, it is

$$A_h = \frac{H}{\sqrt{2f'(R)}}. \tag{5.128}$$

This result deserves some discussion and can be read in two ways. From one side the amplitude of gravitational waves produced during inflation depends on the given theory of gravity that, if different from General Relativity, gives extra degrees of freedom which assume the role of inflaton field in the cosmological dynamics [389]. On the other hand, the Sachs-Wolfe effect related to the Cosmic Microwave Background Radiation temperature anisotropy could constitute a powerful tool to test the true theory of gravity at early epochs, i.e. at very high redshift. This probe, related with data at medium [110] and low redshift [109], could strongly contribute *i*) to reconstruct cosmological dynamics at every scale; *ii*) to further test General Relativity or to rule out it against alternative theories, *iii*) to give constraints on the gravitational waves stochastic background, if $f(R)$ theories are independently probed at other scales.

Chapter 6

Probing the Post Newtonian Limit

6.1. The Problem of Newtonian Limit in Extended Gravity

In this Chapter, we shall study the Newtonian limit of fourth-order gravity theories in which extensions of the Hilbert-Einstein Lagrangian are considered. As we shall see, the generalization to extended Lagrangians means that the standard Newtonian limit has to be corrected. This fact could constitute a major test bed for any theory of gravity also at local scales. We are going to focus on the weak field limit within the metric approach. At this point it is useful to remind that it was already shown in [69] that different variational procedures do not lead to equivalent results in the case of quadratic order Lagrangians.

In principle, any alternative or extended theory of gravity should allow to recover positive results of General Relativity, for example in a weak limit regime, then starting from the Hilbert-Einstein Lagrangian

$$\mathcal{L}_0 = R, \tag{6.1}$$

the following terms

$$\mathcal{L}_1 = R^2, \tag{6.2}$$

$$\mathcal{L}_2 = R_{\alpha\beta}R^{\alpha\beta}, \tag{6.3}$$

$$\mathcal{L}_3 = R_{\alpha\beta\mu\nu}R^{\alpha\beta\mu\nu}, \tag{6.4}$$

and combinations of them, represent the obvious minimal choices for an extended gravity theory with respect to General Relativity. Since the variational derivative of $\mathcal{L}_3$ can be linearly expressed [33, 168, 259, 403] via the variational derivatives of $\mathcal{L}_1$ and $\mathcal{L}_2$, one may omit $\mathcal{L}_3$ in the final Lagrangian of a fourth-order theory without loss of generality. We will consider the Newtonian limit of the combined Lagrangian (6.1)-(6.2)-(6.3), as a straightforward generalization of the Einstein theory which is obviously recovered in low curvature regimes.

6.2. The Field Equations and the Newtonian Limit

We are interested to achieve the correct Newtonian limit of gravity theories with quadratic Lagrangians in the curvature invariants. This result can be achieved under two main hypotheses: *i*) asking for low velocities with respect to the light speed and *ii*) asking for week fields. By these requests, the metric tensor is independent of time and second order perturbation terms can be discarded in the field equations (see also [114] for details). It is worth stressing that the Newtonian limit of any relativistic theory of gravity is related to such hypotheses and it is a misunderstanding to consider only the recovering of the Newtonian potential. In other words, a more general theory of gravity gives rise, in the Newtonian limit, to gravitational potentials which can be very different with respect to the standard Newtonian one. We shall discuss the field equations in the Newtonian limit to show that the fourth-order contributions to the potential cannot be trivially discarded.

6.2.1. General Form of the Field Equations

Let us now come back to the choices displayed in Eqs. (6.1)-(6.2)-(6.3)-(6.4), for which the l.h.s. of the field equations takes the general form

$$
\begin{aligned}
{}^{0}H_{\mu\nu} &= R_{\mu\nu} - \frac{1}{2}g_{\mu\nu}R, && (6.5)\\
{}^{1}H_{\mu\nu} &= 2RR_{\mu\nu} - \frac{1}{2}g_{\mu\nu}R^2 - 2R_{;\mu\nu} + 2g_{\mu\nu}\Box R, && (6.6)\\
{}^{2}H_{\mu\nu} &= 2R_{\mu}{}^{\alpha}R_{\nu\alpha} - \frac{1}{2}g_{\mu\nu}R_{\alpha\beta}R^{\alpha\beta} - 2R_{(\mu|}{}^{\alpha}{}_{;|\nu)\alpha} + \Box R_{\mu\nu} \\
&\quad + g_{\mu\nu}R^{\alpha\beta}{}_{;\alpha\beta}, && (6.7)\\
{}^{3}H_{\mu\nu} &= 2R_{\mu\alpha\beta\gamma}R_{\nu}{}^{\alpha\beta\gamma} - \frac{1}{2}g_{\mu\nu}R_{\alpha\beta\gamma\delta}R^{\alpha\beta\gamma\delta} + 4R_{\mu}{}^{\alpha}{}_{\nu}{}^{\beta}{}_{;(\alpha\beta)}. && (6.8)
\end{aligned}
$$

All of the three expressions in Eqs. (6.6)-(6.7)-(6.8) involve fourth order differential operators. Due to the identity

$$
{}^{1}H_{\mu\nu} - 4\,{}^{2}H_{\mu\nu} + {}^{3}H_{\mu\nu} = 0, \tag{6.9}
$$

which holds in a four-dimensional space-time [259], only two of the expressions in (6.6)-(6.7)-(6.8) are independent, and we are free to use any two independent linear combinations in our analysis. This identity gives rise to the well known Gauss - Bonnet topological invariant which recently acquired a lot of importance in cosmology as a possible source of dark energy [311]. Furthermore, for a Lagrangian comprising a general function of the Ricci scalar, we have, as discussed above

$$
{}^{f(R)}H_{\mu\nu} = \frac{df}{dR}R_{\mu\nu} - \frac{1}{2}g_{\mu\nu}f - \left(\frac{df}{dR}\right)_{;\mu\nu} + g_{\mu\nu}\Box\frac{df}{dR}. \tag{6.10}
$$

With these considerations in mind, let us consider the Newtonian limit of such a theory of gravity.

6.2.2. The Newtonian Limit

Here we are not interested in entering the theoretical discussion on how to formulate a mathematically well sound Newtonian limit of relativistic theories, for this we point the interested reader to [153,171,172,195,249,256,414]. In this section, we provide the explicit form of the field equations for the different admissible choices of Lagrangians collected in Introduction at the lowest, i.e. Newtonian, order. In the language of the post-Newtonian approximation, we are going to consider the field equations up to the order $O\left(c^{-2}\right)$, where c denotes the speed of light.

We only mention, in passing, that there has also been a discussion of a somewhat alternative way to define the Newtonian limit in higher-order theories in the recent literature, see for example [161].

Let us start from a flat background and work out the corresponding field equations and hydrodynamic equations to the Newtonian order. Our conventions are that $g_{\alpha\beta}$, with $\alpha,\beta = 0,1,2,3$, can be transformed to $\eta_{\alpha\beta} = \mathrm{diag}(1,-1,-1,-1)$ along a given curve. Latin indices i,j run from $1,2,3$, the coordinates are labelled by $x^{\alpha} = \left(x^0,x^1,x^2,x^3\right) = \left(ct,x^1,x^2,x^3\right)$. We start with the following ansatz for the metric

$$
\begin{aligned}
g_{00} &= 1-\frac{2U}{c^2}+O(c^{-4}), \\
g_{0a} &= \frac{1}{c^3}h_{0\alpha}+O(c^{-5}), \\
g_{ab} &= -\left(1+\frac{2V}{c^2}\right)\delta_{ab}+O(c^{-4}).
\end{aligned} \tag{6.11}
$$

Apparently, the orders involved in this ansatz for the line element reach beyond the Newtonian order, which we are mainly interested here.

On the matter side, i.e. r.h.s. of the field equations, we start with the general definition of the energy-momentum tensor of a perfect fluid

$$
T_{\alpha\beta} = \left(\rho c^2+\Pi\rho+p\right)u_{\alpha}u_{\beta}-pg_{\alpha\beta}, \tag{6.12}
$$

here Π denotes the internal energy density, ρ the energy density, and p the pressure. Following the procedure outlined in [230], we derive the explicit form of the energy-momentum as follows

$$
T_{00} = \rho c^2+O\left(c^{-2}\right), \tag{6.13}
$$

$$
T_{0a} = c\rho v^a+O\left(c^{-1}\right), \tag{6.14}
$$

$$
T_{ab} = \rho v^a v^b+p\delta_{ab}+O\left(c^{-2}\right). \tag{6.15}
$$

The general form of the field equations is given by

$$
H_{\mu\nu} = \frac{8\pi G}{c^4}T_{\mu\nu}, \tag{6.16}
$$

with a generalized tensor $H_{\mu\nu}$, being a combination of the expressions[1] specified in (6.5)-(6.6)-(6.7)-(6.8), which, in turn, depend on the final form of the Lagrangian.

6.2.3. The Quadratic Lagrangians and the Newtonian Limit of the Field Equations

Let us consider now the field equations, in the Newtonian limit, for the possible quadratic Lagrangians which we compare to the Newtonian limit of the standard Hilbert - Einstein Lagrangian. It is important to stress that the field equations, in Newtonian limit, are considered up to the order $O(c^{-2})$ while the vector component to the order $O(c^{-3})$ is related to the post-Newtonian limit of the theory (see [117] for details). Up to the Newtonian order the left-hand side of the field equations, i.e. (6.5)-(6.6)-(6.7)-(6.8), takes the following form[2] for the metric given in (6.11):

- The Hilbert - Einstein Lagrangian ($\mathcal{L}_0 = R$)

$$
\begin{aligned}
{}^0H_{00} &= -\frac{2}{c^2}\nabla^2 V, && (6.17)\\
{}^0H_{0a} &= 0, && (6.18)\\
{}^0H_{ab} &= \frac{1}{c^2}\left[\left(\nabla^2 V - \nabla^2 U\right)\delta_{ab} - \left(V - U\right)_{,ab}\right]; && (6.19)
\end{aligned}
$$

- The R^2 - Lagrangian ($\mathcal{L}_1 = R^2$)

$$
\begin{aligned}
{}^1H_{00} &= \frac{4}{c^2}\left(2\nabla^4 V - \nabla^4 U\right), && (6.20)\\
{}^1H_{0a} &= 0, && (6.21)\\
{}^1H_{ab} &= \frac{4}{c^2}\left[\left(\nabla^4 U - 2\nabla^4 V\right)\delta_{ab} + \left(2\nabla^2 V - \nabla^2 U\right)_{,ab}\right]; && (6.22)
\end{aligned}
$$

- The $R_{\alpha\beta}R^{\alpha\beta}$ - Lagrangian ($\mathcal{L}_2 = R_{\alpha\beta}R^{\alpha\beta}$)

$$
\begin{aligned}
{}^2H_{00} &= \frac{2}{c^2}\left(\nabla^4 V - \nabla^4 U\right), && (6.23)\\
{}^2H_{0a} &= 0, && (6.24)\\
{}^2H_{ab} &= \frac{1}{c^2}\left[\left(\nabla^4 U - 3\nabla^4 V\right)\delta_{ab} + \left(3\nabla^2 V - \nabla^2 U\right)_{,ab}\right]; && (6.25)
\end{aligned}
$$

- The $R_{\alpha\beta\gamma\delta}R^{\alpha\beta\delta\gamma}$ - Lagrangian ($\mathcal{L}_3 = R_{\alpha\beta\gamma\delta}R^{\alpha\beta\delta\gamma}$)

$$
\begin{aligned}
{}^3H_{00} &= -\frac{4}{c^2}\nabla^4 U, && (6.26)\\
{}^3H_{0a} &= 0, && (6.27)\\
{}^3H_{ab} &= -\frac{4}{c^2}\left(\nabla^4 V\delta_{ab} - \nabla^2 V_{,ab}\right). && (6.28)
\end{aligned}
$$

[1] Obviously ${}^0H_{\mu\nu}$ is the Einstein tensor.

[2] Here we made use of the following operator definition $\nabla^2 := \delta^{ab}\frac{\partial^2}{\partial x^a \partial x^b}$, as well as $\nabla^4 := \nabla^2\nabla^2$.

6.2.4. The Combined Lagrangian

Let us now combine the different terms into the same Lagrangian. This combination is the basis for our investigation. Since terms resulting from R^n with $n \geq 3$ do *not* contribute at the order $O\left(c^{-2}\right)$, the most general choice for the Lagrangian is

$$\mathcal{L} = a_1 R + a_2 R^2 + b_1 R_{\mu\nu}R^{\mu\nu} + c_1 R_{\alpha\beta\mu\nu}R^{\alpha\beta\mu\nu}. \tag{6.29}$$

Due to the identity given in Eq. (6.9), it is sufficient to study

$$\mathcal{L} = a_1 R + a_2 R^2 + b_1 R_{\mu\nu}R^{\mu\nu}, \tag{6.30}$$

in four dimension, where we introduced the constants[3]-[4] a_1, a_2, b_1. If we take into account the results from (6.13)-(6.15) as well as (6.17)-(6.25) the explicit form of the field equations (6.16) up to the Newtonian order, is

$$-2a_1\nabla^2 V + (8a_2 + 2b_1)\nabla^4 V - (4a_2 + 2b_1)\nabla^4 U = 8\pi G\rho\,, \tag{6.31}$$

$$\begin{aligned}&\left[a_1(\nabla^2 V - \nabla^2 U) - (8a_2 + 3b_1)\nabla^4 V + (4a_2 + b_1)\nabla^4 U\right]\delta_{ab}\\ &+\left[(8a_2 + 3b_1)\nabla^2 V - (4a_2 + b_1)\nabla^2 U + a_1(U - V)\right]_{,ab} = 0\end{aligned} \tag{6.32}$$

the equations which we are going to solve.

6.3. Considerations on the Field Equations in the Newtonian Limit

Now, we are going to formulate the problem to solve the field equations (6.31) - (6.32) in the most general way. It is worth noticing that the isotropic coordinates for the metric (6.11) allow to search for solutions independently of the symmetry of the physical system (which can be spherical, cylindrical etc.). The results which we are going to achieve are completely general since we will search for solutions in terms of Green functions. However, being the combined Lagrangian (6.30) built up by various terms, the field equations strictly depend on the coupling constants. As we will see, the value of such coefficients have a crucial role for the validity of the approach since, from a physical viewpoint, we have to obtain the Newtonian limit of General Relativity as soon as the quadratic corrections disappear. This aspect of the problem is not accurately faced in the literature and can lead to wrong conclusions. Our aim is to develop a method which allows to control, step by step, the Newtonian limit in agreement with the results of General Relativity. This is possible at three levels: Lagrangian, field equations and solutions. After, we will analyze the various cases of field equations considering particular values of the coefficients. Specifically, we

[3]Note that $[a_1] = [\text{length}]^0$, $[a_2] = [b_1] = [\text{length}]^2$.
[4]The coefficients of (6.29) are different from ones of (6.30).

will take into account the values where the proposed approach fails. It is interesting to note that any time the Hilbert - Einstein term is absent into the Lagrangian, the field equations are fourth order (see Table 6.1). In other words, if the Hilbert - Einstein term is not present, we do not recover the Laplace/Poisson equations.

6.3.1. The General Approach to Decouple the Field Equations

By introducing two new auxiliary functions (A and B), the Eqs. (6.31)-(6.32) become

$$\nabla^2\left\{\frac{4a_2+b_1}{2a_2+b_1}A+\frac{a_1}{2a_2+b_1}B+\nabla^2\left[\frac{2b_1(3a_2+b_1)}{a_1(2a_2+b_1)}A -\frac{2a_2}{2a_2+b_1}B\right]\right\}=8\pi G\rho\,, \tag{6.33}$$

$$\nabla^2(A+\nabla^2B)\delta_{ab}-(A+\nabla^2B)_{,ab}=0\,, \tag{6.34}$$

where A and B are linked to U and V via

$$A:=a_1(V-U)\,, \tag{6.35}$$

$$B:=(4a_2+b_1)V-(8a_2+3b_1)U\,. \tag{6.36}$$

Obviously we must require $a_1(2a_2+b_1)\neq 0$, which is the determinant of the transformations (6.35)-(6.36). Let us introduce the new function Φ defined as follows:

$$\Phi:=A+\nabla^2B. \tag{6.37}$$

At this point, we can use the new function Φ to decouple the system (6.33)-(6.34). In fact we obtain

$$-\frac{2b_1(3a_2+b_1)}{a_1(2a_2+b_1)}\nabla^6B-\frac{6a_2+b_1}{2a_2+b_1}\nabla^4B+\frac{a_1}{2a_2+b_1}\nabla^2B =8\pi G\rho-\nabla^2\tau_I\,, \tag{6.38}$$

$$\nabla^2\Phi\delta_{ab}-\Phi_{,ab}=0\,, \tag{6.39}$$

where $\tau_I:=\dfrac{4a_2+b_1}{2a_2+b_1}\Phi+\dfrac{2b_1(3a_2+b_1)}{a_1(2a_2+b_1)}\nabla^2\Phi$. We are interested in the solution of (6.38) in terms of the Green function $\mathcal{G}_I(\mathbf{x},\mathbf{x}')$ defined by

$$B(\mathbf{x})=Y_I\int d^3\mathbf{x}'\,\mathcal{G}_I(\mathbf{x},\mathbf{x}')\sigma_I(\mathbf{x}')\,, \tag{6.40}$$

where

$$\sigma_I(\mathbf{x}):=8\pi G\rho(\mathbf{x})-\nabla^2\tau_I(\mathbf{x})\,, \tag{6.41}$$

and Y_I being a constant, which we introduced for dimensional reasons. Then the set of equations (6.31)-(6.32) is equivalent to

$$\frac{2b_1(3a_2+b_1)}{a_1(2a_2+b_1)}\nabla^6_{\boldsymbol{x}}\mathcal{G}_I(\boldsymbol{x},\boldsymbol{x}')+\frac{6a_2+b_1}{2a_2+b_1}\nabla^4_{\boldsymbol{x}}\mathcal{G}_I(\boldsymbol{x},\boldsymbol{x}')$$
$$-\frac{a_1}{2a_2+b_1}\nabla^2_{\boldsymbol{x}}\mathcal{G}_I(\boldsymbol{x},\boldsymbol{x}')=-Y_I^{-1}\delta(\boldsymbol{x}-\boldsymbol{x}')\,, \quad (6.42)$$

$$\nabla^2\Phi(\boldsymbol{x})\delta_{ab}-\Phi(\boldsymbol{x})_{,ab}=0\,, \quad (6.43)$$

where $\delta(\boldsymbol{x}-\boldsymbol{x}')$ is the 3-dimensional Dirac δ-function. The general solutions of equations (6.31)-(6.32) for $U(\boldsymbol{x})$ and $V(\boldsymbol{x})$, in terms of the Green function $\mathcal{G}_I(\boldsymbol{x},\boldsymbol{x}')$ and the function $\Phi(\boldsymbol{x})$, are

$$\begin{aligned}U(\boldsymbol{x}) &= Y_I\frac{(8a_2+3b_1)\nabla^2_{\boldsymbol{x}}-a_1}{2a_1(2a_2+b_1)}\int d^3\boldsymbol{x}'\,\mathcal{G}_I(\boldsymbol{x},\boldsymbol{x}')\Big[8\pi G\rho(\boldsymbol{x}')\\ &\quad -\frac{4a_2+b_1}{2a_2+b_1}\nabla^2_{\boldsymbol{x}'}\Phi(\boldsymbol{x}')-\frac{2b_1(3a_2+b_1)}{a_1(2a_2+b_1)}\nabla^4_{\boldsymbol{x}'}\Phi(\boldsymbol{x}')\Big]\\ &\quad -\frac{8a_2+3b_1}{2a_1(2a_2+b_1)}\Phi(\boldsymbol{x})\,,\end{aligned} \quad (6.44)$$

$$\begin{aligned}V(\boldsymbol{x}) &= Y_I\frac{(4a_2+b_1)\nabla^2_{\boldsymbol{x}}-a_1}{2a_1(2a_2+b_1)}\int d^3\boldsymbol{x}'\,\mathcal{G}_I(\boldsymbol{x},\boldsymbol{x}')\Big[8\pi G\rho(\boldsymbol{x}')\\ &\quad -\frac{4a_2+b_1}{2a_2+b_1}\nabla^2_{\boldsymbol{x}'}\Phi(\boldsymbol{x}')-\frac{2b_1(3a_2+b_1)}{a_1(2a_2+b_1)}\nabla^4_{\boldsymbol{x}'}\Phi(\boldsymbol{x}')\Big]\\ &\quad -\frac{4a_2+b_1}{2a_1(2a_2+b_1)}\Phi(\boldsymbol{x})\,.\end{aligned} \quad (6.45)$$

Eqs. (6.31) - (6.32) represent a coupled set of fourth order differential equations. The total number of integration constants is eight. With the substitution (6.37), it has been possible to decouple the set of equations, but now the differential order is changed. The total differential order is the same, indeed we have one equation of sixth order (6.38), and another equation of second order (6.39), while previously we had two equations of fourth order. Obviously, the number of integration constants is conserved. The possibility to decouple the field equations (6.31) - (6.32) is strictly related to the choice to express the auxiliary field A in terms of B by inverting the relation (6.37), deriving Eq. (6.31) and reducing (6.32) to the second order.

6.3.2. Green Functions for Particular Values of the Coupling Constants

The obtained Green functions deserve some comments. First of all, we have to consider the particular values of the parameters where the general approach developed in Section 6.5.1. does not work. For example, if $b_1=0$, we have only one Yukawa-like correction. The Green function have to satisfy the equation

$$3\,\nabla^4_{\boldsymbol{x}}\mathcal{G}_I(\boldsymbol{x},\boldsymbol{x}')-\frac{a_1}{2a_2}\nabla^2_{\boldsymbol{x}}\mathcal{G}_I(\boldsymbol{x},\boldsymbol{x}')=-Y_I^{-1}\delta(\boldsymbol{x}-\boldsymbol{x}')\,, \quad (6.46)$$

obtained from the (6.42) by setting $b_1 = 0$. In this case, the Green function (Fourier transformed), is:

$$\tilde{\mathcal{G}}_I(\boldsymbol{k}) = -\frac{2a_2 Y_I^{-1}}{6a_2\boldsymbol{k}^4 + a_1\boldsymbol{k}^2}, \tag{6.47}$$

and the Lagrangian becomes: $L = a_1 R + a_2 R^2$. Since at the level of the Newtonian limit, as discussed, the powers of Ricci scalar higher then two do not contribute, we can conclude that (6.47) is the Green function for any $f(R)$ - theory at Newtonian order, if $f(R)$ is some analytical function of the Ricci scalar. The same result is achieved considering a particular choice of the constants in the theory, e.g. $b_1 = -2a_2$. In Table 6.1 (case iv), we provide the field equations for this choice and the related Green function is:

$$\tilde{\mathcal{G}}_{(2a_2\nabla^4 - a_1\nabla^2)}(\boldsymbol{k}) \propto \frac{1}{2a_2\boldsymbol{k}^4 + a_1\boldsymbol{k}^2}. \tag{6.48}$$

The spatial behavior of (6.47) - (6.48) is the same but the coefficients are different since the theories are different. The interpretation of the result is the same than that in Section 6.4.1. since we have to take into account a proper scale length. In fact Eq. (6.47) presents a null pole for $\mathbf{k}^2 = 0$ which gives the standard Newtonian potential, and a pole in $\mathbf{k}^2 = \lambda_2^2$, which gives the Yukawa - like correction. Finally, we need the Green function for the differential operator ∇^4. From Table 6.1, the field equations present always a quadratic Laplacian operator (Case i excluded). This means that the equation to solve is:

$$\nabla_{\boldsymbol{x}}^4 \mathcal{G}_{(\nabla^4)}(\mathbf{x}', \mathbf{x}) = \delta(\mathbf{x} - \mathbf{x}'). \tag{6.49}$$

By introducing the variable $r = |\mathbf{x} - \mathbf{x}'| \neq 0$, we have that Eq. (6.49) becomes

$$\nabla_r^4 \mathcal{G}_{(\nabla^4)}(r) = 0 \tag{6.50}$$

with solution

$$\mathcal{G}_{(\nabla^4)}(r) = K_{I,1} + \frac{K_{I,2}}{r} + K_{I,3}r + K_{I,7}r^2, \tag{6.51}$$

where $K_{I,1}$, $K_{I,2}$, $K_{I,3}$, $K_{I,7}$ are generic integration constants.

Let us now consider the fact that the Green function has to be null at infinity. The only possible physical choice for the squared Laplacian is:

$$\tilde{\mathcal{G}}_{(\nabla^4)}(\boldsymbol{x}, \boldsymbol{x}') \propto \frac{1}{|\boldsymbol{x} - \boldsymbol{x}'|}. \tag{6.52}$$

Considering the last possibility, we will end up with a force law increasing with distance. In conclusion, we have shown the general approach to find solutions of the field equations by using the Green functions. In particular, the vacuum solutions with point-like source have been used to find out directly the potentials, however it remains the most important issue to find out solutions when we consider systems with extended matter distribution.

6.4. Solutions by the Green Functions in Spherically Symmetric Distribution of Matter

Let us now explicitly determine the gravitational potential in the inner and in the outer region of a spherically symmetric matter distribution. This is a delicate problem since the Gauss theorem is not valid for the gravity theories which we are considering. In fact, in the Newtonian limit of General Relativity, the equation for the gravitational potential, generated by a point-like source

$$\nabla_{\boldsymbol{x}}^2 \mathcal{G}_{New.mech.}(\boldsymbol{x},\boldsymbol{x}') = -4\pi\delta(\boldsymbol{x}-\boldsymbol{x}') \qquad (6.53)$$

is not satisfied by the new Green functions developed above. If we consider the flux of force lines $\boldsymbol{F}_{New.mech.}$ defined as

$$\boldsymbol{F}_{New.mech.} := -\frac{GM(\boldsymbol{x}-\boldsymbol{x}')}{|\boldsymbol{x}-\boldsymbol{x}'|^3} = -GM\boldsymbol{\nabla}_{\boldsymbol{x}}\mathcal{G}_{New.mech.}(\boldsymbol{x},\boldsymbol{x}')\,, \qquad (6.54)$$

we obtain, as standard, the Gauss theorem:

$$\int_\Sigma d\Sigma\;\boldsymbol{F}_{New.mech.}\cdot\hat{n} \propto M\,, \qquad (6.55)$$

where Σ is a generic two-dimensional surface and $\hat{n}$ its surface normal. The flux of field $\boldsymbol{F}_{New.mech.}$ on the surface Σ is proportional to the matter content M, inside to the surface independently of the particular shape of surface (Gauss theorem, or Newton theorem for the gravitational field [55]). On the other hand, if we consider the flux defined by the new Green function, its value is not proportional to the enclosed mass but depends on the particular choice of the surface:

$$\int_\Sigma d\Sigma\;\boldsymbol{F}_{New.mech.}\cdot\hat{n} \propto M_\Sigma\,. \qquad (6.56)$$

Hence M_Σ is a mass-function depending on the surface Σ. Then we have to find the solution inside/outside the matter distribution by evaluating the quantity

$$\int d^3x'\,\mathcal{G}_I(\boldsymbol{x},\boldsymbol{x}')\rho(\boldsymbol{x}')\,, \qquad (6.57)$$

and by imposing the boundary condition on the separation surface.

6.4.1. Field Equations for Particular Values of the Coupling Constants

In this subsection, we want to analyze the behavior of the field equations (6.31) - (6.32) for those values of the coupling constants a_1, a_2, b_1 where the transformations (6.35) - (6.36) do not hold. Specifically, in Table 6.1, we display several cases of the field equations (6.31) - (6.32), for different choices of the coupling constants, where the determinant of transformations (6.35) - (6.36) is zero. First of all, we have to note that, for $a_2 = b_1 = 0$ (Case i in the Table 6.1), we trivially obtain the same result of General Relativity in isotropic coordinates. Furthermore, by asking for $U = V$, the spatial equation is satisfied. It is straightforward to derive as solution the Newton potential . Particularly interesting is also

Case iv, where both terms $R_{\alpha\beta}R^{\alpha\beta}$ and R^2 give similar contributions. In other words, we have the same situation of the Lagrangian $\mathcal{L} = a_1 R + a_2 R_{\alpha\beta}R^{\alpha\beta}$ with a redefinition of the couplings. However, this Lagrangian is compatible with the transformations (6.35) - (6.36) and then the above results 6.3.1. hold. In Cases (ii - iii - v - vi - vii), the differential operator ∇^2 never appears as a linear term since the invariants R^2 and $R_{\alpha\beta}R^{\alpha\beta}$ give rise to higher order terms in the field equations. In these cases, the full field equations (not in the weak field regime) give $g_{\mu\nu}\Box R - R_{;\mu\nu}$ for the Lagrangian R^2 and $-2R_{(\mu|}{}^{\alpha}{}_{;|\nu)\alpha} + \Box R_{\mu\nu} + g_{\mu\nu}R^{\alpha\beta}{}_{;\alpha\beta}$ for $R_{\alpha\beta}R^{\alpha\beta}$ which are fourth order equations. In the weak field regime, one obtains the equations reported in Table 6.1.

6.5. Green's Functions for Spherically Symmetric Systems

We are interested in the solutions of field equations (6.16) at order $O(c^{-2})$ by using the method of Green functions. We have to stress that the method of Green's functions does not work in the general case since the field equations are non-linear. However, the Newtonian limit of the theory (based on the hypothesis that metric nonlinear terms can be discarded) allows that also the field equations result linearized). By solving the field equations with the Green function method, one obtains, as a first result, the solution in terms of gravitational potential in the point-mass case. Then by using the equations (6.44)-(6.45), obtained in the weak field limit and then in Newtonian linear approximation for a spatial distribution of matter, we obtain, in principle, the gravitational potential for a given density profile. If the matter possesses a spherical symmetry, also the Green function has to be spherically symmetric. In this case, the correlation between two points has to be a function of the radial coordinate only, that is: $\mathcal{G}(\boldsymbol{x},\boldsymbol{x}') = \mathcal{G}(|\boldsymbol{x}-\boldsymbol{x}'|)$. It is important to stress again the fact that the approach works if and only if we are in the linear approximation, i.e. in the Newtonian limit.

6.5.1. A General Green Function for the Decoupled Field Equations

Let us introduce the radial coordinate $r := |\boldsymbol{x}-\boldsymbol{x}'|$; with this choice, equation (6.42) for $r \neq 0$ becomes

$$2b_1(3a_2+b_1)\nabla_r^6\mathcal{G}_I(r) + a_1(6a_2+b_1)\nabla_r^4\mathcal{G}_I(r) - a_1^2\nabla_r^2\mathcal{G}_I(r) = 0\,, \qquad (6.58)$$

where $\nabla_r^2 = r^{-2}\partial_r(r^{-2}\partial_r)$ is the radial component of the Laplacian in polar coordinates. The solution of (6.58) is:

$$\begin{aligned}\mathcal{G}_I(r) &= K_{I,1} - \frac{1}{r}\left[K_{I,2} + \frac{b_1}{a_1}\left(K_{I,3}e^{-\sqrt{-\frac{a_1}{b_1}}r} + K_{I,4}e^{\sqrt{-\frac{a_1}{b_1}}r}\right)\right.\\ &\left. - \frac{2(3a_2+b_1)}{a_1}\left(K_{I,5}e^{-\sqrt{\frac{a_1}{2(3a_2+b_1)}}r} + K_{I,6}e^{\sqrt{\frac{a_1}{2(3a_2+b_1)}}r}\right)\right] \qquad (6.59)\end{aligned}$$

where $K_{I,1}$, $K_{I,2}$, $K_{I,3}$, $K_{I,4}$, $K_{I,5}$, $K_{I,6}$ are constants. The integration constants $K_{I,i}$ have to be fixed by imposing the boundary conditions at infinity and in the origin. A physically acceptable solution has to satisfy the condition $\mathcal{G}(\boldsymbol{x},\boldsymbol{x}') \to 0$ if $|\boldsymbol{x}-\boldsymbol{x}'| \to \infty$, then the constants

Table 6.1. Explicit form of the field equations for different choices of the coupling constants for which the determinant of the transformations (6.35)-(6.36) vanishes. Cases i, ii, iii are the Lagrangians introduced in the Sec.6.2.3. (R, R^2, $R_{\mu\nu}R^{\mu\nu}$).

Cases	Choices of a_1, a_2, b_1	Corresponding field equations
i	$a_2 = 0$ $b_1 = 0$	$\nabla^2 V = -\frac{4\pi G}{a_1}\rho\,,$ $\nabla^2\left[V-U\right]\delta_{ab}-\left[V-U\right]_{,ab}=0$
ii	$a_1 = 0$ $b_1 = 0$	$\nabla^4(2V-U) = \frac{2\pi G}{a_2}\rho\,,$ $\nabla^2\left[\nabla^2(2V-U)\right]\delta_{ab}-\left[\nabla^2(2V-U)\right]_{,ab}=0$
iii	$a_1 = 0$ $a_2 = 0$	$\nabla^4(U-V) = -\frac{4\pi G}{b_1}\rho\,,$ $\nabla^2\left[\nabla^2(U-3V)\right]\delta_{ab}-\left[\nabla^2(U-3V)\right]_{,ab}=0$
iv	$b_1 = -2a_2$	$2a_2\nabla^4 V - a_1\nabla^2 V = 4\pi G\rho\,,$ $\nabla^2\left[a_1(V-U)-2a_2\nabla^2(V-U)\right]\delta_{ab}$ $-\left[a_1(V-U)-2a_2\nabla^2(V-U)\right]_{,ab}=0$
v	$a_1 = 0$ $b_1 = -4a_2$	$\nabla^4 U = \frac{2\pi G}{a_2}\rho\,,$ $\nabla^2\left[\nabla^2 V\right]\delta_{ab}-\left[\nabla^2 V\right]_{,ab}=0$
vi	$a_1 = 0$ $b_1 = -2a_2$	$\nabla^4 V = \frac{2\pi G}{a_2}\rho\,,$ $\nabla^2\left[\nabla^2(V-U)\right]\delta_{ab}-\left[\nabla^2(V-U)\right]_{,ab}=0;$
vii	$a_1 = 0$ $b_1 = -\frac{8a_2}{3}$	$\nabla^4(2V+U) = \frac{6\pi G}{a_2}\rho\,,$ $\nabla^2\left[\nabla^2 U\right]\delta_{ab}-\left[\nabla^2 U\right]_{,ab}=0$

$K_{I,1}$, $K_{I,4}$, $K_{I,6}$ in equation (6.59) have to vanish. We note that, if $a_2 = b_1 = 0$, the Green function of the Newtonian mechanics is found. In this case, we have the complete analogy with the Electromagnetism. More precisely, when we do not consider higher-order tems than Hilbert - Einstein one in the gravitational Lagrangian, we obtain, in the Newtonian limit, a field equation analog to the electromagnetic one for the scalar component (electric potential). This means that we have the same form of the Green function [239].

To obtain the conditions on the constants $K_{I,2}$, $K_{I,3}$, $K_{I,5}$ we consider the Fourier transform of $\mathcal{G}(\boldsymbol{x},\boldsymbol{x}')$:

$$\mathcal{G}_I(\boldsymbol{x},\boldsymbol{x}') = \int \frac{d^3\boldsymbol{k}}{(2\pi)^{3/2}}\, \tilde{\mathcal{G}}_I(\boldsymbol{k})\, e^{i\boldsymbol{k}\cdot(\boldsymbol{x}-\boldsymbol{x}')} . \tag{6.60}$$

$\mathcal{G}_I(\boldsymbol{x},\boldsymbol{x}')$ depends on the nature of the poles of $|\boldsymbol{k}|$ and on the values of the arbitrary constants a_1, a_2, b_1. If we define two new quantities λ_1, $\lambda_2 \in \mathcal{R}$:

$$\lambda_1^2 := -\frac{a_1}{b_1}, \qquad \lambda_2^2 := \frac{a_1}{2(3a_2+b_1)}, \tag{6.61}$$

we obtain:

$$\mathcal{G}_I(\boldsymbol{x},\boldsymbol{x}') = \sqrt{\frac{\pi}{18}}\frac{Y_I^{-1}}{|\boldsymbol{x}-\boldsymbol{x}'|}\left[\frac{\lambda_2^2-\lambda_1^2}{\lambda_1^2\lambda_2^2} - \frac{e^{-\lambda_1|\boldsymbol{x}-\boldsymbol{x}'|}}{\lambda_1^2} + \frac{e^{-\lambda_2|\boldsymbol{x}-\boldsymbol{x}'|}}{\lambda_2^2}\right]. \tag{6.62}$$

This Green function corresponds to the one in (6.59). Obviously, we have three possibilities for the parameters λ_1 and λ_2. In fact, λ_1 and λ_2 are related to the algebraic signs of a_1, a_2, b_1 and then we can also achieve real values for such parameters. This means that we have three possibilities: both imaginary, one real and one imaginary. In Table 6.2, we provide the complete set of Green functions $\mathcal{G}_I(\boldsymbol{x},\boldsymbol{x}')$, depending on the choices of the coefficients a_2 and b_1 (with a fixed sign of a_1). The various modalities in which we obtain the Green functions are due to the various sign combinations of the arbitrary constants. In general, the parameters $\lambda_{1,2}$ indicate characteristic scale lengths where corrections to the Newtonian potential can be appreciated. It is worth noticing that, thanks to the forms of the Green functions (see Table 6.2), the Newton behavior is always asymptotically recovered. When one considers a point-like source, $\rho \propto \delta(\boldsymbol{x})$, and by setting $\Phi(\boldsymbol{x}) = 0$ the potentials (6.44)-(6.45) are proportional to $\mathcal{G}_I(\boldsymbol{x},\boldsymbol{x}')$. Without losing generality we have:

$$U(\boldsymbol{x}) \sim \frac{U_0}{|\boldsymbol{x}|} + U_1\frac{e^{-\lambda_1|\boldsymbol{x}|}}{|\boldsymbol{x}|} + U_2\frac{e^{-\lambda_2|\boldsymbol{x}|}}{|\boldsymbol{x}|}, \tag{6.63}$$

where U_0, U_1, U_2 are some integration constants. An analogous behavior is obtained for the potential $V(\boldsymbol{x})$. We note that in the vacuum case we found a Yukawa-like corrections to Newtonian mechanics but with two scale lengths related to the quadratic corrections in the Lagrangian (6.30) (see also the above expressions (6.61)). This behavior is strictly linked to the sixth order of (6.42), which depends on the coupled form of the system of equations (6.31)-(6.32). In fact if we consider the Fourier transform of the potentials U and V:

$$U(\boldsymbol{x}) = \int \frac{d^3\boldsymbol{k}}{(2\pi)^{3/2}}\, \tilde{u}(\boldsymbol{k})\, e^{i\boldsymbol{k}\cdot\boldsymbol{x}}, \quad V(\boldsymbol{x}) = \int \frac{d^3\boldsymbol{k}}{(2\pi)^{3/2}}\, \tilde{v}(\boldsymbol{k})\, e^{i\boldsymbol{k}\cdot\boldsymbol{x}}, \tag{6.64}$$

the solutions of equations (6.31)-(6.32) are

$$U(\boldsymbol{x}) = \int \frac{d^3\boldsymbol{k}}{(2\pi)^{3/2}}\frac{4\pi G[a_1+(8a_2+3b_1)\boldsymbol{k}^2]\tilde{\rho}(\boldsymbol{k})e^{i\boldsymbol{k}\cdot\boldsymbol{x}}}{\boldsymbol{k}^2(a_1-b_1\boldsymbol{k}^2)[a_1+2(3a_2+b_1)\boldsymbol{k}^2]}, \tag{6.65}$$

Table 6.2. The complete set of Green functions for equations (6.42). The scale lengths are: $\lambda_1 := |a_1/b_1|^{1/2}$, $\lambda_2 := |a_1/2(3a_2+b_1)|^{1/2}$. It is possible to have a further choice for the scale lengths which turns out to be dependent on the two knows length scales. In fact, if we perform the substitution $\lambda_1 \rightleftharpoons \lambda_2$, we obtain a fourth choice. In addition, for a correct Newtonian component, we assumed $a_1 > 0$. In fact when $a_2 = b_1 = 0$ the field equations (6.31) and (6.32) give us the Newtonian theory of gravity if $a_1 = 1$.

Cases	Choices of a_2, b_1	Green function $\mathcal{G}_I(\boldsymbol{x},\boldsymbol{x}')$
viii	$b_1 < 0$ $3a_2+b_1 > 0$	$\sqrt{\frac{\pi}{18}}\frac{Y_I^{-1}}{\lvert\boldsymbol{x}-\boldsymbol{x}'\rvert}\left[\frac{\lambda_2^2-\lambda_1^2}{\lambda_1^2\lambda_2^2}-\frac{e^{-\lambda_1\lvert\boldsymbol{x}-\boldsymbol{x}'\rvert}}{\lambda_1^2}+\frac{e^{-\lambda_2\lvert\boldsymbol{x}-\boldsymbol{x}'\rvert}}{\lambda_2^2}\right]$
ix	$b_1 > 0$ $3a_2+b_1 < 0$	$\sqrt{\frac{\pi}{18}}\frac{Y_I^{-1}}{\lvert\boldsymbol{x}-\boldsymbol{x}'\rvert}\left[\frac{\lambda_1^2-\lambda_2^2}{\lambda_1^2\lambda_2^2}+\frac{\cos(\lambda_1\lvert\boldsymbol{x}-\boldsymbol{x}'\rvert)}{\lambda_1^2}-\frac{\cos(\lambda_2\lvert\boldsymbol{x}-\boldsymbol{x}'\rvert)}{\lambda_2^2}\right]$
x	$b_1 < 0$ $3a_2+b_1 < 0$	$\sqrt{\frac{\pi}{18}}\frac{Y_I^{-1}}{\lvert\boldsymbol{x}-\boldsymbol{x}'\rvert}\left[\frac{\lambda_1^2+\lambda_2^2}{\lambda_1^2\lambda_2^2}-\frac{e^{-\lambda_1\lvert\boldsymbol{x}-\boldsymbol{x}'\rvert}}{\lambda_1^2}-\frac{\cos(\lambda_2\lvert\boldsymbol{x}-\boldsymbol{x}'\rvert)}{\lambda_2^2}\right]$

$$V(\boldsymbol{x}) = \int \frac{d^3\boldsymbol{k}}{(2\pi)^{3/2}}\frac{4\pi G[a_1+(4a_2+b_1)\boldsymbol{k}^2]\tilde{\rho}(\boldsymbol{k})e^{i\boldsymbol{k}\cdot\boldsymbol{x}}}{\boldsymbol{k}^2(a_1-b_1\boldsymbol{k}^2)[a_1+2(3a_2+b_1)\boldsymbol{k}^2]}, \tag{6.66}$$

where $\tilde{\rho}(\boldsymbol{k})$ is the Fourier transform of the matter density. It is possible to show, by applying the Fourier transform to the potentials U and V, that the poles in Eqs. (6.65) - (6.66) are always three.

Finally, if $\tilde{\rho}(\boldsymbol{k}) = \frac{M}{(2\pi)^{3/2}}$ (the Fourier transform of a point-like source) the solutions (6.65)-(6.66) are similar to (6.63). In fact, if we suppose that $b_1 \neq 0$ and $3a_2+b_1 \neq 0$, the solutions (6.65) - (6.66) are

$$U(\boldsymbol{x}) = \frac{GM}{a_1|\boldsymbol{x}|}\left(1-\frac{4}{3}e^{-\lambda_1|\boldsymbol{x}|}+\frac{1}{3}e^{-\lambda_2|\boldsymbol{x}|}\right), \tag{6.67}$$

$$V(\boldsymbol{x}) = \frac{GM}{a_1|\boldsymbol{x}|}\left(1-\frac{2}{3}e^{-\lambda_1|\boldsymbol{x}|}-\frac{1}{3}e^{-\lambda_2|\boldsymbol{x}|}\right). \tag{6.68}$$

These are the main results of this Section. it is clear that Newtonian potential is nothing else but a particular case related to the specific lagrangian $\mathcal{L} = \mathcal{R}$. As soon as this last condition is relaxed assuming more general theories, corrections emerge. In particular Yukawa-like theories are present.

6.6. Fourth Order Gravity and Experimental Constraints on Eddington Parameters

Starting from the definitions of PPN-parameters in term of a generic analytic function $f(R)$ and its derivatives, we want to now discuss a class of fourth order theories, compatible with data, by means of an inverse procedure which allows to compare PPN-conditions with data. As a matter of fact, it is possible to show that a third order polynomial, in the Ricci scalar, is compatible with observational constraints on PPN-parameters. The degree of deviation from General Relativity depends on the experimental estimate of PPN-parameters.

A useful method to take into account deviation with respect to General Relativity is to develop expansions about the General Relativity solutions up to some perturbation orders. A standard approach as discussed in Chapter 1 is the Parameterized-Post-Newtonian (PPN) expansion of the Schwarzschild metric. In isotropic coordinates, it is

$$ds^2 = \left(\frac{1-\frac{r_g}{4\tilde{r}}}{1+\frac{r_g}{4\tilde{r}}}\right)^2 dt^2 - \left(1+\frac{r_g}{4\tilde{r}}\right)^4 (d\tilde{r}^2+\tilde{r}^2 d\Omega^2) \tag{6.69}$$

where $r_g = 2GM/c^2$ is the Schwarzschild radius. Eddington parameterized deviations with respect to General Relativity, considering a Taylor series in term of $r_g/\tilde{r}$ assuming that in Solar System, the limit $r_g/\tilde{r}) \ll 1$ holds [436]. The resulting metric is

$$ds^2 \simeq \left[1-\alpha\frac{r_g}{\tilde{r}}+\frac{\beta}{2}\left(\frac{r_g}{\tilde{r}}\right)^2+\ldots\right]dt^2 - \left(1+\gamma\frac{r_g}{\tilde{r}}+\ldots\right)\left(d\tilde{r}^2+\tilde{r}^2 d\Omega^2\right) \tag{6.70}$$

where α, β and γ are unknown dimensionless parameters (Eddington parameters) which parameterize deviations with respect to General Relativity. The reason to carry out this expansion up to the order $(r_g/\tilde{r})^2$ in g_{00} and only to the order $(r_g/\tilde{r})$ in g_{ij} is that, in applications to celestial mechanics, g_{ij} always appears multiplied by an extra factor $v^2 \backsim (M/\tilde{r})$. It is evident that the standard General Relativity solution for a spherically symmetric gravitational system in vacuum is obtained for $\alpha = \beta = \gamma = 1$ giving the Schwarzschild solution. Actually, the parameter α can be settled to the unity due to the mass definition of the system itself [436]. As a consequence, the expanded metric (6.70) can be recast in the form :

$$ds^2 \simeq \left[1-\frac{r_g}{r}+\frac{\beta-\gamma}{2}\left(\frac{r_g}{r}\right)^2+\ldots\right]dt^2 - \left[1+\gamma\frac{r_g}{r}+\ldots\right]dr^2 - r^2 d\Omega^2\,, \tag{6.71}$$

where we have restored the standard spherical coordinates by means of the transformation $r = \tilde{r}\left(1+\frac{r_g}{4\tilde{r}}\right)^2$. The two parameters β, γ have a physical interpretation. The parameter γ measures the amount of curvature of space generated by a body of mass M at radius r. In fact, the spatial components of the Riemann curvature tensor are, at post-Newtonian order,

$$R_{ijkl} = \frac{3}{2}\gamma\frac{r_g}{r^3}N_{ijkl} \tag{6.72}$$

independently of the gauge choice, where N_{ijkl} represents the geometric tensor properties (e.g. symmetries of the Riemann tensor and so on). On the other side, the parameter

β measures the amount of non-linearity ($\sim (r_g/r)^2$) in the g_{00} component of the metric. However, this statement is valid only in the standard post-Newtonian gauge.

If one takes into account a more general theory of gravity, the calculation of the PPN-limit can be performed following a well defined pipeline which straightforwardly generalizes the standard General Relativity case [436]. A significant development in this sense has been pursued by Damour and Esposito-Farese [147, 149–152] which have approached to the calculation of the PPN-limit of scalar-tensor gravity by means of a conformal transformation $\tilde{g}_{\mu\nu} = F(\phi)g_{\mu\nu}$ to the standard Einstein frame. In fact, a general scalar-tensor theory

$$\mathcal{A} = \int d^4x\sqrt{-g}\left[F(\phi)R + \frac{1}{2}\phi_{;\mu}\phi^{;\mu} - V(\phi) + \mathcal{L}_m\right], \tag{6.73}$$

where $V(\phi)$ is the self-interaction potential and $F(\phi)$ the non-minimal coupling, can be recast as :

$$\tilde{\mathcal{A}} = \int d^4x\sqrt{-\tilde{g}}\left[-\frac{1}{2}\tilde{R} + \frac{1}{2}\varphi_{;\mu}\varphi^{;\mu} - \tilde{V}(\varphi) + \tilde{\mathcal{L}}_m\right] \tag{6.74}$$

where

$$\left(\frac{d\varphi}{d\phi}\right)^2 = \frac{3}{4}\left(\frac{d\ln F(\phi)}{d\phi}\right)^2 + \frac{1}{2F(\phi)}, \tag{6.75}$$

$8\pi G = 1$, $\tilde{V}(\varphi) = F^{-2}(\phi)V(\phi)$, $\tilde{\mathcal{L}}_m = F^{-2}(\phi)\mathcal{L}_m$. The first consequence of such a transformation is that now the non-minimal coupling is transferred to the ordinary-matter sector. In fact, the Lagrangian $\tilde{\mathcal{L}}_m$ is dependent not only on the conformally transformed metric $\tilde{g}_{\mu\nu}$ and matter fields but it is even characterized by the coupling function $F(\phi)$ [387]. This scheme provides several interesting results up to obtain an intrinsic definition of γ, β in term of the non-minimal coupling function $F(\phi)$. The analogy between scalar-tensor gravity and higher order theories of gravity has been widely demonstrated [373, 403, 423]. Here an important remark is in order. The analogy between scalar-tensor gravity and fourth order gravity, although mathematically straightforward, requires a careful physical analysis. Recasting fourth-order gravity as a scalar-tensor theory, often the following steps, in terms of a generic scalar field ψ, are considered

$$f(R) + \mathcal{L}_m \to F'(\psi)R + F(\psi) - F'(\psi)\psi + \mathcal{L}_m \to F'(\psi)R - V(\psi) + \mathcal{L}_m, \tag{6.76}$$

where, by analogy, $\psi \to R$ and the "potential" is $V(\psi) = F(\psi) - F'(\psi)\psi$. Clearly the kinetic term is not present so that (6.76) is usually referred as a Brans-Dicke description of $f(R)$ gravity where $\omega_{BD} = 0$. This is the so-called O'Hanlon Lagrangian [403]. However, the typical Brans-Dicke action is

$$\mathcal{A} = \int d^4x\sqrt{-g}\left[\psi R - \omega_{BD}\frac{\psi_{;\mu}\psi^{;\mu}}{\psi} + \mathcal{L}_m\right], \tag{6.77}$$

where no scalar field potential is present and ω_{BD} is a constant. In summary, O'Hanlon Lagrangian has a potential but has no kinetic term, while Brans-Dicke Lagrangian has a kinetic term without potential. The most general situation is in (6.73) where we have non-minimal coupling, kinetic term, and scalar field potential. This means that fourth-order gravity and scalar tensor gravity can be "compared" only by means of conformal transformations where kinetic and potential terms are preserved. In particular, it is misleading to state that PPN-limit of fourth order gravity is bad defined since these models provide $\omega_{BD} = 0$ and this is in contrast with observations [316, 317].

Table 6.3. Summary of the three approaches: Scalar-Tensor (ST), Einstein $+\varphi$, $(E+\phi)$ and $f(R)$ and their relations at Lagrangians, field equations and solutions levels. The solutions are in the Einstein frame for the minimally coupled case while they are in Jordan frame for $f(R)$ and ST-gravity. Clearly, $f(R)$ and ST theories can be rigorously compared only recasting them in the Einstein frame.

$\mathcal{L}_{ST}$	$\longrightarrow$	$E+\phi$	$\longleftarrow$	$\mathcal{L}_{f(R)}$
$\downarrow$		$\downarrow$		$\downarrow$
ST -Eqs.	$\longrightarrow$	Einstein Eqs.	$\longleftarrow$	$f(R)$-Eqs.
$\downarrow$		$\downarrow$		$\downarrow$
J-frame Sol.	$\longrightarrow$	E-frame Sol.	$\longleftarrow$	J-frame Sol.

Scalar-tensor theories and $f(R)$ theories can be rigorously compared, after conformal transformations, in the Einstein frame where both kinetic and potential terms are present. With this consideration in mind, $F(\phi)$ and $f'(R)$ can be considered analogous quantities in Jordan frame and then the PPN limit can be developed [5].

Starting from this analogy, the PPN results for scalar-tensor gravity can be extended to fourth order gravity [119]. In fact, identifying $\phi \rightarrow R$ [423], it is possible to extend the definition of the scalar-tensor PPN-parameters [147, 149, 372] to the case of fourth order gravity :

$$\gamma - 1 = -\frac{f''(R)^2}{f'(R) + 2f''(R)^2}\,, \qquad \beta - 1 = \frac{1}{4}\left(\frac{f'(R)\cdot f''(R)}{2f'(R) + 3f''(R)^2}\right)\frac{d\gamma}{dR}\,. \tag{6.78}$$

In [119], these definitions have been confronted with the observational upper limits on γ and β coming from Mercury Perihelion Shift [365] and Very Long Baseline Interferometry [366]. Actually, it is possible to show that data and theoretical predictions from Eqs. (6.78) agree in the limits of experimental measures for several classes of fourth order theories. Such a result tells us that extended theories of gravity are not ruled out from Solar System experiments but a more careful analysis of theories against experimental limits has to be performed. A possible procedure could be to link the analytic form of a generic fourth order theory with experimental data. In fact, the matching between data and theoretical predictions, found in [119], holds provided some restrictions for the model parameters but gives no general constraints on the theory. In general, the function $f(R)$ could contain an infinite number of parameters (i.e. it can be conceived as an infinite power series [373]) while, on the contrary, the number of useful relations is finite (in our case we have only two relations). An attempt to deduce the form of the gravity Lagrangian can be to consider

[5] To be precise, conformal transformations should be operated "before" performing PPN-limit and results discussed in the same frame. A back conformal transformation, after PPN limit, could be misleading due to gauge troubles.

the relations (6.78) as differential equations for $f(R)$, so that, taking into account the experimental results, one could constrain, in principle, the model parameters by the measured values of γ and β.

6.7. Fourth Order Theories Compatible with Experimental Limits on γ and β

The idea is supposing the relations for γ and β as differential equations. This hypothesis is reasonable if the derivatives of $f(R)$ function are smoothly evolving with the Ricci scalar. Formally, one can consider the r.h.s. of the definitions (6.78) as differential relations which have to be matched with values of PPN-parameters. In other words, one has to solve the equations (6.78) where γ and β are two parameters. Based on such an assumption, on can try to derive the largest class of $f(R)$ theories compatible with experimental data. In fact, by the integration of Eqs. (6.78), one obtains a solution parameterized by β and γ which have to be confronted with the experimental quantities β_{exp} and γ_{exp}.

Assuming $f'(R)+2f''(R)^2 \neq 0$ and defining $A = \left|\frac{1-\gamma}{2\gamma-1}\right|$, we obtain from (6.78) a differential equation for $f(R)$:

$$[f''(R)]^2 = Af'(R)\,. \tag{6.79}$$

The general solution of such an equation is a third order polynomial $f(R) = aR^3 + bR^2 + cR + d$ whose coefficients have to satisfy the conditions : $a = b = c = 0$ and $d \neq 0$ (trivial solution) or $a = \frac{A}{12}, b = \pm\frac{\sqrt{Ac}}{2}$, with $c, d \neq 0$. Thus, the general solution for the non-trivial case, in natural units, reads

$$f(R) = \frac{1}{12}\left|\frac{1-\gamma}{2\gamma-1}\right|R^3 \pm \frac{\sqrt{c}}{2}\sqrt{\left|\frac{1-\gamma}{2\gamma-1}\right|}R^2 + cR + d\,. \tag{6.80}$$

It is evident that the integration constants c and d have to be compatible with General Relativity prescriptions and, eventually, with the presence of a cosmological constant. Indeed, when $\gamma \to 1$, which implies $f(R) \to cR + d$, the General Relativity - limit is recovered. As a consequence the values of these constants remain fixed ($c=1$ and $d=\Lambda$, where Λ is the cosmological constant). Therefore, the fourth order theory provided by Eq. (6.80) becomes

$$f_{\pm}(R) = \frac{1}{12}\left|\frac{1-\gamma}{2\gamma-1}\right|R^3 \pm \frac{1}{2}\sqrt{\left|\frac{1-\gamma}{2\gamma-1}\right|}R^2 + R + \Lambda\,, \tag{6.81}$$

where we have formally displayed the two branch form of the solution depending on the sign of the coefficient entering the second order term. Since the constants a, b, c, d of the general solution satisfy the relation $3ac - b^2 = 0$, one can easily verify that it gives :

$$\left.\frac{d\gamma}{dR}\right|_{f_{\pm}(R)} = -\frac{d}{dR}\left.\frac{f''(R)^2}{f'(R)+2f''(R)^2}\right|_{f_{\pm}(R)} = 0\,, \tag{6.82}$$

where the subscript $_{f_{\pm}(R)}$ refers the calculation to the solution (6.81). This result, compared with the second differential equation Eq. (6.78), implies $4(\beta - 1) = 0$, which means the compatibility of the solution even with this second relation.

6.8. Comparing with Experimental Measurements

Up to now we have discussed a family of fourth order theories (6.81) parameterized by the PPN-quantity γ; on the other hand, for this class of Lagrangians, the parameter β is compatible with General Relativity value being unity.

Table 6.4. A schematic resume of recent experimental constraints on the PPN-parameters. They are the perihelion shift of Mercury [365], the Lunar Laser Ranging [437], the upper limit coming from the Very Long Baseline Interferometry [366] and the results obtained by the estimate of the Cassini spacecraft delay into the radio waves transmission near the Solar conjunction [52].

Mercury Perihelion Shift	$\lvert 2\gamma-\beta-1\rvert < 3\times 10^{-3}$
Lunar Laser Ranging	$4\beta-\gamma-3 = -(0.7\pm 1)\times 10^{-3}$
Very Long Baseline Interf.	$\lvert\gamma-1\rvert = 4\times 10^{-4}$
Cassini Spacecraft	$\gamma-1 = (2.1\pm 2.3)\times 10^{-5}$

Now, the further step directly characterizes such a class of theories by means of the experimental estimates of γ. In particular, by fixing γ to its observational estimate γ_{exp}, we will obtain the weight of the coefficients relative to each of the non-linear terms in the Ricci scalar of the Lagrangian (6.81). In such a way, since General Relativity predictions require exactly $\gamma_{exp} = \beta_{exp} = 1$, in the case of fourth order gravity, one could to take into account small deviations from this values as inferred from experiments. Some plots can contribute to the discussion of this argument. In Fig.6.1, the Lagrangian (6.81) is plotted. It is parameterized for several values of γ compatible with the experimental bounds coming from the Mercury perihelion shift (see Table.1 and [365]). The function is plotted in the range $R \geq 0$. Since the property $f_+(R) = -f_-(-R)$ holds for the function (6.81), one can easily recover the shape of the plot in the negative region. As it is reasonable, the deviation from General Relativity becomes remarkable when scalar curvature is large.

In order to display the differences between the theory (6.81) and Hilbert-Einstein one, the ratio $f(R)/R$ is plotted in Fig.6.2. Again it is evident that the two Lagrangians differ significantly for great values of the curvature scalar. It is worth noting that the formal difference between the PPN-inspired Lagrangian and the General Relativity expression can be related to the physical meaning of the parameter γ which is the deviation from the Schwarzschild-like solution. It measures the spatial curvature of the region which one is investigating, then the deviation from the local flatness can be due to the influence of higher order contributions in Ricci scalar. On the other hand, one can reverse the argument and notice that if such a deviation is measured, it can be recast in the framework of fourth order gravity, and in particular its "amount" indicates the deviation from General Relativity. Furthermore, it is worth considering that, in the expression (6.81), the modulus of the coefficients in γ (i.e. the strength of the term) decreases by increasing the degree of R. In

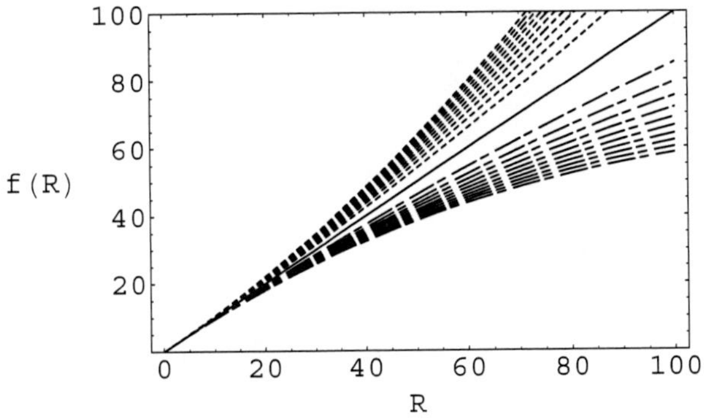

Figure 6.1. Plot of the two branch solution provided in Eq. (6.81). The $f_+(R)$ (dotted line) branch family is up to General Relativity solution (straight line), while the one indicated with $f_-(R)$ (dotted-dashed line) remains below this line. The different plots for each family refer to different values of γ fulfilling the condition $|\gamma - 1| \leq 10^{-4}$ and increased by step of 10^{-5}.

particular, the highest values of cubic and squared terms in R are, respectively, of order 10^{-4} and 10^{-2} (see Fig.6.3) then General Relativity remains a viable theory at short distances (i.e. Solar System) and low curvature regimes.

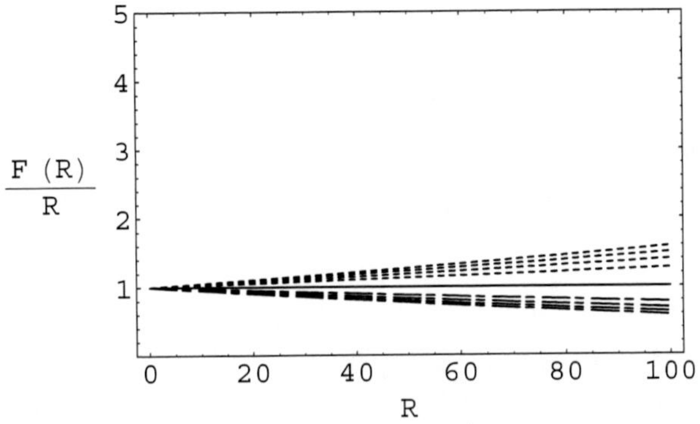

Figure 6.2. The ratio $f(R)/R$. It is shown the deviation of the fourth order gravity from General Relativity considering the PPN-limit. Dotted and dotted-dashed lines refer to the $f_+(R)$ and $f_-(R)$ branches plotted with respect to several values of γ (the step in this case is 2.5×10^{-5}).

A remark is in order at this point. The class of theories which we have discussed is a third order function of the Ricci scalar R parameterized by the experimental values of the PPN parameter γ. In principle, any analytic $f(R)$ can be compared with the Lagrangian (6.81) provided suitable values of the coefficients. However, more general results can be achieved relaxing the condition $\beta = 1$ which is an intrinsic feature for (6.81) (see for example [119]). These considerations suggest to take into account, as physical theories, functions of the Ricci scalar which slightly deviates from General Relativity, i.e. $f(R) = f_0 R^{(1+\varepsilon)}$ with ε a small parameter which indicates how much the theory deviates from General Rela-

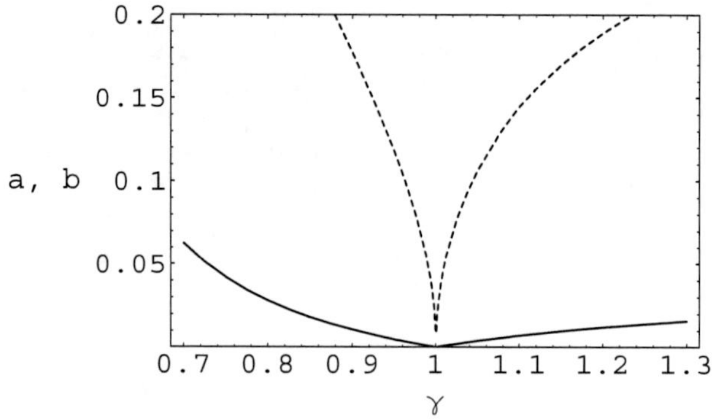

Figure 6.3. Plot of the modulus of coefficients: a (R^3) (line) and b (R^2) (dashed line). The choice of plotting the modulus of coefficients is the consequence of two solution for $f(R)$.

tivity [36]. In fact, supposing ε sufficiently small, it is possible to approximate this expression

$$f_0|R|^{(1+\varepsilon)} \simeq f_0|R|\left(1+\varepsilon\ln|R|+\frac{\varepsilon^2\ln^2|R|}{2}+\ldots\right). \tag{6.83}$$

This relation can be easily confronted with the solution (6.81) since, also in this case, the corrections have very small "strength".

6.9. $f(R)$ Viable Models

Let us consider now a class of $f(R)$ models which do not contain cosmological constant and are explicitly designed to satisfy cosmological and Solar-System constraints in given limits of the parameter space. In practice, we choose a class of functional forms of $f(R)$ capable of matching, in principle, observational data (see [110, 334] for the general approach). Firstly, the cosmological model should reproduce the Cosmic Microwave Background Radiation constraints in the high-redshift regime (which agree with the presence of an effective cosmological constant). Secondly, it should give rise to an accelerated expansion, at low redshift, according to the ΛCDM model. Thirdly, there should be sufficient degrees of freedom in the parameterization to encompass low redshift phenomena (e.g. the large scale structure) according to the observations [319]. Finally, small deviations from General Relativity should be consistent with Solar System tests. All these requirements suggest that we can assume the limits

$$\lim_{R\to\infty} f(R) = \text{constant}, \tag{6.84}$$

$$\lim_{R\to 0} f(R) = 0, \tag{6.85}$$

which are satisfied by a general class of broken power law models, proposed in [228], which are

$$f_I(R) = -m^2\frac{c_1\left(\frac{R}{m^2}\right)^n}{c_2\left(\frac{R}{m^2}\right)^n+1} \tag{6.86}$$

or otherwise written as

$$f_I(R) = R - \lambda R_c \frac{\left(\frac{R}{R_c}\right)^{2n}}{\left(\frac{R}{R_c}\right)^{2n} + 1} \tag{6.87}$$

where m is a mass scale and $c_{1,2}$ are dimensionless parameters.

Besides, another viable class of models was proposed in [391]

$$f_{II}(R) = R + \lambda R_c \left[\left(1 + \frac{R^2}{R_c^2}\right)^{-p} - 1 \right]. \tag{6.88}$$

Since $f(R=0) = 0$, the cosmological constant has to disappear in a flat space-time. The parameters $\{n, p, \lambda, R_c\}$ are constants which should be determined by experimental bounds.

Other interesting models with similar features have been studied in [24, 139, 307, 312, 314, 417]. In all these models, a de-Sitter stability point, responsible for the late-time acceleration, exists for $R = R_1$ (> 0), where R_1 is derived by solving the equation $R_1 f_{,R}(R_1) = 2f(R_1)$ [36]. For example, in the model (6.88), we have $R_1/R_c = 3.38$ for $\lambda = 2$ and $p = 1$. If λ is of the unit order, R_1 is of the same order of R_c. The stability conditions, $f_{,R} > 0$ and $f_{,RR} > 0$, are fulfilled for $R > R_1$ [391,417]. Moreover the models satisfy the conditions for the cosmological viability that gives rise to the sequence of radiation, matter and accelerated epochs [77,417].

In the region $R \gg R_c$ both classes of models (6.86) and (6.88) behave as

$$f_{III}(R) \simeq R - \lambda R_c \left[1 - (R_c/R)^{2s}\right], \tag{6.89}$$

where s is a positive constant. The model approaches ΛCDM in the limit $R/R_c \to \infty$.

Finally, let also consider the class of models [21, 185, 267]

$$f_{IV}(R) = R - \lambda R_c \left(\frac{R}{R_c}\right)^q. \tag{6.90}$$

Also in this case λ, q and R_c are positive constants (note that n, p, s and q have to converge toward the same values to match the observations). We do not consider the models whit negative q, because they suffer for instability problems associated with negative $f_{,RR}$ [16, 43, 155, 166]. In Fig.(6.4), we have plotted some of the selected models as function of $\frac{R}{R_c}$ for suitable values of $\{p, n, q, s, \lambda\}$.

Let us now estimate m_σ for the models discussed above [94]

$$m_\sigma^2 \equiv \frac{1}{2}\frac{d^2V(\sigma)}{d\sigma^2} = \frac{1}{2}\left[\frac{R}{f'(R)} - \frac{4f(R)}{(f'(R))^2} + \frac{1}{f''(R)}\right], \tag{6.91}$$

which, in the weak field limit, could induce corrections to the Newton law.

For Model I [228], when the curvature is large, we find

$$f_I(R) \sim -\frac{m^2 c_1}{c_2} + \frac{m^{2+2n} c_1}{c_2^2 R^n} + \cdots, \tag{6.92}$$

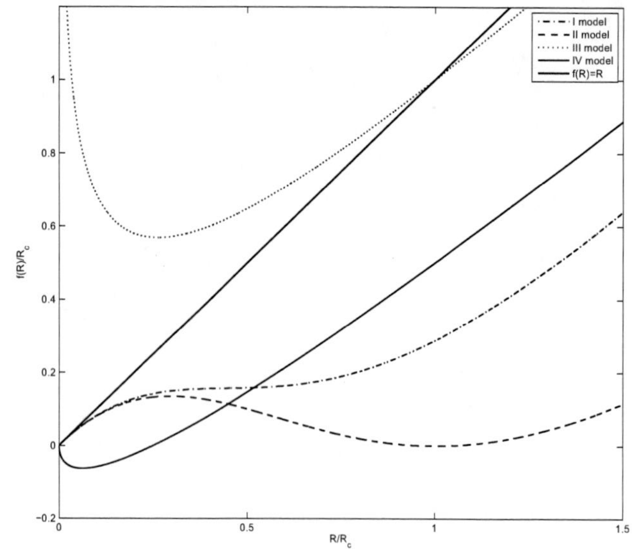

Figure 6.4. Plots of four different $f(R)$ models as function of $\frac{R}{R_c}$. Model I in Eq. (6.86) with $n = 1$ and $\lambda = 2$ (dashed line). Model II in Eq. (6.88) with $p = 2$, $\lambda = 0.95$ (dashdot line). Model III in Eq. (6.89) with $s = 0.5$ and $\lambda = 1.5$ (dotted). Model IV in Eq. (6.90) with $q = 0.5$ and $\lambda = 0.5$ (solid line). We also plot $f(R) = R$ (solid thick line) to see whether or not the stability condition $f_{,R} > 0$ is violated.

and obtain the following expression:

$$m_\sigma^2 \sim \frac{m^2 c_2^2}{2n(n+1)c_1}\left(\frac{R}{m^2}\right)^{n+2} . \tag{6.93}$$

Here the order of the mass-dimensional parameter m^2 should be $m^2 \sim 10^{-64}\,\mathrm{eV}^2$. Then in Solar System, where $R \sim 10^{-61}\,\mathrm{eV}^2$, the mass is given by $m_\sigma^2 \sim 10^{-58+3n}\,\mathrm{eV}^2$ while on the Earth atmosphere, where $R \sim 10^{-50}\,\mathrm{eV}^2$, it has to be $m_\sigma^2 \sim 10^{-36+14n}\,\mathrm{eV}^2$. The order of the radius of the Earth is $10^7\,\mathrm{m} \sim \left(10^{-14}\,\mathrm{eV}\right)^{-1}$. Therefore the scalar field σ is enough heavy if $n \gg 1$ and the correction to the Newton law is not observed, being extremely small. In fact, if we choose $n = 10$, the order of the Compton length of the scalar field σ becomes that of the Earth radius. On the other hand, in the Earth atmosphere, if we choose $n = 10$, for example, we find that the mass is extremely large:

$$m_\sigma \sim 10^{43}\,\mathrm{GeV} \sim 10^{29} \times M_{\mathrm{Planck}} . \tag{6.94}$$

Here M_{Planck} is the Planck mass. Hence, the Newton law correction should be extremely small.

In Model II

$$f_{II}(R) = -\lambda R_0 \left[1 - \left(1 + \frac{R^2}{R_0^2}\right)^{-p}\right] , \tag{6.95}$$

if R is large compared with R_0, whose order of magnitude is that of the curvature in the present universe, we find

$$f_{II}(R) = -\lambda R_0 + \lambda \frac{R_0^{2p+1}}{R^{2p}} + \cdots . \tag{6.96}$$

By comparing Eq. (6.96) with Eq. (6.92), if the curvature is large enough when compared with R_0 or m^2, as in the Solar System or on the Earth, we can set the following identifications:

$$\lambda R_0 \leftrightarrow \frac{m^2 c_1}{c_2} , \quad \lambda R_0^{2p+1} \leftrightarrow \frac{m^{2+2n} c_1}{c_2^2} , \quad 2p \leftrightarrow n . \tag{6.97}$$

We have $41m^2 \sim R_0$. Then, if p is large enough, there is no correction to the Newton law as in Model I given by Eq. (6.87).

Let us now discuss the instability of fluid matter proposed in [166], which may appear if the matter-energy density (or the scalar curvature) is large enough when compared with the average density the Universe, as it is inside the Earth. Considering the trace of the above field equations and with a little algebra, one obtains

$$\Box R + \frac{f^{(3)}(R)}{f^{(2)}(R)} \nabla_\rho R \nabla^\rho R + \frac{f'(R)R}{3f^{(2)}(R)} - \frac{2F(R)}{3F^{(2)}(R)} = \frac{\kappa^2}{6f^{(2)}(R)} T . \tag{6.98}$$

Here T is the trace of the matter energy-momentum tensor: $T \equiv T_\rho^{(m)\rho}$. We also denote the derivative $d^n f(R)/dR^n$ by $f^{(n)}(R)$. Let us now consider the perturbation of the Einstein gravity solutions. We denote the scalar curvature, given by the matter density in the Einstein gravity, by $R_b \sim (\kappa^2/2)\rho > 0$ and separate the scalar curvature R into the sum of R_b (background) and the perturbed part R_p as $R = R_b + R_p$ ($|R_p| \ll |R_b|$). Then Eq. (6.98) leads to the perturbed equation:

$$\begin{aligned} 0 \;=\; & \Box R_b + \frac{f^{(3)}(R_b)}{f^{(2)}(R_b)} \nabla_\rho R_b \nabla^\rho R_b + \frac{f'(R_b)R_b}{3f^{(2)}(R_b)} - \frac{2F(R_b)}{3f^{(2)}(R_b)} - \frac{R_b}{3f^{(2)}(R_b)} \\ & + \Box R_p + 2\frac{f^{(3)}(R_b)}{f^{(2)}(R_b)} \nabla_\rho R_b \nabla^\rho R_p + U(R_b) R_p . \end{aligned} \tag{6.99}$$

Here the potential $U(R_b)$ is given by

$$\begin{aligned} U(R_b) \;\equiv\; & \left(\frac{f^{(4)}(R_b)}{f^{(2)}(R_b)} - \frac{f^{(3)}(R_b)^2}{f^{(2)}(R_b)^2} \right) \nabla_\rho R_b \nabla^\rho R_b + \frac{R_b}{3} - \frac{F^{(1)}(R_b) f^{(3)}(R_b) R_b}{3f^{(2)}(R_b)^2} \\ & - \frac{f^{(1)}(R_b)}{3f^{(2)}(R_b)} + \frac{2f(R_b) f^{(3)}(R_b)}{3f^{(2)}(R_b)^2} - \frac{f^{(3)}(R_b) R_b}{3f^{(2)}(R_b)^2} \end{aligned} \tag{6.100}$$

It is convenient to consider the case where R_b and R_p are uniform and do not depend on the spatial coordinates. Hence, the d'Alembert operator can be replaced by the second derivative with respect to the time, that is: $\Box R_p \to -\partial_t^2 R_p$. Eq. (6.100) assumes the following structure:

$$0 = -\partial_t^2 R_p + U(R_b) R_p + \text{const} \,. \tag{6.101}$$

If $U(R_b) > 0$, R_p becomes exponentially large with time, i.e. $R_p \sim \mathrm{e}^{\sqrt{U(R_b)}t}$, and the system becomes unstable.

In the $1/R$-model, considering the background values, we find

$$U(R_b) = -R_b + \frac{R_b^3}{6\mu^4} \sim \frac{R_0^3}{\mu^4} \sim \left(10^{-26}\text{sec}\right)^{-2} \left(\frac{\rho_m}{\text{g cm}^{-3}}\right)^3 ,$$
$$R_b \sim \left(10^3 \text{sec}\right)^{-2} \left(\frac{\rho_m}{\text{g cm}^{-3}}\right) . \tag{6.102}$$

Here the mass parameter μ is of the order

$$\mu^{-1} \sim 10^{18}\text{sec} \sim \left(10^{-33}\text{eV}\right)^{-1} . \tag{6.103}$$

Eq. (6.102) tells us that the model is unstable and it would decay in 10^{-26} sec (considering the Earth size). In Model I, however, $U(R_b)$ is negative:

$$U(R_0) \sim -\frac{(n+2)m^2 c_2^2}{c_1 n(n+1)} < 0 \,. \tag{6.104}$$

Therefore, there is no matter instability.

For Model (6.95), as it is clear from the identifications (6.97), there is no matter instability too.

In order to study the stability of the de Sitter solution, let us proceed as follows. From the field equations, we obtain the trace

$$\Box f'(R) = \frac{1}{3}\left[R - f'(R)R + 2f(R) + \kappa^2 T\right] . \tag{6.105}$$

Here, as above, $T \equiv g^{\mu\nu} T_{\mu\nu}^{(m)}$.

Now we consider the (in)stability around the de Sitter solution, where $R = R_0$, and therefore $f(R_0)$ and $f'(R_0)$, are constants. Then since the l.h.s. in Eq. (6.105) vanishes for $R = R_0$, we find

$$R_0 - f'(R_0)R_0 + 2f(R_0) + \kappa^2 T_0 = 0 \,. \tag{6.106}$$

Let us expand both sides of (6.106) around $R = R_0$ as

$$R = R_0 + \delta R \,. \tag{6.107}$$

One obtains

$$f''(R_0)\Box\delta R = \frac{1}{3}\left(1 - f''(R_0)R_0 + f'(R_0)\right)\delta R \,. \tag{6.108}$$

Since

$$\Box\delta R = -\frac{d^2\delta R}{dt^2} - 3H_0\frac{d\delta R}{dt}\,, \tag{6.109}$$

in the de Sitter background, if

$$C(R_0) \equiv \lim_{R\to R_0} \frac{1 - f''(R)R + f'(R)}{f''(R)} > 0\,, \tag{6.110}$$

the de Sitter background is stable but, if $C(R_0) < 0$, the de Sitter background is unstable. The expression for $C(R_0)$ could be valid even if $f''(R_0) = 0$. More precisely, the solution of (6.108) is given by

$$\delta R = A_+ \mathrm{e}^{\lambda_+ t} + A_- \mathrm{e}^{\lambda_- t}\,. \tag{6.111}$$

Here $A_\pm$ are constants and

$$\lambda_\pm = \frac{-3H_0 \pm \sqrt{9H_0^2 - C(R_0)}}{2}\,. \tag{6.112}$$

Then, if $C(R_0) < 0$, λ_+ is always positive and the perturbation grows up. This leads to the instability. We have also to note that, when $C(R_0)$ is positive, if $C(R_0) > 9H_0^2$, δR oscillates and the amplitude becomes exponentially small being:

$$\delta R = (A\cos\omega_0 t + B\sin\omega_0 t)\,\mathrm{e}^{-3H_0 t/2}\,, \quad \omega \equiv \frac{\sqrt{C(R_0) - 9H_0^2}}{2}\,. \tag{6.113}$$

Here A and B are constant. On the other hand, if $C(R_0) < 9H_0^2$, there is no oscillation in δR.

Let us now consider the case where the matter contribution T can be neglected in the de Sitter background and assume $f'(R) = 0$ in the same background. We can assume that there are two de Sitter background solutions satisfying $f'(R) = 0$, for $R = R_1$ and $R = R_2$ as it could be the physical case if one asks for an inflationary and a dark energy epoch. We also assume $f'(R) \neq 0$ if $R_1 < R < R_2$ or $R_2 < R < R_1$. In the case $C(R_1) < 0$ and $C(R_2) > 0$, the de Sitter solution, corresponding to $R = R_1$, is unstable but the solution corresponding to $R = R_2$ is stable. Then there should be a solution where the (nearly) de Sitter solution corresponding to R_1 transits to the (nearly) de Sitter solution R_2. Since the solution corresponding to R_2 is stable, the universe remains in the de Sitter solution corresponding to R_2 and there is no more transition to any other de Sitter solution.

As an example, we consider Model I. For large curvature values, we find

$$f_I(R) = -\Lambda + \frac{\alpha}{R^{2n+1}}\,. \tag{6.114}$$

Here Λ and α are positive constants and n is a positive integer. Then we find

$$C(R) \sim \frac{1}{f''(R)} \sim \frac{R^{2n+2}}{2n(2n+1)\alpha} > 0\,. \tag{6.115}$$

This means that the de Sitter solution in Model I can be stable. We have also to note that $C(R_0) \sim H_0^{4n+4}/m^{4n+2}$. Here m^2 is the mass scale introduced in [228] and $m^2 \ll H_0^2$: this means that $C(R_0) \gg 9H_0^2$ and therefore there could be no oscillation.

We may also consider the model proposed in [312](here Model V):

$$f_V(R) = \frac{\alpha R^{2n} - \beta R^n}{1 + \gamma R^n}. \tag{6.116}$$

Here α, β, and γ are positive constants and n is a positive integer. In Fig.6.5, we show the behavior of Model V and of its first derivative. When the curvature is large ($R \to \infty$), $f(R)$ behaves as a power law. Since the derivative of $f(R)$ is given by

$$f_V'(R) = \frac{nR^{n-1}\left(\alpha\gamma R^{2n} - 2\alpha R^n - \beta\right)}{\left(1 + \gamma R^n\right)^2}, \tag{6.117}$$

we find that the curvature R_0 in the present universe, which satisfies the condition $f'(R_0) = 0$, is given by

$$R_0 = \left[\frac{1}{\gamma}\left(1 + \sqrt{1 + \frac{\beta\gamma}{\alpha}}\right)\right]^{1/n}, \tag{6.118}$$

and

$$f(R_0) \sim -2\tilde{R}_0 = \frac{\alpha}{\gamma^2}\left(1 + \frac{(1 - \beta\gamma/\alpha)\sqrt{1 + \beta\gamma/\alpha}}{2 + \sqrt{1 + \beta\gamma/\alpha}}\right). \tag{6.119}$$

As shown in [312], the magnitudes of the parameters is given by

$$\alpha \sim 2\tilde{R}_0 R_0^{-2n}, \; \beta \sim 4\tilde{R}_0^2 R_0^{-2n} R_I^{n-1}, \; \gamma \sim 2\tilde{R}_0 R_0^{-2n} R_I^{n-1}. \tag{6.120}$$

Here R_I is the curvature in the inflationary epoch and we have assumed $f(R_I) \sim (\alpha/\gamma)R_I^n \sim R_I$.

$C(R_0)$ in (6.110) is given by

$$C(R_0) \sim \frac{1}{f''(R_0)} = \frac{1 + \gamma R_0^n}{2n^2 \alpha R_0^{2n-2}\left(\gamma R_0^n - 1\right)}. \tag{6.121}$$

By using the relations (6.120), we find

$$C(R_0) \sim \frac{R_0^2}{4n^2 \tilde{R}_0}, \tag{6.122}$$

which is positive and therefore the de Sitter solution is stable. We notice that $C(R_0) < 9H_0^2$ and therefore, there could occur oscillations as in (6.113).

Furthermore, we can take into account the following model [139] (Model VI):

$$\begin{aligned} f_{VI}(R) &= -\alpha\left[\tanh\left(\frac{b(R - R_0)}{2}\right) + \tanh\left(\frac{bR_0}{2}\right)\right] \\ &= -\alpha\left[\frac{e^{b(R-R_0)} - 1}{e^{b(R-R_0)} + 1} + \frac{e^{bR_0} - 1}{e^{bR_0} + 1}\right] \end{aligned} \tag{6.123}$$

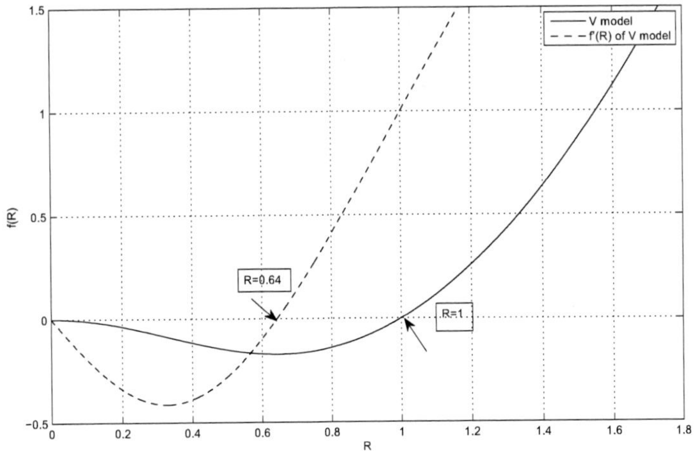

Figure 6.5. Plots of Model V (6.116) (solid line) and its first derivative (dashed line). Here $n = 2$ and α, β, γ are assumed as in (6.120) with the value of R_0 taken in the Solar System. $f'(R)$ is negative for $0 < R < 0.64$. $f(R)$ is given in the range $0 < R < 1$ where we have adopted suitable units.

where α and b are positive constants. When $R \to 0$, we find that

$$f_{VI}(R) \to -\frac{\alpha b R}{2\cosh^2\left(\frac{bR_0}{2}\right)}, \tag{6.124}$$

and thus $f(0) = 0$. On the other hand, when $R \to +\infty$,

$$f_{VI}(R) \to -2\Lambda_{\text{eff}} \equiv -\alpha\left[1 + \tanh\left(\frac{bR_0}{2}\right)\right]. \tag{6.125}$$

If $R \gg R_0$, in the present universe, Λ_{eff} plays the role of the effective cosmological constant. We also obtain

$$f'_{VI}(R) = -\frac{\alpha b}{2\cosh^2\left(\frac{b(R-R_0)}{2}\right)}, \tag{6.126}$$

which has a minimum when $R = R_0$, that is:

$$f'_{VI}(R_0) = -\frac{\alpha b}{2}. \tag{6.127}$$

Then in order to avoid anti-gravity, we find

$$0 < 1 + f'_{VI}(R_0) = 1 - \frac{\alpha b}{2}. \tag{6.128}$$

Beside the above model, we can consider a model which is able to describe, in principle, both the early inflation and the late acceleration epochs. The following two-step model [139] (Model VII):

$$\begin{aligned} f_{VII}(R) &= -\alpha_0 \left[\tanh\left(\frac{b_0(R-R_0)}{2}\right) + \tanh\left(\frac{b_0 R_0}{2}\right)\right] \\ &\quad -\alpha_I \left[\tanh\left(\frac{b_I(R-R_I)}{2}\right) + \tanh\left(\frac{b_I R_I}{2}\right)\right], \end{aligned} \tag{6.129}$$

could be useful to this goal. Let us assume

$$R_I \gg R_0\,, \quad \alpha_I \gg \alpha_0\,, \quad b_I \ll b_0\,, \tag{6.130}$$

and

$$b_I R_I \gg 1\,. \tag{6.131}$$

When $R \to 0$ or $R \ll R_0 \ll R_I$, $f_{VII}(R)$ behaves as

$$f_{VII}(R) \to -\left[\frac{\alpha_0 b_0}{2\cosh^2\left(\frac{b_0 R_0}{2}\right)} + \frac{\alpha_I b_I}{2\cosh^2\left(\frac{b_I R_I}{2}\right)}\right] R\,, \tag{6.132}$$

and we find again $f_{VII}(0) = 0$. When $R \gg R_I$, we find

$$\begin{aligned} f(R)_{VII} \to -2\Lambda_I &\equiv -\alpha_0 \left[1 + \tanh\left(\frac{b_0 R_0}{2}\right)\right] - \alpha_I \left[1 + \tanh\left(\frac{b_I R_I}{2}\right)\right] \\ &\sim -\alpha_I \left[1 + \tanh\left(\frac{b_I R_I}{2}\right)\right]. \end{aligned} \tag{6.133}$$

On the other hand, when $R_0 \ll R \ll R_I$, we find

$$\begin{aligned} f_{VII}(R) \to -\alpha_0 \left[1 + \tanh\left(\frac{b_0 R_0}{2}\right)\right] - \frac{\alpha_I b_I R}{2\cosh^2\left(\frac{b_I R_I}{2}\right)} &\sim -2\Lambda_0 \\ \equiv -\alpha_0 \left[1 + \tanh\left(\frac{b_0 R_0}{2}\right)\right]. \end{aligned} \tag{6.134}$$

Here, we have assumed the condition (6.131). We also find

$$f'_{VII}(R) = -\frac{\alpha_0 b_0}{2\cosh^2\left(\frac{b_0(R-R_0)}{2}\right)} - \frac{\alpha_I b_I}{2\cosh^2\left(\frac{b_I(R-R_I)}{2}\right)}, \tag{6.135}$$

which has two minima for $R \sim R_0$ and $R \sim R_I$. When $R = R_0$, we obtain

$$f'_{VII}(R_0) = -\alpha_0 b_0 - \frac{\alpha_I b_I}{2\cosh^2\left(\frac{b_I(R_0-R_I)}{2}\right)} > -\alpha_I b_I - \alpha_0 b_0\,. \tag{6.136}$$

On the other hand, when $R = R_I$, we get

$$f'_{VII}(R_I) = -\alpha_I b_I - \frac{\alpha_0 b_0}{2\cosh^2\left(\frac{b_0(R_0 - R_I)}{2}\right)} > -\alpha_I b_I - \alpha_0 b_0 \,. \tag{6.137}$$

Then, in order to avoid the anti-gravity behavior, we find

$$\alpha_I b_I + \alpha_0 b_0 < 1 \,. \tag{6.138}$$

Let us now investigate the correction to the Newton potential and the matter instability issue related to Models VI and VII. In the Solar System domain, on or inside the Earth, where $R \gg R_0$, $f(R)$ in Eq. (6.123) can be approximated by

$$f_{VI}(R) \sim -2\Lambda_{\rm eff} + 2\alpha e^{-b(R-R_0)} \,. \tag{6.139}$$

On the other hand, since $R_0 \ll R \ll R_I$, by assuming Eq. (6.131), $f(R)$ in (6.129) can be also approximated by

$$f_{VII}(R) \sim -2\Lambda_0 + 2\alpha e^{-b_0(R-R_0)} \,, \tag{6.140}$$

which has the same expression, after having identified $\Lambda_0 = \Lambda_{\rm eff}$ and $b_0 = b$. Then, we may check the case of (6.139) only. In this case, the effective mass has the following form

$$m_\sigma^2 \sim \frac{e^{b(R-R_0)}}{4\alpha b^2} \,, \tag{6.141}$$

which could be again very large. In fact, in the Solar System, we find $R \sim 10^{-61}\,\mathrm{eV}^2$. Even if we choose $\alpha \sim 1/b \sim R_0 \sim \left(10^{-33}\,\mathrm{eV}\right)^2$, we find that $m_\sigma^2 \sim 10^{1{,}000}\,\mathrm{eV}^2$, which is, ultimately, extremely heavy. Then, there will be no appreciable correction to the Newton law. In the Earth atmosphere, $R \sim 10^{-50}\,\mathrm{eV}^2$, and even if we choose $\alpha \sim 1/b \sim R_0 \sim \left(10^{-33}\,\mathrm{eV}\right)^2$ again, we find that $m_\sigma^2 \sim 10^{10{,}000{,}000{,}000}\,\mathrm{eV}^2$. Then, a correction to the Newton law is never observed in such models. In this case, we find that the effective potential $U(R_b)$ has the form

$$U(R_e) = -\frac{1}{2\alpha b}\left(2\Lambda + \frac{1}{b}\right) e^{-b(R_e - R_0)} \,, \tag{6.142}$$

which could be negative, what would suppress any instability.

In order that a de Sitter solution exists in $f(R)$-gravity, the following condition has to be satisfied:

$$R = Rf'(R) - 2f(R) \,. \tag{6.143}$$

For the model (6.123), the r.h.s of (6.143) has the following form:

$$R = -\frac{b\alpha R}{2\cosh^2\left(\frac{b(R-R_0)}{2}\right)} + 2\alpha\left[\tanh\left(\frac{b(R-R_0)}{2}\right) + \tanh\left(\frac{bR_0}{2}\right)\right] . \tag{6.144}$$

For large R, the r.h.s. behaves as

$$-\frac{b\alpha R}{2\cosh^2\left(\frac{b(R-R_0)}{2}\right)}+2\alpha\left[\tanh\left(\frac{b(R-R_0)}{2}\right)+\tanh\left(\frac{bR_0}{2}\right)\right]\to 2\alpha\,, \tag{6.145}$$

although the l.h.s. goes to infinity. On the other hand, when R is small, the r.h.s. behaves as

$$-\frac{b\alpha R}{2\cosh^2\left(\frac{b(R-R_0)}{2}\right)}+2\alpha\left[\tanh\left(\frac{b(R-R_0)}{2}\right)+\tanh\left(\frac{bR_0}{2}\right)\right] \to \frac{b\alpha R}{2\cosh^2\left(\frac{bR_0}{2}\right)}\,. \tag{6.146}$$

Then if

$$\frac{b\alpha}{2\cosh^2\left(\frac{bR_0}{2}\right)}>1\,, \tag{6.147}$$

there is a de Sitter solution. Combining Eq. (6.147) with Eq. (6.128), we find

$$2>\alpha b>\frac{1}{2\cosh^2\left(\frac{bR_0}{2}\right)}\,. \tag{6.148}$$

The stability, as above, is given by $C(R_{dS})$, where R_{dS} is the solution of (6.144). The expression is given by

$$C(R_{dS})=-R_{dS}+\frac{2\cosh^3\left(\frac{b(R_{dS}-R_0)}{2}\right)}{\alpha b^2\sinh\left(\frac{b(R_{dS}-R_0)}{2}\right)}-\frac{1}{b\tanh\left(\frac{b(R_{dS}-R_0)}{2}\right)}\,. \tag{6.149}$$

Let us now rewrite Eq. (6.144) as follows,

$$R_{dS}=2\alpha\left[\tanh\left(\frac{b(R_{dS}-R_0)}{2}\right)+\tanh\left(\frac{bR_0}{2}\right)\right] \times\left[1+\frac{\alpha b}{2\cosh^2\left(\frac{b(R_{dS}-R_0)}{2}\right)}\right]^{-1}\,. \tag{6.150}$$

Then by using (6.150), we may rewrite (6.149) in the following form:

$$C(R_{dS})=\frac{-\alpha^2b^2\left(1-x^2\right)\left[(x-x_0)^2+1-x_0^2\right]+4}{\alpha b^2x\left(1-x^2\right)\left[2+\alpha b\left(1-x^2\right)\right]}\,, \tag{6.151}$$

where

$$x=\tanh\left(\frac{b(R_{dS}-R_0)}{2}\right)\,,\quad x_0=-\tanh\left(\frac{bR_0}{2}\right)\,, \tag{6.152}$$

and therefore we have

$$-1 < x_0 \leq x < 1 \,, \quad x_0 < 0 \,. \tag{6.153}$$

Let us now consider (6.144) in order to find a de Sitter solution. Since Eq. (6.144) is difficult to solve in general, we assume $0 < R_{\mathrm{dS}} \ll R_0$. Then we find

$$R_{\mathrm{dS}} = \frac{\varepsilon}{b x_0} \,, \quad \varepsilon \equiv 1 - \frac{2\cosh^2\left(\frac{bR_0}{2}\right)}{\alpha b} = 1 - \frac{2}{\alpha b\left(1 - x_0^2\right)} \,. \tag{6.154}$$

Eq. (6.147) tells that the parameter ε is positive and, by assumption, very small: $0 < \varepsilon \ll 1$. Since ε is small, by using Eqs. (6.152), we find

$$x = x_0 + \frac{\left(1 - x_0^2\right)}{2x_0}\varepsilon + o\left(\varepsilon^2\right) \,. \tag{6.155}$$

Then by using the expression (6.151) for $C(R_{\mathrm{dS}})$, we find

$$C(R_{\mathrm{dS}}) \sim \frac{-\alpha^2 b^2 \left(1 - x_0^2\right)^2 + 4}{\alpha b^2 x_0 \left(1 - x_0^2\right)\left[2 + \alpha b\left(1 - x_0^2\right)\right]} \,. \tag{6.156}$$

From the definition of ε in (6.154), we find

$$\alpha b\left(1 - x_0^2\right) = 2 + 2\varepsilon + o\left(\varepsilon^2\right) \,, \tag{6.157}$$

and then, from Eq. (6.157), Eq. (6.156) can be written as follows;

$$C(R_{\mathrm{dS}}) \sim -\frac{\varepsilon}{b x_0} \,. \tag{6.158}$$

Since $x_0 < 0$ in the condition (6.153), we find $C(R_{\mathrm{dS}}) > 0$ and therefore the de Sitter solution is stable.

In Fig. 6.6, we have plotted the two models (6.123) and (6.129) written in the form $F(R) = R + f(R)$. We have used the inequalities (6.130) assuming, $R_I \sim \rho_g \sim 10^{-24}$ g/cm^3 for the Galactic density in the Solar vicinity and $R_0 \sim \rho_g \sim 10^{-29}$ g/cm^3 for the present cosmological density. .

Our task is now to find reliable experimental bounds for such models working at small and large scales. To this goal, we shall take into account constraints coming from Solar System experiments (which, at present, are capable of giving upper limits on the PPN parameters) and constraints coming from interferometers, in particular those giving limits on the (eventual) scalar components of gravitational waves. If constraints (and in particular the ranges of model parameters given by them) are comparable, this could constitute, besides other experimental and observational probes, a good hint to achieve a self-consistent $f(R)$ theory at very different scales.

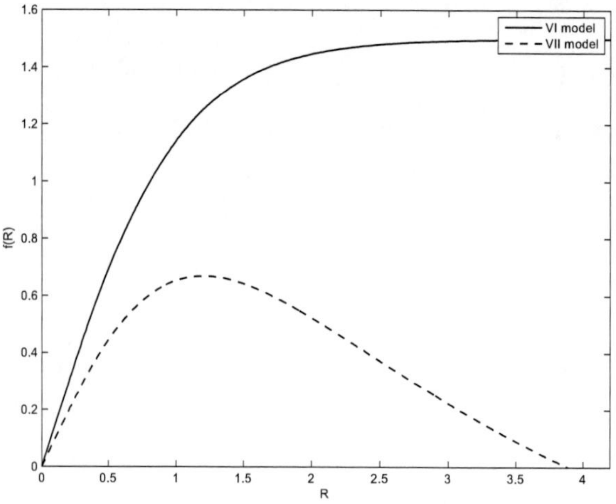

Figure 6.6. Plots of Model VI (6.123) (solid line) and Model VII (6.129) (dashed line). Here $b = 2$ and $b_I = 0.5$ with $\alpha = 1.5$ and $\alpha_I = 2$. The value of R_I is taken in the Solar System while R_0 corresponds to the present cosmological value.

6.10. Constraining $f(R)$-models by PPN Parameters

The above models can be constrained at Solar System level by considering the PPN formalism. This approach is extremely important in order to test gravitational theories and to compare them with General Relativity [78, 119, 120]. Let us take into account before the $f(R)$-models (6.87)-(6.90). Specifically, we want to investigate the values or the ranges of parameters in which they match the Solar-System experimental constraints in Table 6.4. In other words, we use these models to search under what circumstances it is possible to significantly address cosmological observations by $f(R)$-gravity and, simultaneously, evade the local tests of gravity.

By integrating the equations for β and γ, one obtains $f(R)$ solutions depending on β and γ which has to be confronted with β_{exp} and γ_{exp} [120]. If we plug into such equations the models (6.87)-(6.90) and the experimental values of PPN parameters, we will obtain algebraic constraints for the phenomenological parameters $\{n, p, q, \lambda, s\}$. This is the issue which we want to take into account in this section.

Assuming $f'(R) + 2f''(R)^2 \neq 0$ and defining $A = \left|\frac{1-\gamma}{2\gamma-1}\right|$, we obtain

$$\left[f''(R)\right]^2 - Af'(R) = 0\,. \tag{6.159}$$

The general solution of such an equation is a polynomial function [120].

Considering Model II given by (6.88), we obtain

$$\left[1-\frac{2pR\left(\frac{R^2}{R_c^2}+1\right)^{-p-1}\lambda}{R_c}\right]\left|\frac{\gamma-1}{2\gamma-1}\right| -\frac{4p^2\left(\frac{R^2}{R_c^2}+1\right)^{-2p}R_c^2\left(R_c^2-(2p+1)R^2\right)^2\lambda^2}{(R^2+R_c^2)^4}=0. \tag{6.160}$$

Our issue is now to find the values of λ, p, and R/R_c for which the Solar System experimental constraints are satisfied. Some preliminary considerations are in order at this point. Considering the de Sitter solution achieved from (6.88), we have $R = const = R_1 = x_1 R_c$, and $x_1 > 0$. It is straightforward to obtain

$$\lambda = \frac{x_1\left(1+x_1^2\right)^{p+1}}{2\left[\left(1+x_1^2\right)^{p+1}-1-(p+1)x_1^2\right]}. \tag{6.161}$$

On the other hand, the stability conditions $F_{,R} > 0$ and $F_{,RR} > 0$ give the inequality

$$\left(1+x_1^2\right)^{p+2} > 1+(p+2)x_1^2+(p+1)(2p+1)x_1^4, \tag{6.162}$$

which has to be satisfied. In particular, for $p=1$, it is $x_1 > \sqrt{3}$ and then $\lambda > \frac{8}{3\sqrt{3}} = 1.5396$. In addition, the value of x_1 satisfying the relation (6.162) is also the point where $\lambda(x_1)$, in Eq. (6.161), reaches its minimum.

To determine values of R compatible with PPN constraints, let us consider the trace of the field equations and explicit solutions, given the density profile $\rho(r)$, in the Solar vicinity. One can set the boundary condition considering $f_{,R_\infty} = f_{R_g}$

$$f_{,R_g} = f_{,R}(R = k^2\rho_g) \tag{6.163}$$

where $\rho_g \sim 10^{-24}$ g/cm^3 is the observed Galactic density in the Solar neighborhoods. At this point, we can see when the relation (6.160) satisfies the constraints for very Long Baseline Interferometer ($\gamma - 1 = 4\times 10^{-4}$) and Cassini Spacecraft ($\gamma - 1 = 2.1\times 10^{-}5$). This allows to find out suitable values for p.

An important remark is in order at this point. These constraint equations work if stability conditions hold. In the range

$$0 < \frac{R}{R_c} < \frac{1}{\sqrt{2p+1}} \tag{6.164}$$

$f_{,RR}$ is negative for the model (6.88) and then stability conditions are violated. To avoid this range, we need, at least, $\frac{R}{R_c} > 1$. For example, we can choose $\frac{R}{R_c} = 3.38$, corresponding to de Sitter behavior. Then we have $p = 1$ and $\lambda = 2$. On the other hand, for $0.944 < \lambda < 0.966$, we have $p = 2$ and $\frac{R}{R_c} = \sqrt{3}$; finally, for $R >> R_c$, we have $\lambda = 2$ and $p = 1.5$. For these values of parameters, the Solar System tests are evaded.

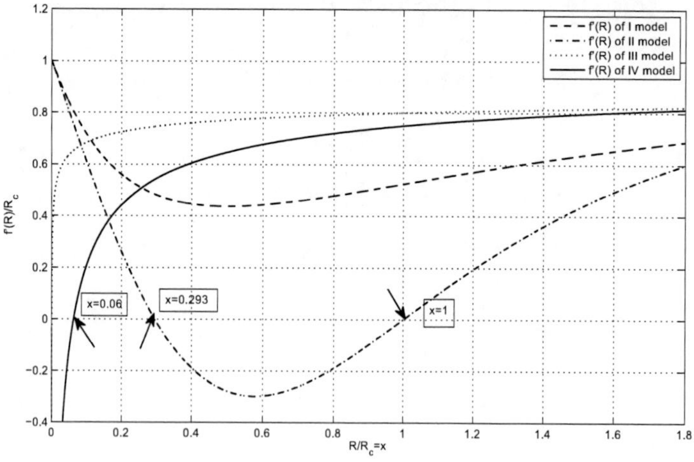

Figure 6.7. Plots of the first derivatives of four different models as function of $x = \frac{R}{R_c}$. Model I (dashed) is drawn for $n = 1$ and $\lambda = 2$. Model II (dashdot), for $p = 2$, $\lambda = 0.95$. Model III (dotted), for $s = 0.5$ and $\lambda = 1.5$. Model IV (solid) is for $q = 0.5$ and $\lambda = 0.5$. The labelled values of x indicate where the derivative changes its sign.

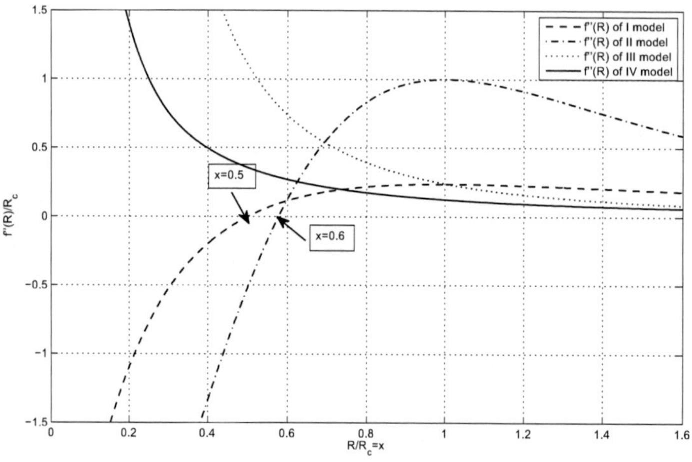

Figure 6.8. As above for the second derivatives of the models.

Let us consider now Model I, given by (6.86). Inserting it into the relation (6.159), we get

$$R^3\left[\left(\frac{R}{R_c}\right)^{2n}+1\right]^4\left[R\left(\left(\frac{R}{R_c}\right)^{2n}+1\right)^2-2n\left(\frac{R}{R_c}\right)^{2n}R_c\lambda\right]$$
$$\left|\frac{\gamma-1}{2\gamma-1}\right|-4n^2\left[(2n+1)\left(\frac{R}{R_c}\right)^{2n}-2n+1\right]^2$$
$$\left(\frac{R}{R_c}\right)^{4n}R_c^2\lambda^2\times R^4\left[\left(\frac{R}{R_c}\right)^{2n}+1\right]^6=0$$

Using the same procedure as above, λ is related to the de Sitter behavior. This means

$$\lambda=\frac{\left(1+x_1^{2n}\right)^2}{x_1^{2n-1}\left(2+2x_1^{2n}-2n\right)}, \tag{6.165}$$

while, from the stability conditions, we get

$$2x_1^4-(2n-1)(2n+4)x_1^{2n}+(2n-1)(2n-2)\geq 0\,. \tag{6.166}$$

For $n=1$, one obtains $x_1>\sqrt{3}$, $\lambda>\frac{8}{3\sqrt{3}}$. In this model, $f_{,RR}$ is negative for

$$0<\frac{R}{R_c}<\left(\frac{2n-1}{2n+1}\right)^{\frac{1}{2n}}. \tag{6.167}$$

The VLBI constraint is satisfied for $n=1$ and $\lambda=2$, while, for $n=1$ and $\lambda=1.5$, Cassini constraint holds.

By inserting Model III, given by Eq. (6.89), into the relation (6.159), we obtain

$$\frac{R^3\left[R-2sR_c\left(\frac{R_c}{R}\right)^{2s}\lambda\right]\left|\frac{\gamma-1}{2\gamma-1}\right|-4\left(2s^2+s\right)^2R_c^2\left(\frac{R_c}{r}\right)^{4s}\lambda^2}{R^4}=0\,. \tag{6.168}$$

The de-Sitter point corresponds to

$$\lambda=\frac{x_1^{2s+1}}{2(x_1^{2s}-s-1)}. \tag{6.169}$$

while the stability condition is $x_1^{2s}>2s^2+3s+1$. VLBI and Cassini constraints are satisfied by the sets of values: $s=1$, $\lambda=1.53$, for $\frac{R}{R_c}\sim 1$; $s=2$, $\lambda=0.95$, for $\frac{R}{R_c}=\sqrt{3}$, ; $s=1$, $\lambda=2$, for $\frac{R}{R_c}=3.38$.

Finally let us consider Model VI, given by Eq. (6.123), and Model VII, given by

Eq. (6.129). Using Eq. (6.159) for (6.123), we get

$$-\frac{1}{4}b\alpha\mathrm{sech}^2\left(\frac{1}{2}b(R-R_0)\right)$$
$$\times\left[b^3\alpha\mathrm{sech}^2\left(\frac{1}{2}b(R-R_0)\right)\tanh^2\left(\frac{1}{2}b(R-R_0)\right)-2\left|\frac{\gamma-1}{2\gamma-1}\right|\right]=0\,. \qquad (6.170)$$

As above, considering the stability conditions and the de Sitter behavior, we get the parameter ranges $0<b<2$ and $0<\alpha\leq 2$ which satisfy both VLBI and Cassini constraints. Inserting now Model VII in (6.159), we have

$$\frac{1}{2}\left|\frac{\gamma-1}{2\gamma-1}\right|\left[b\alpha\mathrm{sech}^2\left(\frac{1}{2}b(R-R_0)\right)-b_I\alpha_I\mathrm{sech}^2\left(\frac{1}{2}b_I(R-R_I)\right)+2\right]$$
$$-\frac{1}{4}\left[b^2\alpha\mathrm{sech}^2\left(\frac{1}{2}b(R-R_0)\right)\tanh\left(\frac{1}{2}b(R-R_0)\right)\right.$$
$$\left.-b_I^2\alpha_I\mathrm{sech}^2\left(\frac{1}{2}b_I(R-R_I)\right)\tanh\left(\frac{1}{2}b_I(R-R_I)\right)\right]^2=0\,. \qquad (6.171)$$

From the stability condition, we have that $f_{,R}>0$ for $R>0$, (see Fig.6.9) and $f_{,RR}<0$ for $0<R<2.35$ in suitable units (see Fig.6.10). Observational constraints from VLBI and Cassini experiments are fulfilled for

$$R_I\gg R_0\,,\quad \alpha_I\gg\alpha\,,\quad b_I\ll b\,. \qquad (6.172)$$

Plots for $b=2$, $b_I=0.5$, $\alpha=1.5$ and $\alpha_I=2$, verifying the constraints, are reported in Figs. 6.9 and 6.10.

Considering now the relation for β, one can easily verify that it is

$$\frac{d\gamma}{dR}=-\frac{d}{dR}\left[\frac{f''(R)^2}{f'(R)+2f''(R)^2}\right]=0\,, \qquad (6.173)$$

and this result implies

$$4(\beta-1)=0\,. \qquad (6.174)$$

This means the complete compatibility of the $f(R)$ solutions between the PPN-parameters β and γ.

Now we want to see if the parameter values, obtained for these models, are compatible with bounds coming from the stochastic background of gravitational waves achieved by interferometric experiments.

6.11. Further Experimental Constrainsts

As we discussed above, also the stochastic background of gravitational waves can be taken into account in order to constrain models. This approach could reveal very interesting because production of primordial gravitational waves could be a robust prediction for any model attempting to describe the cosmological evolution at primordial epochs. However,

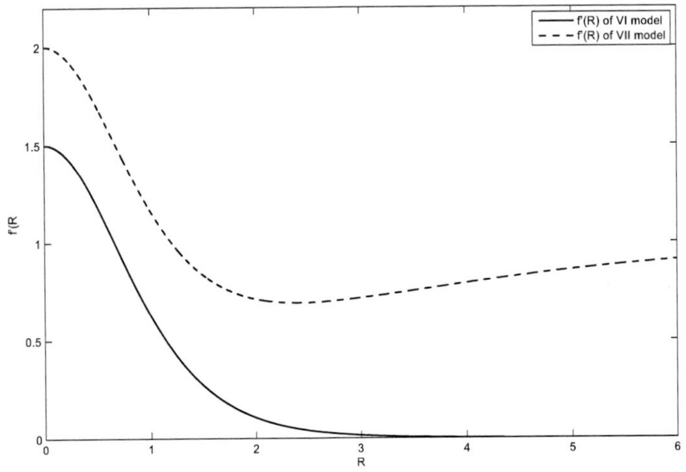

Figure 6.9. Plots represent the first derivatives of functions (6.128) (solid line) and (6.129) (dashed line). Here, $b = 2$, $b_I = 0.5$, $\alpha = 1.5$ and $\alpha_I = 2$ with R_I with the Solar System value and R_0 the today cosmological value. It is $f_{,R} > 0$ for $R > 0$.

bursts of gravitational radiation emitted from a large number of unresolved and uncorrelated astrophysical sources generate a stochastic background at more recent epochs, immediately following the onset of galaxy formation. Thus, astrophysical backgrounds might overwhelm the primordial one and their investigation provides important constraints on the signal detectability coming from the very early Universe, up to the bounds of the Planck epoch and the initial singularity [13, 14, 204, 288].

It is worth stressing the unavoidable and fundamental character of such a mechanism. It directly derives from the inflationary scenario [206, 426], which well fits the WMAP data with particular good agreement with almost exponential inflation and spectral index ≈ 1, [45, 380].

The main characteristics of the gravitational backgrounds produced by cosmological sources depend both on the emission properties of each single source and on the source rate evolution with redshift. It is therefore interesting to compare and contrast the probing power of these classes of $f(R)$-models at hight, intermediate and zero redshift [104, 106–108].

To this purpose, let us take into account the primordial physical process which gave rise to a characteristic spectrum Ω_{sgw} for the early stochastic background of relic scalar gravitational waves by which we can recast the further degrees of freedom coming from fourth-order gravity. This approach can greatly contribute to constrain viable cosmological models. The physical process related to the production has been analyzed, for example, in [14, 15, 204] but only for the first two tensorial components due to standard General Relativity. Actually the process can be improved considering also the third scalar-tensor component strictly related to the further $f(R)$ degrees of freedom [104, 106–108].

At this point, using the above LIGO, VIRGO and LISA upper bounds, calculated for the characteristic amplitude of GW scalar component, let us test the $f(R)$-gravity models,

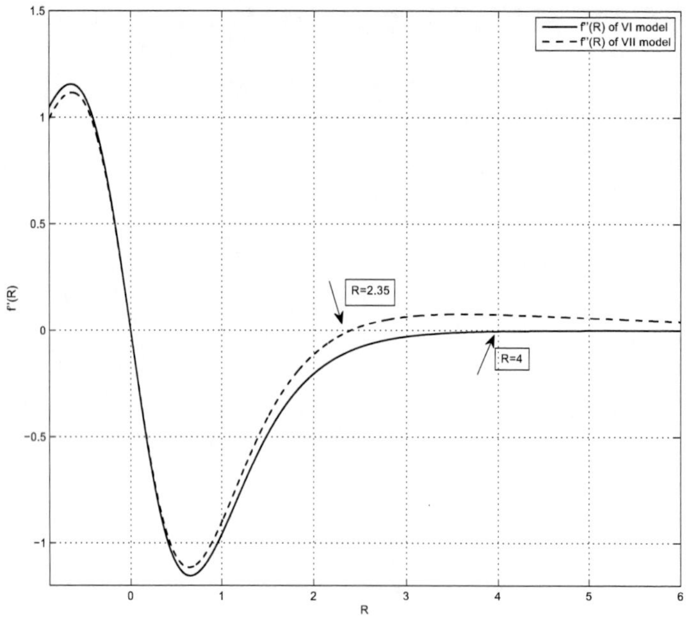

Figure 6.10. Second derivatives of Model VI (solid line) and VII (dashed line). Here $f_{,RR}$ is negative in the range $0 < R < 4$ for Model VI and in the range $0 < R < 2.35$ for Model VII. As above, we have used $b = 2$, $b_I = 0.5$, $\alpha = 1.5$ and $\alpha_I = 2$ with the value of R_I taken in the Solar System and R_0 for the today cosmological value.

considered in the previous sections, to see whether they are compatible both with the Solar System and GW stochastic background.

As above, for the considered models, we have to determine the values of the characteristic parameters which are compatible with both Solar System and gravitational waves stochastic background.

Let us start, for example, with the model (6.89). It is straightforward to derive the scalar component amplitude

$$\Phi_{III} = \frac{s(2s+1)\left(\frac{R_c}{R}\right)^{2s+1}\lambda}{\left[sR_c\left(\frac{R_c}{R}\right)^{2s}\lambda - R\right]\log\left[2 - 2s\left(\frac{R_c}{R}\right)^{2s+1}\lambda\right]}. \tag{6.175}$$

Such an equation satisfies the constraints in Table.3.26 for the values $s = 0.5$, $\frac{R}{R_c} \sim 1$, $\lambda = 1.53$ and $s = 1$, $\frac{R}{R_c} \sim 1$, $\lambda = 0.95$ (LIGO); $s = 2$, $\frac{R}{R_c} = \sqrt{3}$, $\lambda = 2$ (VIRGO); $s = 1$, $\lambda = 2$ and $\frac{R}{R_c} = 3.38$ (LISA).

It is important to stress the nice agreement with the figures achieved from the PPN constraints. In this case, we have assumed $R_c \sim \rho_c \sim 10^{-29}$ g/cm^3, where ρ_c is the present day cosmological density.

Considering the model (6.86), we obtain

$$\Phi_I = -\frac{n\left[(2n+1)\left(\frac{R}{R_c}\right)^{2n} - 2n+1\right]\left(\frac{R}{R_c}\right)^{2n-1}\lambda}{\left[\left(\frac{R}{R_c}\right)^{2n}+1\right]\left\{R\left[\left(\frac{R}{R_c}\right)^{2n}+1\right]^2 - n\left(\frac{R}{R_c}\right)^{2n} R_c\lambda\right\}\log\left(1-\frac{2n\left(\frac{R}{R_c}\right)^{2n-1}\lambda}{\left(\left(\frac{R}{R_c}\right)^{2n}+1\right)^2}\right)}. \quad (6.176)$$

The expected constraints for gravitational waves scalar amplitude are fulfilled for $n=1$ and $\lambda=2$ and for $n=1$ and $\lambda=1.5$ when $0.3 < \frac{R}{R_c} < 1$.

Furthermore, considering the model (6.88), one gets

$$\Phi_I = -\frac{2p\left(1+\frac{R^2}{R_c^2}\right)^{-p} R_c\left((1+2p)R^2 - R_c^2\right)\lambda}{(R^2-R_c^2)^2\left[2-\frac{2p\left(1+\frac{R^2}{R_c^2}\right)^{-1-p}\lambda}{R_c}\right]\ln\left[2-\frac{2pR\left(1+\frac{R^2}{R_c^2}\right)^{-1-p}\lambda}{R_c}\right]}. \quad (6.177)$$

The LIGO upper bound is fulfilled for $p=1$, $\frac{R}{R_c} > \sqrt{3}$, $\lambda > \frac{8}{3\sqrt{3}}$; the VIRGO one for $p=1$, $\frac{R}{R_c} = 3.38$, $\lambda = 2$; finally, for LISA, we have $p=2$, $\frac{R}{R_c} = \sqrt{3}$ and $0.944 < \lambda < 0.966$. Besides, considering LISA in the regime $R >> R_c$, we have $\lambda = 2$ and $p = 1.5$.

Finally, let us consider Models VI and VII. We have

$$\Phi_{VI} = \frac{b^2\alpha\tanh\left[\frac{1}{2}b(R-R_0)\right]}{\left[b\alpha+\cosh(b(R-R_0))+1\right]\ln\left[\frac{b\alpha}{\cosh(b(R-R_0))+1}\right]}, \quad (6.178)$$

and

$$\begin{aligned}\Phi_{VII} = &\log\left[0.5\left(b\alpha\mathrm{sech}^2(0.5b(R-R_0)) - b_I\alpha_I\mathrm{sech}^2(0.5b_I(R-R_I))+2\right)\right]\\ &\times\left[b\alpha\mathrm{sech}^2(0.5b(R-R_0)) - b_I\alpha_I\mathrm{sech}^2(0.5b_I(R-R_I))+4\right]\\ &\times\left[b^2\alpha\mathrm{sech}^2(0.5b(R-R_0))\tanh(0.5b(R-R_0))\right.\\ &\left.- b_I^2\alpha_I\mathrm{sech}^2(0.5b_I(R-R_I))\tanh(0.5b_I(R-R_I))\right]. \end{aligned} \quad (6.179)$$

These equations satisfy the constraints for VIRGO, LIGO and LISA for $b=2$, $b_I=0.5$, $\alpha=1.5$ and $\alpha_I=2$ with R_I valued at Solar System scale and R_0 at cosmological scale.

As concluding remarks, it is worth that several class of extended model can address the so called "Chameleon approach" [77] which essentially consists in evading the Solar system tests and matching the astrophysical observations. This is extremely important to avoid the puzzling issues related to Dark Energy and Dark Matter. However, despite of this encouraging results, a degeneracy problem remains since a final, fully self self-consistent model, working at all scales, is not available yet.

Chapter 7

Future Perspectives and Conclusions

7.1. A Brief Summary

Before concluding this book and discussing future perspectives, let us shortly summarize the results presented so far. The motivations coming from astrophysics, cosmology and quantum field theory have been thoroughly outlined in Chapter 1. It has been stressed that extending the Einstein-Hilbert scheme is not forbidden, in principle, and could be a useful approach to address several shortcomings of modern physics ranging from inflation to Dark Energy and Dark Matter.

In the Chapter 2, after a summary of the bundle approach to the gauge theories with a discussion, in particular, of the bundle structure of gravitation, a nonlinearly realized representation of the local conformal-affine group has been determined. Before the physical applications, we have reviewed, in details, the mathematical tools to show that gravity and spin are the results of the local conformal-affine group so then it is possible to deal with an Invariance Induced Gravity. It has been found that the nonlinear Lorentz transformations contain contributions coming from the linear Lorentz parameter as well as conformal and shear contributions via the nonlinear 4-boosts and symmetric GL_4 parameters. We have identified the pullback of the nonlinear translational connection coefficient to M as a space-time coframe. In this way, the frame fields of the theory are obtained from the (nonlinear) gauge prescription. The mixed index coframe component (tetrad) is used to convert from Lie algebra indices into space-time indices. The space-time metric is a secondary object constructed (induced!) from the constant H group metric and the tetrads. The gauge fields $\overset{\circ}{\Gamma}{}^{\alpha\beta}$ are the analogues of the Christoffel connection coefficients of General Relativity and serve as the gravitational gauge potentials used to define covariant derivative operators. The gauge fields ϑ, Φ, and Υ encode information regarding special conformal, dilatonic and deformational degrees of freedom of the bundle manifold [91]. The space-time geometry is therefore determined by gauge field interactions as in the so called *Emergent Gravity* [370]. Furthermore, the bundle curvature and the Bianchi identities have been determined and then the gauge Lagrangian density have been modelled after the boundary topological invariants have been defined. As a consequence of this approach, no mixed field strength terms involving different components of the total curvature arise in the action. The analogue of the Einstein equations contains a non-trivial torsion contribution which is directly related to the spin fields of the theory (see also [100]). The Einstein-like three-form includes symmetric GL_4 as well as special conformal contributions. A mixed translational-conformal cosmological constant term arises due to the structure of the generalized curvature of the manifold. We also obtain a Yang-Mills-like equation that represents the generalization of the Gauss torsion-free equation. Variation of I with respect to Υ_α^β leads to a constraint equation relating the GL_4 deformation gauge field to the translational and special conformal field strengths. The gravi-scalar field equation has non-vanishing translational and special conformal contributions. As a concluding remark, we can say that gravity (and in general any gauge field) can be derived as the nonlinear realization of a local conformal-affine symmetry group and then gravity can be considered an interaction induced from invariance properties. This approach can be adopted also to generalized theories of gravity [88, 308, 387].

In Chapter 3, we have proposed a novel definition of space-time metric deformations parameterizing them in terms of scalar field matrices. The main result is that deformations can be described as extended conformal transformations. This fact gives a straightforward physical interpretation of conformal transformations: conformally related metrics can be seen as the "background" and the "perturbed" metrics. In other words, the relations between the Jordan frame and the Einstein frame can be directly interpreted through the action of the deformation matrices contributing to solve the issue of what the true physical frame is [9]. Besides, space-time metric deformations can be immediately recast in terms of perturbation theory allowing a completely covariant approach to the problem of gravitational waves.

A general discussion about Extended Theories of Gravity has been exposed in Chapter 4. We have outlined what one should intend for Extended Theories of Gravity in the metric and in the Palatini approaches. In particular, we have discussed the higher-order and the scalar-tensor theories of gravity showing the relations between them and their connection to General Relativity via the conformal transformations.

In the so called Einstein frame, any Extended Theories of Gravity can be reduced to the Hilbert-Einstein action plus one or more than one scalar field(s). The physical meaning of conformal transformations can be particularly devised in the Palatini approach. In this framework, the conformal transformation is not only a mathematical tool capable of disentangling matter from gravitational degrees of freedom, but it is related to the bi-metric structure of space-time where chronological structure and geodesic structure are, a priori, independent. These physical properties have been shown to be relevant at cosmological scales. In space-time regions interesting for standard observations, the conformal factor is slowly varying and cannot be distinguished from a constant, which ca be normalized to 1.

While in Hilbert-Einstein gravity the affine connections can be assumed in any case Levi-Civita, this is not true in the Palatini approach, being the fields g and Γ independent. Due to this fact, the ambiguities to work out a given theory in the Einstein or in the Jordan frame are assume an immediate and different physical meaning, since in the Palatini formalism, field equations and, first of all the structural equation of space-time, give, at the same time, information on both frames. In other words, discussing if "Jordan" or "Einstein" is the true physical frame results to be a minor problem in the Palatini approach, where both metric and connection are, a priori, independent fields. These results become crucial in cosmology since, by them, it is possible to show that solutions taken into account as different ones are the same in the Palatini approach. For example, the recently observed acceleration of the Hubble fluid [154, 332, 341, 411, 419] is an evidence that some form of "dark energy" should be present in the cosmic dynamics. Despite of this general result, such an accelerated dynamics can be achieved in several ways (cosmological constant [352], scalar fields dynamics [396], curvature quintessence [94, 193, 313]) but no definite answer, up to now, has been given about its nature. In what we have discussed, it has been shown that cosmic dynamics ruled by the cosmological constant in the Einstein frame becomes ruled by a non-minimally coupled, self-interacting scalar field (evolving in time) in the Jordan frame. Consequently, matching the data against a solution in the Einstein frame or in the Jordan frame could lead to highly misleading results and interpretation. The shortcoming is completely overcome in the Palatini approach which give, at the same time, dynamics and conformal structure of the given Extended Theories of Gravity avoiding such ambiguities. These considerations have to be further developed considering concretely the matching with the data. The results here obtained, however, do not provide a definitive answer on the very important question: which frame is the physical one in the purely metric formalism? The results obtained surely provide some hints and tools in this long-time discussion which is at the moment still open, very interesting and surely deserves further studies and investigations. Moreover both theoretical and experimental tests to establish which is the physical frame will be necessary and we have here given some hints in the direction of a different interpretation of the problem.

Chapters 5 and 6 have bee devoted to the analysis of post-Minkowskian and post-Newtonian limits of Extended Theories of Gravity. Such an analysis is essential in order to test experimentally the models. The main results are that further gravitational modes (massive and ghost) and Yukawa-like corrections emerges in such a weak field approximations. Our analysis covers extended gravity models with a generic class of Lagrangian density with higher order and terms of the form $f(R,P,Q)$, where $P \equiv R_{ab}R^{ab}$ and $Q \equiv R_{abcd}R^{abcd}$. We have linearized the field equations for this class of theories around a Minkowski background and found that, besides a massless spin-2 field (the graviton), the theory contains also spin-0 and spin-2 massive modes with the latter being, in general, ghosts. Then, we have investigated the detectability of additional polarization modes of a stochastic gravitational waves with ground-based laser-interferometric detectors and space-interferometers. Such polarization modes, in general, appear in the extended theories of gravitation and can be utilized to constrain the theories beyond General Relativity in a model-independent way.

However, a point has to be discussed in detail. If the interferometer is directionally sensitive and we also know the orientation of the source (and of course if the source is coherent) the situation is straightforward. In this case, the massive mode coming from the simplest extension, $f(R)$-gravity, would induce longitudinal displacements along the direction of propagation which should be detectable and only the amplitude due to the scalar mode would be the true, detectable, "new" signal [104, 106–108]. But even in this case, we could have a second scalar mode inducing a similar effect, coming from the massive ghost, although with a minus sign. So in this case, one has deviations from the prediction of $f(R)$-gravity, even if only the massive modes are considered as new signal.

On the other hand, in the case of the stochastic background, there is no coherent source and no directional detection of the gravitational radiation. What the interferometer picks is just an averaged signal coming from the contributions of all possible modes from (uncorrelated) sources all over the celestial sphere. Since we expect the background to be isotropic, the signal will be the same regardless of the orientation of the interferometer, no matter how or on which plane it is rotated, it would always record the characteristic amplitude h_c. So there is intrinsically no way to disentangle any of the modes in the background, being h_c related to the total energy density of the gravitational radiation, which depends on the number of modes available. Every mode, essentially, contributes in the same manner, at least in the limit where the mass for the massive and ghost modes are very small (as they should be). So, it should be the number of the modes available that makes the difference, not their origin.

Again, even if this does not hold, one should still get into consideration at least the massive ghost mode to get a constraint. This is the why we have considered only h_{GR}, h_{HOG} and h_s in the above cross-correlation analysis without giving further fine details coming from polarization. For the situation considered here, we find that the massive modes are certainly of interest for direct attempts at detection with the LISA experiment. It is, in principle, possible that massive gravitational waves modes could be produced in more significant quantities in cosmological or early astrophysical processes in alternative theories of gravity, being this possibility still unexplored. This situation should be kept in mind when looking for a signature distinguishing these theories from General Relativity, and seems to deserve further investigation. As an example, we have shown that the amplitudes of tensor gravitational waves are conformally invariant and their evolution depends on the cosmological background. Such a background is tuned by a conformal scalar field which is not present in the standard General Relativity. Assuming that primordial vacuum fluctuations produce stochastic gravitational waves, beside scalar perturbations, kinematical distortions and so on, the initial amplitude of these ones is a function of the specific Extended Theories of Gravity model and then the stochastic background can be, in a certain sense "tuned" by the theory. Vice versa, data coming from the Sachs-Wolfe effect could contribute to select a suitable Extended Theories of Gravity which can be consistently matched with other observations. However, further and accurate studies are needed in order to test the relation between Sachs-Wolfe effect and Extended Theories of Gravity. This goal could, in principle, be achieved very soon through the forthcoming space (LISA) and ground-based (VIRGO, LIGO) interferometers.

In Chapter 6, we have formally developed the Newtonian limit of theories containing higher-order curvature invariants. Two main achievements have to be stressed. The Poisson equation has to be substituted with equations containing higher-order Laplace operators. Besides, Newtonian potential results corrected in any case.

Furthermore, we have investigated the possibility that some viable Extended Theories of Gravity models could be constrained considering both Solar System experiments and upper bounds on the stochastic background of gravitational radiation. Such bounds come from interferometric ground-based (VIRGO and LIGO) and space (LISA) experiments. The underlying philosophy is to show that the Extended Theories of Gravity approach, in order to describe consistently the observed Universe, should be tested at very different scales, that is at very different redshifts. In other words, such a proposal could partially contribute to remove the unpleasant degeneracy affecting the wide class of dark energy models, today on the ground.

Beside the request to evade the Solar System tests, new methods have been recently proposed to investigate the evolution and the power spectrum of cosmological perturbations [319]. The investigation of stochastic background, in particular of the scalar component of gravitational waves coming from the Extended Theories of Gravity additional degrees of freedom, could acquire, if revealed by the running and forthcoming experiments, a fundamental importance to discriminate among the various gravity theories [104,106–108]. These data (today only upper bounds coming from simulations) if combined with Solar System tests, Cosmic Microwave Background Radiation anisotropies, LSS, etc. could greatly help to achieve a self-consistent cosmology bypassing the shortcomings of ΛCDM model.

Specifically, we have taken into account some broken power law $f(R)$ models fulfilling the main cosmological requirements which are to match the today observed accelerated expansion and the correct behavior in early epochs. In principle, the adopted parametrization allows to fit data at extragalactic and cosmological scales [77,228]. Furthermore, such models are constructed to evade the Solar System experimental tests. Beside these broken power laws, we have considered also two models capable of reproducing the effective cosmological constant, the early inflation and the late acceleration epochs [78, 139]. These $f(R)$-functions are combinations of hyperbolic tangents.

We have discussed the behavior of all the considered models. In particular, the problem of stability has been addressed determining suitable and physically consistent ranges of parameters. Then we have taken into account the results of the main Solar System current experiments. Such results give upper limits on the PPN parameters which any self-consistent theory of gravity should satisfy at local scales. Starting from these, we have selected the $f(R)$ parameters fulfilling the tests. As a general remark, all the functional forms chosen for $f(R)$ present sets of parameters capable of matching the two main PPN quantities, that is γ_{exp} and β_{exp}. This means that, in principle, extensions of General Relativity are not *a priori* excluded as reasonable candidates for gravity theories.

The interesting feature, is that such sets of parameters are not in conflict with bounds coming from the cosmological stochastic background of gravitational waves. In particular, some sets of parameters reproduce quite well both the PPN upper limits and the constraints on the scalar component amplitude of gravitational waves.

Far to be definitive, these preliminary results indicate that self-consistent models could be achieved comparing experimental data at very different scales without extrapolating results obtained only at a given scale.

7.2. Concluding Remarks

To conclude, even though some significant progress has been made by developing alternative theories of gravity, it is still unclear how to relate principles and experiments in practice, in order to formulate self-consistent viability criteria. The inability to express these criteria in a representation-invariant model plays a crucial role in the lack of a theory of gravity at foundamental level. This is a critical obstacle to overcome if we want to go beyond a mere trial-and-error approach in developing alternative theories of gravity.

In our opinion such an approach should be one of the future goals to achieve a self-consistent theory of gravity. This is not to say, of course, that efforts to search for reliable toy-models, such as $f(R)$-gravity, should be abandoned. These theories proved, so far, to be excellent tools to this goal, but there are still a lot of unexplored corners to be considered.

The motivations to modifies General Relativity, coming from High Energy Physics, Cosmology and Astrophysics, are definitely strong. Even though modifying gravity might not be the only way to address the problems mentioned in Chapter 1, it is our hope that the reader is by now convinced that it should be considered seriously as one of the possible solutions. The path to the final answer is probably long. However, this has never been a good enough reason for scientists to be discouraged.

If I have ever made any valuable discoveries, it has been owing more to patient attention, than to any other talent.

Isaac Newton

References

[1] Abbott L. F., Farhi E., Wise M. B., *Phys. Lett. B* **117**, 29 (1982).

[2] Abbott B. *et al.* (the LIGO Scientific Collaboration), *Phys. Rev. D* **72**, 042002 (2005).

[3] Abbott B. *et al.*, *Phys. Rev D* **76**, 082003 (2007).

[4] Acernese F. *et al.* (the Virgo Collaboration), *Class. Quant. Grav.* **24**, 19, S381-S388 (2007).

[5] Albrecht A., Steinhardt P. J.,Turner M. S., Wilczek F., *Phys. Rev. Lett.* **48**, 1437 (1982).

[6] Albrecht A., Steinhardt P.J., *Phys. Rev. Lett.* **48**, 1220 (1982).

[7] Ali S.A., Capozziello S., *Int. Jou. Geom. Meth. Mod. Phys.* **4**, 1041 (2007).

[8] Allemandi G., Borowiec A., Francaviglia M., *Phys. Rev. D* **70**, 103503 (2004).

[9] Allemandi G., Capone M., Capozziello S., Francaviglia M., *Gen. Rel. Grav.* **38**, 33 (2006).

[10] Adams F.C., Freese K., *Phys. Rev. D* **43**, 353 (1991).

[11] Adams F., Freese K., Guth A., *PHys. Rev. D* **43**, 965 (1991).

[12] Allen S. W., *et al.*, *Mon. Not. Roy. Astr. Soc.* **342**, 287 (2003).

[13] Allen B., Ottewill A.C., *Phys. Rev. D* **56**, 545–563 (1997).

[14] Allen B. -Proceedings of the Les Houches School on Astrophysical Sources of Gravitational Waves, eds. Jean-Alain Marck and Jean-Pierre Lasota, Cambridge University Press, Cambridge, England (1998).

[15] Allen B., *Phys. Rev. D* **307**, 2078 (1988).

[16] Amarzguioui M., Elgaroy O., Mota D. F., Multamaki T., Astron. Astrophys. **454**, 707 (2006).

[17] Amendola L., Capozziello S., Litterio M., Occhionero F., *Phys. Rev. D* **45**, 417 (1992).

[18] Amendola L., Battaglia Mayer A., Capozziello S., Gottlöber S., Müller V., Occhionero F., Schmidt H.J., *Class. Quantum Grav.* **10**, L43 (1993).

[19] Amendola L., Polarski D., Tsujikawa S., *Phys. Rev. Lett.* **98**, 131302 (2007).

[20] Amendola L., Gannouji R., Polarski D., Tsujikawa S., *Phys. Rev. D* **75**, 083504 (2007).

[21] Amendola L., Tsujikawa S., *Phys. Lett. B* **660**, 125–132 (2008).

[22] Anderson J.D. *et al.*, *Phys. Rev. D* **65**, 082004 (2002).

[23] Ando M. and the TAMA Collaboration - *Class. Quant. Grav.* **197**, 1615–1621 (2002).

[24] Appleby S. A., Battye R. A., *Phys. Lett. B* **654**, 7 (2007).

[25] Appelquist T., Chodos A., Freund P.G.O., *Modern Kaluza-Klein Theories*, Addison-Wesley, Reading, 1987.

[26] Armendariz-Picon C., Mukhanov V. F., Steinhardt P. J., *Phys. Rev. Lett.* **85**, 4483 (2000).

[27] Arnold V.I., *Mathematical Methods of Classical Mechanics*, Springer-Verlag, Berlin, 1978.

[28] Arnowitt R., Deser S., Misner C.W., *Phys. Rev.* **117**, 1595 (1960).

[29] Arnowitt R., Deser S., Misner C.W., in *Gravitation: An Introduction to Current Research*, edited by L. Witten, John Wiley and Sons, New York, 1962.

[30] Ashtekar A., Lewandowski J., *Class. Quant. Grav.* **21**, R53 (2004).

[31] Ashtekar A., *Lectures on Non-Perturbative Canonical Gravity*, World Scientific, Singapore, 1991.

[32] Astier P. *et al.*, *Astron. Astroph.* **447**, 3 (2006).

[33] Bach R., *Math. Zeit.* **9**, 110 (1921).

[34] Bahcall N. A., Ostriker J. P., Perlmutter S., Steinhardt P. J., *Science* **284**, 1481 (1999).

[35] Bardeen J., *Phys. Rev.* D **22**, 1882 (1980).

[36] Barrow J., Ottewill A.C., *J. Phys. A: Math. Gen.* **16**, 2757 (1983).

[37] Barrow J. D., Tipler F. J. , *The Anthropic Cosmological Principle*, Clarendon Press, Oxford, 1986.

[38] Basini G., Capozziello S., *Gen. Rel. Grav.* **35**, 2217 (2003).

[39] Basini G., Capozziello S., *Int. Jou. Mod. Phys. D* **15**, 583 (2006).

[40] Basini G., Capozziello S., *Mod. Phys. Lett. A* **20**, 251 (2005).

[41] Bassett B.A., Kunz M., Parkinson D., Ungarelli C., *Phys. Rev. D* 68, 043504 (2003).

[42] Battaglia-Mayer A., Schmidt H.-J., *Class. Quantum Grav.* **10**, 2441 (1993).

[43] Bean R.,Bernat D., Pogosian L., Silvestri A., Trodden M., *Phys. Rev. D* **75**, 064020 (2007).

[44] Bellucci S.,Capozziello S., De Laurentis M., Faraoni V., *Phys. Rev. D* **79**, 104004 (2009).

[45] Bennet C.L. *et al.*, ApJS **148**, 1 (1996).

[46] Bennett C. L. *et al.*, *Astrophys. J.* **464**, L1 (1996).

[47] Bergmann P.G., *Int. Journ. Theor. Phys.* **1**, 25 (1968).

[48] Berkin A., Maeda K., *Phys. Lett. B* **245**, 348 (1990).

[49] Bernabei R. *et al.*, *Riv. Nuovo Cim.* **26**, 1, 1 (2003).

[50] Bertolami O., Böhmer Ch.G., Lobo F.S.N., *Phys. Rev. D* **75**, 104016 (2007).

[51] Bertone G. , Hooper D., Silk J., *Phys. Rept.* **405**, 279 (2005).

[52] Bertotti B., Iess L., Tortora P., *Nature* **425**, 374 (2003).

[53] Blasone M., Capolupo A., Capozziello S., Carloni S., Vitiello G., *Phys. Lett. A* **323**, 182 (2004).

[54] Blasone M., Capolupo A., Capozziello S., Vitiello G., Nuclear Instruments and Methods in Physics Research A 588 (NIMA), pp. 272-275, (2008).

[55] Binney J., Tremaine S., *Galactic Dynamics*, Princeton Univ. Press, Princeton, (1987).

[56] Birrel N.D., Davies P.C.W., *Quantum Fields in Curved Space*, Cambridge Univ. Press, Cambridge, (1982).

[57] Blagojevic M., *SFIN* **A1**, 147 (2003).

[58] Bogdanos C., Capozziello S., De Laurentis M., Nesseris S., arXiv:0911.3094.

[59] Bondi H., *Cosmology*, Cambridge Univ. Press, Cambridge, 1952.

[60] Borisov A.B., Ogievetskii V.I., *Theor. Mat. Fiz.* **21**, 329 (1994).

[61] Brans C., Dicke R.H., *Phys. Rev.* **124**, 925 (1961).

[62] Brill D. R., Gowdy R. H. , *Rep. Prog. Phys.* **33**, 413 (1970).

[63] Brill D. R. *et al.*, *Rev. Mod. Phys.* **29**, 465 (1957).

[64] Bosma A., *The distribution and kinematics of neutral hydrogen in spiral galaxies of various morphological types*, PhD Thesis, Rejksuniversiteit Groningen, (1978).

[65] Bosma A., lectures at *"Chaotic Dynamics of Gravitational Systems"*, in Celestial Mechanics*preprint* arXiv: astro-ph/9812015.

[66] Brout R., Englert F., Gunzig E., *Gen. Rel. Grav.* **10**, 1 (1979).

[67] Buchbinder I. L., Odintsov S. D., Shapiro I. L., *Effective Action in Quantum Gravity*, IOP Publishing, Bristol, (1992).

[68] Buchdahl H., *Acta Math.* **85**, 63 (1951).

[69] Buchdahl H.A., *J. Phys. A* **12**, 8, 1229 (1979).

[70] Buchert T. , *AIPConf. Proc.* **910**, 361–380 (2007).

[71] Buonanno A., *TASI Lectures on Gravitational Waves from the Early Universe*, arXiv:gr-qc/0303085.

[72] Burles S., Nollett K. M., Turner M. S., *Astrophys. J. Lett.* **552**, L1 (2001).

[73] Caldwell R. R., Dave R., Steinhardt P. J., *Phys. Rev. Lett.* **80**, 1582 (1998).

[74] Callan C. G., Coleman S., Jackiw R., *Ann. Phys. (N.Y.)* **59**, 42 (1970).

[75] Callan C. G.,Coleman S., Wess J., Zumino B., *Phys. Rev.* **117**, 2247 (1969).

[76] Capolupo A., Capozziello S., Vitiello G., *Phys. Lett. A* **363**, 53–56 (2007).

[77] Capozziello S., Tsujikawa S., *Phys. Rev. D* **77**, 107501-4 (2008).

[78] Capozziello S., De Laurentis M., Nojiri S., Odintsov S.D., *General Relativity and Gravitation* **41**, 10, 2313 (2009).

[79] Capozziello S., De Laurentis M., Francaviglia M., Mercadante S., *Foundations of Physics* **39** 1161–1176 (2209).

[80] Capozziello S., In: Quantum Gravity Research Trends ISBN 1-59454-324-0 Editor: Albert Reimer, pp. 227-276, Nova Science Publishers, Inc., (2005).

[81] Capozziello S., de Ritis R., Rubano C., Scudellaro P., *La Rivista del Nuovo Cimento* **19**, no. 4 (1996).

[82] Capozziello S., Cianci R., Stornaiolo C., Vignolo S., *Class. Quantum Grav.* **24**, 6417 (2007).

[83] Capozziello S., Cianci R., Stornaiolo C., Vignolo S., *Int. J. Geom. Methods Mod. Phys.* **5**, 765 (2008).

[84] Capozziello S., De Laurentis M., Nojiri S., Odintsov S.D., *Phys. Rev. D* **79**, 124007 (2009).

[85] Capozziello S., Occhionero F., Amendola L., *Int. Journ. Mod. Phys.* D **1**, 615 (1993).

[86] Capozziello S., De Laurentis M., *Int. Jou. Geom. Meth. in Mod. Phys.* **6**, 1 (2009).

[87] Capozziello S., De Laurentis M., to appear in *Foundations of Physics*, 2009, arXiv:0910.2881.

[88] Capozziello S., M. Francaviglia M., *Gen. Rel. Grav. Special Issue on Dark Energy*, **40**, 357 (2008).

[89] Capozziello S., de Ritis R. *Phys. Lett.* A **195**, 48 (1994).

[90] Capozziello S., Nojiri S., Odintsov S.D., *Phys. Lett. B* **634**, 93 (2006).

[91] Capozziello S., Stornaiolo C., *Int. Jou. Geom. Methods in Mod. Phys.* **5**, 185 (2008).

[92] Capozziello S., de Ritis R., Marino A.A., *Gen. Relativ. Grav.* **30**, 1247 (1998).

[93] Capozziello S., Cardone V.F., Elizalde E., Nojiri S., Odintsov S.D., *Phys. Rev. D* **73**, 043512 (2006).

[94] Capozziello S., *Int. J. Mod. Phys. D* **11**, 483 (2002).

[95] Cardone V.F., Troisi A., Capozziello S., *Phys. Rev. D* **69**, 083517 (2004).

[96] Capozziello S., Cardone V.F., Piedipalumbo E., Sereno M., Troisi A., *Int. J. Mod. Phys. D* **12**, 381 (2003).

[97] Capozziello S., Cardone V.F., Carloni S., Troisi A., *Int. J. Mod. Phys. D* **12**, 1969 (2003).

[98] Capozziello S., de Ritis R., Rubano C., Scudellaro P., *Int. Journ. Mod. Phys.* D **5**, 85 (1996).

[99] Capozziello S., de Ritis R., *Class. Quantum Grav.* **11**, 107 (1994).

[100] Capozziello S., Lambiase G., Stornaiolo C., *Ann. Phys. (Leipzig)* **10**, 713 (2001).

[101] Capozziello S., Elizalde E., Nojiri, S., Odintsov S. D., *Phys. Lett. B* **671**, 193 (2009).

[102] Capozziello S., Carloni S., Troisi A., *Rec. Res. Devel. Astronomy. & Astrophys.* **1**, 625 (2003).

[103] Capozziello S., Cardone V. F., Carloni S., Troisi A., *Int. J. Mod. Phys. D* **12**, 1969 (2003).

[104] Capozziello S., De Laurentis M., Corda C., *Phys. Lett. B* **699**, 255 (2008).

[105] Capozziello S., Corda C., *Int. J. Mod. Phys. D* **15**, 1119 (2006).

[106] Capozziello S., De Laurentis M., Francaviglia M., *Astrop. Phys.* **29**, 2, 125–129 (2008).

[107] Capozziello S., De Laurentis M., Corda C., *Modern Physics Letters A* **22**, 35 (2007).

[108] Capozziello S., Corda C., De Laurentis M., *Modern Physics Letters A* **22**, 15, 1097–1104. (2007).

[109] Capozziello S., Cardone V.F., Funaro M., Andreon S., *Phys. Rev. D* **70**, 123501 (2004).

[110] Capozziello S., Cardone V.F., Troisi A., *Phys. Rev. D* **71**, 043503 (2005).

[111] Capozziello S., Nojiri S., Odintsov S.D., Troisi A., *Phys. Lett. B* **639**, 135 (2006).

[112] Carloni S., Dunsby P., Capozziello S., Troisi A, *Class. Quant. Grav.* **22**, 4839 (2005).

[113] Carloni S., Troisi A., Dunsby P.K.S. 2007, ArXiv: 0706.0452 [gr-qc]

[114] Capozziello S., Stabile A., Troisi A., *Phys. Rev. D* 76, 104019 (2007).

[115] Capozziello S., Stabile A., Troisi A., *Class. Quant. Grav.* **25**, 085004 (2008).

[116] Capozziello S., Stabile A., Troisi A ., *Class. Quant. Grav.* **24**, 2153 (2007).

[117] Capozziello S., De Laurentis M., Garufi F., Milano L. *Physica Scripta* **79**, 025901 (2009).

[118] Capozziello S., De Filippis E., Salzano V., (2008) arXiv:0809.1882 [astro-ph], to appear in MNRAS.

[119] Capozziello S., Troisi A., *Phys. Rev. D* 72 (2005).

[120] Capozziello S., Stabile A., Troisi A., Mod. Phys. Lett. A **21**, 2291 (2006).

[121] Carroll S. M., Kaplinghat M., *Phys. Rev. D* **65**, 063507 (2002).

[122] Carroll S. M., Press W. H., Turner E. W., *Annu. Rev. Astron. Astrophys.* **30**, 499 (1992).

[123] Carroll S. M., *Living Rev. Relativity* **4**, 1 (2001).

[124] Carroll S. M., *EFI*, 27 (2001).

[125] Carroll S. M., *Phys. Rev. Lett.* **81**, 3067 (1998).

[126] Carroll S.M., Duvvuri V., Trodden M., Turner M.S., *Phys. Rev. D* **70**, 043528 (2004).

[127] Carroll S. *et al.*, *Phys. Rev. D* **50**, 3867 (1994).

[128] Carroll S. M., De Felice A., Duvvuri V., Easson D. A., Trodden M., Turner M. S., *Phys. Rev. D* **71**, 063513 (2005).

[129] Carroll S. M., arXiv:gr-qc/9712019.

[130] Cartan E., *Ann. Ec. Norm.* **42**, 17 (1925).

[131] Carter B. , in *International Astronomical Union Symposium 63: Confrontation of Cosmological Theories with Observational Data*, edited by M. S. Lorgair, Dordrecht, Reidel, 1974.

[132] Carter B., in *The Constants of Physics, Proceedings of a Royal Society Discussion Meeting, 1983*, edited by W. H. McCrea and M. J. Rees, Cambrige University Press, Cambridge, (1983).

[133] Chang L. N., Macrae K. I., Mansouri F., *Phys. Rev.D* **13**, 235 (1976).

[134] Chang L. N. *et al.*, *Phys. Rev. D* **17**, 3168 (1978).

[135] Chen G., Rathra B., *ApJ* **582**, 586 (2003).

[136] Podariu S., Daly R.A., Mory M.P., Rathra B., *ApJ* **584**, 577 (2003).

[137] Chiba T., *JCAP* **0503**, 008 (2005).

[138] Clowe D., Bradac M., Gonzalez A. H., Markevitch M., Randall S. W., Jones C., Zaritsky D. , *Astrophys.J.* **648**, L109-L113 (2006).

[139] Cognola G., Elizalde E., Nojiri S., Odintsov S.D., Sebastiani L., Zerbini, S., *Phys. Rev. D* **77**, 046009 (2008).

[140] Coleman S., Weinberg E.J., *Phys. rev, D* **7**, 1888 (1973).

[141] Coleman S.,Wess J., Zumino B., *Phys. Rev.* **117**, 2239 (1969).

[142] Coll B., *A universal law of gravitational deformation for general relativity*, Proc. of the Spanish Relativistic Meeting, EREs, Salamanca Spain, (1998).

[143] Coll B., Llosa J., Soler D., *Gen. Rel. Grav.* **34**, 269 (2002).

[144] Copeland E. J.,Liddle A. R., Wands D., *Phys. Rev. D* **f 57**, 4686 (1998).

[145] Copeland E.J.,Sami M., Tsujikawa S., *Int. Jou. Mod. Phys. D* **15**, 1753 (2006).

[146] Corda C., *JCAP* **0704**, 009 (2007).

[147] Damour T., Gibbons G.W., Gundlach C., *Phys. Rev. Lett.* **64**, 123 (1990).

[148] Damour T., Gibbons G.W., Taylor J.H., *Phys. Rev. Lett.* **61**, 1151 (1988).

[149] Damour T., Esposito-Farèse G., *Class. Quantum Grav.* **9**, 2093 (1992).

[150] Damour T., Esposito-Farese G., *Phys. Rev. Lett.* **70**, 2220 (1993).

[151] Damour T., Esposito-Farese G. *Phys. Rev. D* **54**, 1474 (1996).

[152] Damour T., Esposito-Farese G. *Phys. Rev. D* **58**, 042001 (1998).

[153] Dautcourt G., *Acta Phys. Polon.* **25**, 637 (1964).

[154] de Bernardis P. et al., *Nature* **404**, 955 (2000).

[155] De Felice A., Mukherjee P., Wang Y., arXiv:0706.1197 [astro-ph].

[156] Delamotte B., *Am. J. Phys.* **72**, 170 (2004).

[157] Deser S., *Gen. Rel. Grav.* **1**, 181 (1970).

[158] de Sitter, W. *Mon. Not. R. Astron. Soc.* **78**, 3 (1917).

[159] De Witt B. S., *Phys. Rev. Lett.* **13**, 114 (1964).

[160] De Witt B. S., *"Dynamical theory of groups and . . . elds* (Les Houches Lectures 1963), Relativity, Groups and Topology", Gordon and Breach Science Publishers, New York, (1965).

[161] Dick R., *Gen. Rel. Grav.* **36**, 217 (2004).

[162] Dicke R.H., *The many faces of Mach* in *Gravitation and Relativity* eds. H.Y. Chiu and W.F. Hofmann, Benjamin, New York, 1964.

[163] Dicke R.H., *Phys. Rev. D* **125**, 2163 (1962).

[164] Dicke R. H., E. Peebles P. J., in *General Relativity: An Einstein Centenary Survey*, edited by S. W. Hawking and W. Israel, Cambridge University Press, Cambridge, (1979).

[165] Dickey J.O. *et al.*, *Science*, 265, (1994) (and reference therein).

[166] Dolgov A. D., Kawasaki, M. *Phys. Lett. B* **573**, 1 (2003).

[167] Duff M. J., *Int. J. Mod. Phys. A* **11**, 5623 (1996).

[168] Duruisseau J.P., Kerner R., *Gen. Rel. Grav.* **15**, 797-807 (1983).

[169] Dvali G.R., Gabadadze G., Porrati M., *Phys. Lett. B* **485**, 208 (2000).

[170] Eddington A. S., *The Mathematical Theory of Relativity*, Cambridge University Press, Cambridge, (1923).

[171] Ehlers J., *Ann. N. Y. Acad. Scien.* **336**, 279 (1980).

[172] Ehlers J., *Grundlagenprobleme der modernen Physik*, Eds. J. Nitsch, J. Pfarr, E.W. Stachow, B.I.-Wissenschaftsverlag, Mannheim, 65, (1981).

[173] Ellis G. F. R., *Gen. Rel. and Grav.*, GR10 Conf. Rep. Ed. B. Bertotti, Reidel, Dordrecht p. 215, (1984).

[174] Ellis J. R. , *preprint* arXiv: astro-ph/0204059.

[175] Einstein A., *Ann. der Physik* **49**, 769 (1916).

[176] Einstein A., *Sitzungber. Preuss. Akad. Wiss. Phys.-Math. Kl.*, 142 (1917).

[177] Einstein A., *Sitzung-ber. Preuss. Akad. Wiss.*, 414 (1925).

[178] Eisenhart L.P., *Riemannian Geometry* Princeton Univ. Press, Princeton, (1955).

[179] Eisenstein D. J. *et al.* [SDSS Collaboration], *Astrophys. J.* **633**, 560 (2005).

[180] Emparan R., Garriga J., *JHEP* **0603**, 028 (2006).

[181] Faraoni V., *Class. Quantum Grav.* **22**, 32352 (2005).

[182] Faraoni V. *Cosmology in Scalar-Tensor Gravity*, Kluwer Academic, Dordrecht, (2004).

[183] Capozziello S., De Laurentis M., Faraoni V., The Special Issue in Cosmology, The Open Astronomy Journal, 2009, arXiv:0909.4672.

[184] Farmer A.J., Phinney E.S., *Mon. Not. Roy. Astron. Soc.* **346**, 1197 (2003).

[185] Faulkner T., Tegmark M., Bunn E. F., Mao Y., *Phys. Rev. D* **76**, 063505 (2007).

[186] Feldman H., Brandenberger R., *Phys. Lett.* B **227**, 359 (1989).

[187] Ferraris M., Francaviglia M., Magnano G., *Class. Quantum Grav.* **5**, L95 (1988).

[188] Ferraris M., Francaviglia M., Reina C., *Gen. Relativ. Grav.* **14**, 243 (1982).

[189] Ferraris M., Francaviglia M., Volovich I., *Class. Quantum Grav.* **11**, 1505 (1994).

[190] Ferraris M., Francaviglia M., in: *Mechanics, Analysis and Geometry: 200 Years after Lagrange*; Editor: M. Francaviglia, Elsevier Science Publishers B:V:, (1991).

[191] Ferreira P. G., Joyce M., *Phys. Rev. D* **58**, 023503 (1998).

[192] Finkelstein R., *Ann. Phys.* **12**, 200 (1961).

[193] Flanagan E.E., *Phys. Rev. Lett.* **92**, 071101 (2004).

[194] Freese K., Lewis M., *Phys. Lett. B* **540**, 1 (2002).

[195] Friedrichs K., *Math. Ann.* **98**, 566 (1927).

[196] Fulling S.A., *Aspects of Quantum Field Theory in Curved Space Times*, Cambridge Univ. Press, Cambridge, (1989).

[197] Gasperini M., Veneziano G., *Phys. Lett. B* **277**, 256 (1992).

[198] Gautreau R., *Phys. Rev. D* **29**, 198-206 (1984).

[199] Giachetta G., *J. Math. Phys.* **40**, 939 (1999).

[200] Goldwirth D.S., Piran T., *Phys. Rep.* **214**, 223 (1992).

[201] Gottlöber S., Schmidt H.-J., Starobinsky A.A., *Class. Quantum Grav.* **7**, 893 (1990).

[202] Green M.B., Schwarz J.H., Witten E., *Superstring Theory*, Cambridge Univ. Press, Cambridge, (1987).

[203] Grignani G., Nardelli G., *Phys. Rev. D* **45**, 2719 (1992).

[204] Grishchuk L. *et al.*, *Phys. Usp.* **44**, 1-51 (2001).

[205] Grishchuk L. *et al.*, *Usp. Fiz. Nauk* **171** 3-59 (2001).

[206] Guth A. H., *Phys. Rev. D* **23**, 347 (1981).

[207] Hanany S. *et al.*, *Astrophys. J.* **545**, L5 (2000).

[208] Hawking S. W., *Proc. R. Soc. A* **300**, 187 (1967).

[209] Hawking S. W., Ellis G. F. R., *The Large Scale Structure of Space-Time*, Cambridge University Press, Cambridge, (1973).

[210] Hehl F. W., Datta B.K., *J. Math. Phys.* **12**, 1334 (1971).

[211] Hehl F. W., von der Heyde P., Kerlick G. D., Nester J. M., *Rev. Mod. Phys.* **48**, 393 (1976).

[212] Hehl F. W. *et al.Phys. Rep.* **258**, 1 (1995).

[213] Hehl F. W., McCrea J. D., *Found. of Phys.* **16**, 267 (1986).

[214] Hinshaw G. *et al. Ap. J.* **148**, 135 (2003).

[215] Horvat R., *Mod. Phys. Lett. A* **14**, 2245 (1999).

[216] http://map.gsfc.nasa.gov/

[217] http://www.rssd.esa.int/index.php?project=Planck

[218] http://www.mso.anu.edu.au/2dFGRS

[219] http://www.sdss.org/

[220] http://www.onera.fr/prix-en/brun2002/edmond-brun2002.html

[221] http://astro.estec.esa.nl/astrogen/COSPAR/step/

[222] http://lpsc.in2p3.fr/mayet/dm.html

[223] http:// lhc.web.cern.ch/lhc/

[224] http:// www.ligo.org/pdf public/camp.pdf.; http://www.ligo.org/pdf public/hough02.pdf.

[225] http:// www.virgo.infn.it

[226] http:// www.lisa.nasa.gov; www.lisa.esa.int

[227] Hu W., White M., *Astrophys. J.* **471**, 30 (1996).

[228] Hu W., Sawicki I., *Phys. Rev. D* **76**, 064004 (2007).

[229] Hulse R.A., Taylor J.H., *Ap. J.* **195**, L51 (1975).

[230] Hwang J., Noh H., Puetzfeld D. astro-ph/0507085, (2005).

[231] Isham C. J., in *Quantum Gravity 2: A Second Oxford Symposium*, edited by Isham, Penrose C. J. R., Sciama D. W., Clarendon Press, Oxford, (1981).

[232] Isham C. J., Salam A., Strathdee J., *Phys. Rev. D* **3**, 1805 (1971).

[233] Inomata A., Trikala M., *Phys. Rev. D* **19**, 1665 (1978).

[234] Ivanov E. A., Ogievetskii, V. I. *Gauge theories as theories of spontaneous breakdown*, Preprint of the Joint Institute of Nuclear Research, E2-9822, 1976, 3-10.

[235] Ivanov E. A., Niederle J., *Phys. Rev. D* **25**, 976 (1982).

[236] Ivanov E. A., Niederle J., *Phys. Rev. D* **25**, 988.

[237] Ivanenko D., Sardanashvily G. A., *Phys. Rep.* **94**, 1 (1983).

[238] Itzykson C., Zuber J.C., *Quantum Field Theory*, McGraw-Hill, New York, (1980).

[239] Jackson J.D., *Classical Electrodynamics*, Academic Press, New York, (1998).

[240] Julve J. et. al., *Class. Quantum Grav.* **12**, 1327 (1995).

[241] Kahn F., Waltjer L., *Astrophys. J.* , **130**, 705 (1959).

[242] Kaku M., *Quantum Field Theory*, Oxford Univ. Press, Oxford, 1993.

[243] Kamenshchik A., Moschella U., Pasquier V., *Phys. Lett. B* **511**, 265 (2001).

[244] Kazanas D., *Astrophys. J.* **241**, L59 (1980).

[245] Khriplovich I. B., *Sov. J. Nucl. Phys.* **3**, 415 (1966).

[246] Klein O., *New Theories in Physics*, 77, Intern.Inst. of Intellectual Co-operation, League of Nations, (1938).

[247] Kibble T.W., *J. Math. Phys.* **2**, 212 (1960).

[248] Kibble T.W., Lorentz Invariance and the Gravitational Field, *J. Math. Phys.* **2**, 212 (1961).

[249] Kilmister C.W., *J. Math. Phys.* **12**, 1 (1963).

[250] Kofman L., Linde A. D., Starobinsky A. A., *Phys. Rev. Lett.* **73**, 3195 (1994).

[251] Kofman L., Linde A. D., Starobinsky A. A., *Phys. Rev. D* **56**, 3258 (1997).

[252] Kolb E.W., Turner M.S., *The Early Universe*, Ed. Addison–Wesley, New York, (1990).

[253] Kold E. W., Turner M. S., *The Early Universe*, Addison-Wesley, California, 1990.

[254] Komatsu E. *et al.*, *Astrophys. J. Suppl.* , **148**, 119 (2003).

[255] Krauss L.M., White M., *Ap. J.* **397**, 357 (1992).

[256] Kuenzle H.P., *Gen. Rel. Grav.* **7**, 445 (1976).

[257] Kung J., Brandenberger R., *Phys. Rev.* D **42**, 1008 (1990).

[258] La D., Steinhardt P.J., *Phys. Rev. Lett.* **62**, 376 (1989).

[259] Lanczos C., *Ann. Math.* **39**, 842 (1938).

[260] Landau L., Lifschitz E.M., *Théorie des Champs*, ed. Mir, Moscow, (1970).

[261] Lemaître G., *Ann. Soc. Sci. Bruxelles A* **47**, 49 (1933).

[262] Lemaître G., *Gen. Rel. Grav.* **29**, 641 (1997).

[263] Levi-Civita T., *The Absolute Differential Calculus*, Blackie and Son, London, 1929.

[264] Li B., Barrow J.D., *Phys. Rev. D* **75**, 084010 (2007).

[265] Nojiri S., Odintsov S.D., hep-th/06012113 (2006).

[266] Li B., Barrow J.D., Mota D.F., submitted to *Phys. Rev. D*, gr-qc/0705.3795.

[267] Li B., Barrow J. D., Phys. Rev. D **75**, 084010 (2007).

[268] Li B., Chu M.C., *Phys. Rev. D* **74**, 104010 (2006).

[269] Li B., Chan K.C., Chu M.C., Phys. Rev. D,in press, astro-ph/0610794 (2006).

[270] Liddle A. R., Lyth D. H., *Phys. Lett. B* **291**, 391 (1992).

[271] Liddle A. R., Lyth D. H.,, *Phys. Rep.* **231**, 1 (1993).

[272] Liddle A. R., Lyth D. H., *Cosmological Inflation and Large Scale Structure*, Cambridge University Press, Cambridge, 2000.

[273] Liddle A. R., Scherrer R. J., *Phys. Rev. D* **59**, 023509 (1999).

[274] Linde A., *Particle Physics and Inflationary Cosmology*, Harwood Academic Publishers, Switzerland, 1990.

[275] Linde A.,*Phys. Lett. B* **108**, 389 (1982).

[276] Linde A., *Phys. Lett. B* **114**, 431 (1982).

[277] Linde A.,*Phys. Lett. B* **129**, 177 (1983).

[278] Linde A.D., in *The Very Early Universe* eds. S.W. Hawking, G. Gibbons and S. Siklos, Cambridge Univ. Press, Cambridge, (1983).

[279] Llosa J., Soler D., *Class. Quant. Grav.* **21**, 3067 (2004).

[280] Lord A., Goswami P., *J. Math. Phys.* **27**, 3051 (1986).

[281] Lord A., Goswami P., *J. Math. Phys.* **29**, 258 (1987).

[282] Lopez-Pinto A., Tiemblo A., Tresguerres R., *Class. Quant. Grav.* **12**, 1503 (1995).

[283] Lucchin F., Matarrese S., *Phys. Rev.* D **32**, 1316 (1985).

[284] Lue A., Scoccimarro R., Starkman G., *Phys. Rev. D* **69**, 044005 (2004).

[285] Lyth D.H., *Phys. Rev.* D **31**, 1792 (1985).

[286] Lyth D.H., Stewart E.D., *Phys. Lett.* B **274**, 168 (1992).

[287] Maeda K., *Phys. Rev.* D **39**, 3159 (1989).

[288] Maggiore M., *Phys. Rep.* **331**, 283-367 (2000).

[289] Magnano G., SokoLowski L.M., *Phys. Rev. D* **50**, 5039 (1994).

[290] Magnano G., Ferraris M., Francaviglia M., *Class. Quantum Grav.* **7**, 557 (1990).

[291] Mannheim P.D., Kazanas D., *Ap. J.* **342**, 635 (1989).

[292] Mansouri F., Chang L.N., *Phys. Rev.* **D13**, 3192 (1976).

[293] Mansouri F., *Phys. Rev. Lett.* **42**, 1021 (1979).

[294] McVittie G.C., MNRAS **93**, 325-3 (1933).

[295] Meng X., Wang P., *GRG* **36** (8), 1947 (2004); *GRG* **36** (12), 2673 (2004).

[296] Miller A. D. *et al.*, *Astrophys. J.* **524**, L1 (1999).

[297] Mijiić M.B., Morris M.S., Suen W.M., *Phys. Rev.* D **34**, 2934 (1986).

[298] Misner C. W., *Astrophys. J.* **151**, 431 (1968).

[299] Misner C. W., *Phys. Rev. Lett.* **22**, 1071 (1969).

[300] Misner C. W., Thorne K. S., Wheeler J. A., *Gravitation*, W.H. Feeman and Company, (1973).

[301] Moore B., *preprint* arXiv: astro-ph/0103100.

[302] Mukhanov V. F., *preprint* arXiv: astro-ph/0303077.

[303] Mukhanov V., *Physical Foundations of Cosmology*, Cambridge University Press, Cambridge, (2005).

[304] Nakahara M., *Geometry, Topology and Physics*, Second Edition, Graduate Student Series in Physics, Institute of Physics, Bristol, (2003).

[305] Ne'eman Y., T. Regge, *Riv. Nuovo Cimento* **1**, 1 (1978).

[306] Ne'eman Y., Sijacki D., *Gravity from Symmetry Breakdown of a Gauge Affine Theory*, The Center for Particle Theory, University of Texas at Austin, D6-87/40, (1987).

[307] Nojiri S., Odintsov S.D., *Phys. Rev. D* **68**, 123512 (2003).

[308] Nojiri S., Odintsov S.D., *Int. J. Meth. Mod. Phys.* **4**, 115 (2007).

[309] Nojiri S., Odintsov S. D., *Phys. Lett. B* **576**, 5 (2003).

[310] Nojiri S., Odintsov S.D., *Phys. Lett. B* **576**, 5 (2003).

[311] Nojiri S., Odintsov S.D., *Phys. Rev. D* **71**, 123509 (2005).

[312] Nojiri S., Odintsov S. D., *Phys. Rev. D* **77**, 026007 (2008).

[313] Nojiri S., Odintsov S.D., *GRG* **36** 1765 (2004).

[314] Nojiri S., Odintsov S.D., *Phys. Lett. B* **657**, 238 (2008).

[315] Nunez A., Solganik S., *Phys. Lett. B* **608**, 189 (2005), [arXiv:hep-th/0411102].

[316] Olmo G.J., *Phys. Rev. Lett.* **95**, 261102 (2005).

[317] Olmo G.J., *Phys. Rev. D* **72**, 083505 (2005).

[318] Ostriker J. P., Steinhardt P. J., *Nature* **377**, 600 (1995).

[319] Oyaizu H., arXiv:0807.2449 [astro-ph]; H. Oyaizu, M. Lima, W. Hu, arXiv:0807.2462 [astro-ph].

[320] Pais A., *'Subtle is the Lord...': The Science and the Life of Albert Einstein*, Oxford University Press, New York, 1982.

[321] Palatini A., *Rend. Circ. Mat. Palermo* **43**, 203 (1919).

[322] Padmanabhan T., *Phys. Rept.* **380**, 235 (2003).

[323] Padmanabhan T., *Phys. Rev. D* **66**, 021301 (2002).

[324] Pauli W., *Theory of Relativity*, Pergamon Press, London, 1967.

[325] Peebles P. J. E., Ratra B., *Rev. Mod. Phys.* **75**, 559 (2003).

[326] Peebles P. J. E., *Principles of Physical Cosmology*, Princeton University Press, Princeton, 1993.

[327] Peebles P. J. E., Yu J. T., *Astrophys. J.* **162**, 815 (1970).

[328] Peebles P. J., Ratra R., *Astrophys. J.* **325**, L17 (1988).

[329] Peiris H. V. *et al.*, *Astrophys. J. Suppl.* **148**, 213 (2003).

[330] Penrose R., *Proc. R. Soc.* A **284**, 159 (1965).

[331] Penzias A. A., Wilson R. W., *Astrophys. J.* **142**, 419 (1965).

[332] Perlmutter S. *et al.*, *ApJ* **517**, 565 (1999); R.A. Knop *et al.*, *ApJ* **598**, 102 (2003).

[333] Polchinski J., *String Theory*, Cambridge University Press, New York, (1998).

[334] Pogosian L., Silvestri A., *Phys. Rev.* D **77**, 023503, (2008).

[335] Pound R.V., Rebka G.A., *Phys. Rev. Lett.* **4**, 337 (1960).

[336] Puetzfeld D. *Comp. Phys. Comm.* **175**, 497 (2006).

[337] Quant I., Schmidt H.-J., *Astron. Nachr.* **312**, 97 (1991).

[338] Ramond P., *Field Theory: A modern Primer*, ed. Addison–Wesley Pub. Co. Menlo Park (CA), (1988).

[339] Ratra R., Peebles P. J., *Phys. Rev. D* **37**, 3406 (1988).

[340] Refregier A., *Ann. Rev. Astron. Astrophys.* **41**, 645 (2003).

[341] Riess A.G. *et al.*, *ApJ* **116**, 1009 (1998).

[342] Riess A.G. *et al.*, *ApJ* **607**, 665 (2004).

[343] Riess A. G.*et al.* [Supernova Search Team Collaboration], *Astrophys. J.* **607**, 665 (2004).

[344] Riess A.G. *et al.*, *AJ* **116**, 1009 (1998); J.L. Tonry *et al.*, *ApJ* **594**, 1 (2003).

[345] Rovelli C., *Living Rev. Relativity* **1**, 1 (1998).

[346] Rovelli C., *Quantum Gravity*, Cambridge University Press, New York, (2004).

[347] Rubano C., Scudellaro P., *Gen. Relativ. Grav.* **37**, 521 (2005).

[348] Rubin V., Ford W. K., Jr, *Astrophys. J.* **159**, 379 (1970).

[349] Rubin V., Ford W. K., Jr, Thonnard N., *Astrophys. J.* **238**, 471 (1980).

[350] Ruzmaikina T.V., Ruzmaikin A.A., *JETP*, **30**, 372 (1970).

[351] Sachs R., Wolfe A., *Astrophys. J.* **147**, 73 (1967).

[352] Sahni V., Starobinsky A., *Int. J. Mod. Phys. D* **9**, 373 (2000).

[353] Sahni V., Wang L., *Phys. Rev. D* **62**, 103517 (2000).

[354] Sanders R.H., *Ann. Rev. Astr. Ap.* **2**, 1 (1990).

[355] Sahni V., Starobinski A., *Int. J. Mod. Phys. D* **9**, 373 (2000).

[356] Salam A., Strathdee J., *Phys. Rev.* **184**, 1750 (1969); *Phys. Rev.* **184**, 1760 (1969).

[357] Sardanashvily G., arXiv: gr-qc/0201074 (2002).

[358] Sato K., *Phys. Lett. B* **33**, 66 (1981).

[359] Scherk J., Schwarz H.J., *Nucl. Phys.* B **81**, 118 (1974).

[360] Schneider P., Ehlers J., Falco E.E., *Gravitational Lenses*, Springer–Verlag, Berlin, (1992).

[361] Schrödinger E., *Space-Time Structure*, Cambridge Univ. Press, Cambridge, (1960).

[362] Schwarz A., *Topology for Physicists*, Springer-Verlag, Berlin-Heidelberg, (1994).

[363] Schwinger J., *Phys. Rev.* **130**, 800 (1963).

[364] Shapiro I. L., *Phys. Rept.* **357**, 113 (2002).

[365] Shapiro I.I., in *General Relativity and Gravitation* 12, Ashby N., *et al.*, Eds. Cambridge University Press (1993).

[366] Shapiro S.S., *et al.*, *Phys. Rev. Lett. D* **92**, 121101 (2004).

[367] Sciama D.W., in *Recent Developments in General Relativity*, 415, Pergamon Press, Oxford, 1962.

[368] Sciama D.W., *Mon. Not. R. Astron. Soc.* **113**, 34 (1953).

[369] Sciama D.W., *On the analog between charge and spin in General Relativity,* in Recent Developments in General Relativity, Festschrift for Leopold Infeld, 415, Pergamon Press, New York, (1962).

[370] Seiberg N., *"Emergent space-time"*, Rapporteur talk at the 23rd Solvay Conference in Physics, December, (2005), arXiv:hep-th/0601234.

[371] Shtanov Y.,Traschen J. H., Brandenberger R. H., *Phys. Rev. D* **51**, 5438 (1995).

[372] Schimd C., Uzan J. P., Riazuelo A., *Phys. Rev. D* **71**, 083512 (2005).

[373] Schmidt H.J., *Class. Quant. Grav.* **7**, 1023 (1990).

[374] Schmidt H.J., *Class. Quantum Grav.* **7**, 1023 (1990).

[375] Sigg D., (for the LIGO Scientific Collaboration) www.ligo.org/pdf public/P050036.pdf

[376] Smoot G. F. *et al.*, *Astrophysic. J.* **396**, L1 (1992).

[377] Spergel D. N. *et al.*, *Astrophys. J. Suppl.* **148**, 175 (2003).

[378] Spergel D.N. *et al.*, *ApJS* **148**, 175 (2003).

[379] Spergel D.N. *et al.*, astro-ph/0603449 (2006).

[380] Spergel D.N. *et al.*, *ApJS* **148**, 195.

[381] Springel V. *et al.*, *Nature* **435**, 629 (2005).

[382] Sokolowski L. M., *Class. Quantum Grav.* **6**, 2045 (1989).

[383] Sotiriou T.P., Faraoni V., arXiv:0805.1726 [gr-qc], at press in *Phys. Rep.* 2009.

[384] Sotiriou T., Faraoni V. arXIv:0805.1726 [gr-qc]

[385] Sotiriou, T. P., *Class. Quant. Grav.* **23**, 1253 (2006).

[386] Sotiriou, T. P., *Class. Quant. Grav.* **23**, 5117 (2006).

[387] Sotiriou T.P., Faraoni V., arXiv:0805.1726 [gr-qc], at press in *Phys. Rep.* 2009.

[388] Stabile A., *TheWeak Field Limit of Higher Order Gravity*, arXiv:0809.3570.

[389] Starobinsky A. A., *Phys. Lett. B* **91**, 99 (1980).

[390] Starobinsky A.A., astro-ph/0706.2041 (2007).

[391] Starobinsky A. A., JETP Lett. **86**, 157 (2007).

[392] Stelle K. S., *Gen. Rel. Grav.* **9**, 353 (1978).

[393] Stelle K. S. *et al.*, *Phys. Rev. D* **21**, 1466 (1980).

[394] Stelle K. S., *Phys. Rev. D* **16**, 953 (1977).

[395] SteinhardtP. J., Wang L., Zlatev I., *Phys. Rev. D* **59**, 123504 (1999).

[396] Steinhardt P.J., Wang L., Zlatev I., *Phys. Rev.* D **59**, 123504 (1999).

[397] Stompor R. *et al.*, *ApJ* **561**, L7 (2001).

[398] Sumner T. J., *Living Rev. Relativity* **5**, 4 (2002).

[399] Sunyaev R. A., in *Large Scale Structure of the Universe*, edited by M. S. Longair and J. Einasto, Dordrecht, Reidel, 1978.

[400] Susskind L., *preprint* arXiv: hep-th/0302219.

[401] Tatsumi D., Tsunesada Y. and the TAMA Collaboration, *Class. Quant. Grav.* **21** 5 S451 (2004).

[402] Taylor J.H., Weinberg J.M., *Ap. J.* **235**, 908 (1982).

[403] Teyssandier P., Tourranc P., *J. Math. Phys.* **24**, 2793 (1983).

[404] Teyssandier P., Tourrenc Ph., *J. Math. Phys.* **24**, 2793 (1983).

[405] Thiemann T., *Lect. Notes Phys.* **631**, 41 (2003).

[406] Tiemblo A., Tresguerres R., Gravitational contribution to fermion masses, arXiv: gr-qc/0506034 *Eur. Phys. J. C* **42**, 437 (2005).

[407] Tiemblo A., Tresguerres R., *Recent Res. Devel. Phys.* **5**, 1255 (2004).

[408] Tiemblo A., Tresguerres R., arXiv: gr-qc/9607066

[409] Tseytlin A.A., *Int. Journ. Mod. Phys.* A **4**, 1257 (1989),

[410] Tseytlin A.A., Vafa C., *Nucl. Phys.* B **372**, 443 (1992).

[411] Tonry J.L., Schmidt B.P., Barris B. *et al. Ap. J.* **594**, 1 (2003).

[412] Tolman R. C., *Relativity, Thermodynamics, and Cosmology*, Clarendon Press, Oxford, (1934).

[413] Tourrenc Ph., *General Relativity and Gravitational Waves*, in Proc. of the Int. Summer School on Experimental Physics of Grav. Waves, September 6-18, 1999, Urbino (Italy), Eds. M. Barone et al., World Scientific (Singapore), (2000).

[414] Trautman A., *Comp. Rend. Heb, Sean.* **257**, 617 (1963).

[415] Tresguerres R., Mielke E. W., *Phys. Rev. D* **62**, 044004 (2000).

[416] Tresguerres R., *Phys. Rev.* **66**, 064025 (2002).

[417] Tsujikawa S., arXiv:0709.1391 [astro-ph], to appear in *Physical Review D.*

[418] van Dam H., Veltman M. J. G., *Nucl. Phys. B* **22**, 397 (1970).

[419] Verde L. *et al., MNRAS* **335**, 432 (2002).

[420] Vilkovisky G., *Class. Quantum Grav.* **9**, 895 (1992).

[421] Vollik D.N. *Phys. Rev. D* **68** 063510 (2003).

[422] Wald R. M., *General Relativity*, University of Chicago Press, United States of America, 1984.

[423] Wands D., *Class. Quant. Grav.* **11**, 269 (1994).

[424] Wang L. M., Caldwell R. R., Ostriker J. P., Steinhardt P. J., *Astrophys. J.* **530**, 17 (2000).

[425] Wang X., Tegmark M., Zaldarriaga M., *Phys. Rev. D* **65**, 123001 (2002).

[426] Watson G.S., *An exposition on inflationary cosmology*, North Carolina University Press, (2000).

[427] Weinberg S., *Phys. Rev. Lett.* **59**, 2607 (1987).

[428] Weinberg S., *Phys. Rev, D* **9**, 3357 (1974).

[429] Weinberg S., *Gravitation and Cosmology*, John Wiley & Sons, United States of America, (1972).

[430] Weinberg S., *Rev. Mod. Phys.* **61**, 1 (1989).

[431] Weinberg E., *Phys. Rev.* D **40**, 3950 (1989).

[432] Wetterich C., *Nucl. Phys. B* **302**, 668 (1988).

[433] Wheeler J. A., in *Relativity, Groups and Topology*, edited by B. S. DeWitt, C. M. DeWitt, (Gordon and Breach, New York, 1964).

[434] White S. D. M., Frenk C., Davis M., *Astrophys. J. Lett.* **274**, L1 (1983).

[435] Will C.M., *Theory and Experiments in Gravitational Physics*, Cambridge Univ. Press, Cambridge, (1993).

[436] Will C.M., *Living Rev. Rel.* **4**, 4 (2001).

[437] Williams J.G., *et al.*, *Phys. Rev. D* **53**, 6730 (1996).

[438] Willke B. *et al.*, *Class. Quant. Grav.* **23**, 8S207-S214 (2006)

[439] Vassilevich D. V., *Phys. Rept.* **388**, 279 (2003).

[440] Vilkovisky G. A., *Class. Quant. Grav.* **9**, 895 (1992).

[441] Volovik G. E., *Int. J. Mod. Phys. D* **15**, 1987 (2006).

[442] Yang C.N., Mills R.L., *Phys. Rev.* **96**, 191 (1954).

[443] Utiyama R., *"Invariant Theoretical Interpretation of Interaction"*, *Phys. Rev.* **101** (1956) 1597.

[444] Utiyama R., De Witt B. S., *J. Math. Phys.* **3**, 608 (1962).

[445] Zlatev I., Wang L., Steinhardt P. J., *Phys, Rev. Lett.* **82**, 896 (1999).

[446] Zlatev I., Steinhardt P. J., *Phys. Lett. B* **459**, 570 (1999).

[447] Zwicky F., *Helvetica Physica Acta* **6**, 110 (1933).

[448] Zwicky F., *Astrophys. J.* **86**, 217 (1937).

Index

E

F

G

H

I

Q

R

S

T

U

V

W

X

Y